T0335757

# Scientific Basis for Nuclear Waste Management XXXVII

MATERIALS RESEARCH SOCIETY
SYMPOSIUM PROCEEDINGS VOLUME 1665

# Scientific Basis
# for Nuclear Waste Management
# XXXVII

Symposium held September 29 – October 3, 2013, Barcelona, Spain

**EDITORS**

## Lara Duro
AMPHOS 21
Barcelona, Spain

## Javier Giménez
Universitat Politècnica de Catalunya – BarcelonaTech
Dept. of Chemical Engineering
Barcelona, Spain

## Ignasi Casas
Universitat Politècnica de Catalunya – BarcelonaTech
Dept. of Chemical Engineering
Barcelona, Spain

## Joan de Pablo
Universitat Politècnica de Catalunya – BarcelonaTech
Dept. of Chemical Engineering
Barcelona, Spain

Materials Research Society
Warrendale, Pennsylvania

# CAMBRIDGE
## UNIVERSITY PRESS

University Printing House, Cambridge CB2 8BS, United Kingdom

One Liberty Plaza, 20th Floor, New York, NY 10006, USA

477 Williamstown Road, Port Melbourne, VIC 3207, Australia

314-321, 3rd Floor, Plot 3, Splendor Forum, Jasola District Centre, New Delhi - 110025, India

79 Anson Road, #06-04/06, Singapore 079906

Cambridge University Press is part of the University of Cambridge.

It furthers the University's mission by disseminating knowledge in the pursuit of education, learning and research at the highest international levels of excellence.

www.cambridge.org
Information on this title: www.cambridge.org/9781605116426

Materials Research Society
506 Keystone Drive, Warrendale, PA 15086
http://www.mrs.org

First published 2014

CODEN: MRSPDH

*A catalogue record for this publication is available from the British Library*

ISBN 978-1-60511-642-6 Hardback

# CONTENTS

## RADIONUCLIDES SOLUBILITY, SPECIATION, SORPTION AND MIGRATION

# CORROSION STUDIES OF ZIRCALOY, CONTAINER AND CARBON STEEL

# HIGH LEVEL WASTE

# CERAMIC AND ADVANCED MATERIALS

# PREFACE

The 37th International Symposium on the Scientific Basis for Nuclear Waste Management (Materials Research Society Symposium Proceedings Volume 1665) was held in Barcelona (Catalonia, Spain), September 30–October 3, 2013. The symposium was officially opened by Dr. Antoni Gurgui, commissioner of Consejo Seguridad Nuclear (Nuclear Safety Council) in Spain. About 80 attendees from 12 countries listened to 51 presentations and discussed 29 posters during the three and a half days of scientific sessions.

The symposium covered the following topics: national and international programs; performance assessment/geological disposal; radionuclide solubility, speciation, sorption and migration; corrosion studies of zircaloy, container and carbon steel; high level waste and ceramic and advanced materials.

Special posters highlighted the First Nuclides EC project and the EC Pooled Facilities called TALISMAN.

Lara Duro
Ignasi Casas
Javier Giménez
Joan de Pablo

December 2013

# Acknowledgments

The Editors would like to formally acknowledge the following people:

- The invited speakers: Jordi Bruno (AMPHOS 21), Pablo Zuloaga (ENRESA), Bernhard Kienzler (INE-KIT), Borje Torstenfelt (SKB), Rod Ewing (University of Michigan), Satoru Suzuki (NUMO, Japan), Mikhail Ojovan (IAEA), Monica Regalbuto (Fuel Cycle Technologies, DOE, USA), Gérald Ouzounian (ANDRA, France), Julio Astudillo (ENRESA), and Bernd Grambow (Subatech)

- The local organizing committee: Alexandra Espriu, Isabel Rojo, Albert Martínez-Torrents, and Frederic Clarens for continuous advice and help

- Krimson Events, in particular J. Prado Middleton, for taking care of the practical organization of the symposium and for much advice

- Marc Torrentsgenerós, who prepared and updated the symposium webpage

- Amanda de Pablo for her assistance in the preparation of the proceedings

- Sessions chairs: P. Zuluoaga, B. Grambow, J. Giménez, L. Duro, J.M. Soler, E. Wieland, R.C. Ewing, D. Bosbach, N. Hyatt, J. Sánchez, K. Spahiu, J. de Pablo, M. Jonsson, M. Grivé, D. Serrano-Purroy, L. Zetterström- Evins, H. Tanabe, M.I. Ojovan, M.R. Gilbert, and N. Dacheux

- Reviewers of papers: B. Kienzler, R.C. Ewing, E. Wieland, M. García-Gutiérrez, T. Missana, H. Tanabe, I. Casas, D. Serrano-Purroy, J.M. Soler, J. de Pablo, F. Clarens, I. Rojo, I. Azcárate, N. Hyatt, J. Giménez, K. Spahiu, G. Deissmann, J. Cobos, M.R. Gilbert, N. Dacheux, M.I. Ojovan, M. Grivé, L. Duro, T. Schaeffer, W. Von Lensa, E. Giffaut, J. Bruno, L- Zetterström-Evins, C. Ayora, J. Cama, E. González-Robles, V. Metz, L. Johnson, and D. Arcos.

Financial Support from UPC-Barcelona Tech, Fundació CTM Centre Tecnòlogic, and AMPHOS 21 is gratefully acknowledged.

# MATERIALS RESEARCH SOCIETY SYMPOSIUM PROCEEDINGS

# MATERIALS RESEARCH SOCIETY SYMPOSIUM PROCEEDINGS

Volume 1634E – Diamond Electronics and Biotechnology—Fundamentals to Applications VII, 2014, J.C. Arnault, C.L. Cheng, M. Nesladek, G.M. Swain, O.A. Williams, ISBN 978-1-60511-611-2

Volume 1635 – Compound Semiconductor Materials and Devices, 2014, F. Shahedipour-Sandvik, L.D. Bell, K.A. Jones, A. Clark, K. Ohmori, ISBN 978-1-60511-612-9

Volume 1636E – Magnetic Nanostructures and Spin-Electron-Lattice Phenomena in Functional Materials, 2014, A. Petford-Long, ISBN 978-1-60511-613-6

Volume 1638E – Next-Generation Inorganic Thin-Film Photovoltaics, 2014, C. Kim, C. Giebink, B. Rand, A. Boukai, ISBN 978-1-60511-615-0

Volume 1639E – Physics of Organic and Hybrid Organic-Inorganic Solar Cells, 2014, P. Ho, M. Niggemann, G. Rumbles, L. Schmidt-Mende, C. Silva, ISBN 978-1-60511-616-7

Volume 1640E – Sustainable Solar-Energy Conversion Using Earth-Abundant Materials, 2014, S. Jin, K. Sivula, J. Stevens, G. Zheng, ISBN 978-1-60511-617-4

Volume 1641E – Catalytic Nanomaterials for Energy and Environment, 2014, J. Erlebacher, D. Jiang, V. Stamenkovic, S. Sun, J. Waldecker, ISBN 978-1-60511-618-1

Volume 1642E – Thermoelectric Materials—From Basic Science to Applications, 2014, Q. Li, W. Zhang, I. Terasaki, A. Maignan, ISBN 978-1-60511-619-8

Volume 1643E – Advanced Materials for Rechargeable Batteries, 2014, T. Aselage, J. Cho, B. Deveney, K.S. Jones, A. Manthiram, C. Wang, ISBN 978-1-60511-620-4

Volume 1644E – Materials and Technologies for Grid-Scale Energy Storage, 2014, B. Chalamala, J. Lemmon, V. Subramanian, Z. Wen, ISBN 978-1-60511-621-1

Volume 1645 – Advanced Materials in Extreme Environments, 2014, M. Bertolus, H.M. Chichester, P. Edmondson, F. Gao, M. Posselt, C. Stanek, P. Trocellier, X. Zhang, ISBN 978-1-60511-622-8

Volume 1646E – Characterization of Energy Materials In-Situ and Operando, 2014, I. Arslan, L. Mai, E. Stach, ISBN 978-1-60511-623-5

Volume 1647E – Surface/Interface Characterization and Renewable Energy, 2014, R. Opila, P. Sheldon, ISBN 978-1-60511-624-2

Volume 1648E – Functional Surfaces/Interfaces for Controlling Wetting and Adhesion, 2014, D. Beysens, ISBN 978-1-60511-625-9

Volume 1649E – Bulk Metallic Glasses, 2014, S. Mukherjee, ISBN 978-1-60511-626-6

Volume 1650E – Materials Fundamentals of Fatigue and Fracture, 2014, A.A. Benzerga, E.P. Busso, D.L. McDowell, T. Pardoen, ISBN 978-1-60511-627-3

Volume 1651E – Dislocation Plasticity, 2014, J. El-Awady, T. Hochrainer, G. Po, S. Sandfeld, ISBN 978-1-60511-628-0

Volume 1652E – Advances in Scanning Probe Microscopy, 2014, T. Mueller, ISBN 978-1-60511-629-7

Volume 1653E – Neutron Scattering Studies of Advanced Materials, 2014, J. Lynn, ISBN 978-1-60511-630-3

Volume 1654E – Strategies and Techniques to Accelerate Inorganic Materials Innovation, 2014, S. Curtarolo, J. Hattrick-Simpers, J. Perkins, I. Tanaka, ISBN 978-1-60511-631-0

Volume 1655E – Solid-State Chemistry of Inorganic Materials, 2014, S. Banerjee, M.C. Beard, C. Felser, A. Prieto, ISBN 978-1-60511-632-7

Volume 1656 – Materials Issues in Art and Archaeology X, 2014, P. Vandiver, W. Li, C. Maines, P. Sciau, ISBN 978-1-60511-633-4

Volume 1657E – Advances in Materials Science and Engineering Education and Outreach, 2014, P. Dickrell, K. Dilley, N. Rutter, C. Stone, ISBN 978-1-60511-634-1

Volume 1658E – Large-Area Graphene and Other 2D-Layered Materials—Synthesis, Properties and Applications, 2014, editor TBD, ISBN 978-1-60511-635-8

Volume 1659 – Micro- and Nanoscale Systems – Novel Materials, Structures and Devices, 2014, J.J. Boeckl, R.N. Candler, F.W. DelRio, A. Fontcuberta i Morral, C. Jagadish, C. Keimel, H. Silva, T. Voss, Q.H. Xiong, ISBN 978-1-60511-636-5

Volume 1660E – Transport Properties in Nanocomposites, 2014, H. Garmestani, H. Ardebili, ISBN 978-1-60511-637-2

Volume 1661E – Phonon-Interaction-Based Materials Design—Theory, Experiments and Applications, 2014, D.H. Hurley, S.L. Shinde, G.P. Srivastava, M. Yamaguchi, ISBN 978-1-60511-638-9

Volume 1662E – Designed Cellular Materials—Synthesis, Modeling, Analysis and Applications, 2014, K. Bertoldi, ISBN 978-1-60511-639-6

# MATERIALS RESEARCH SOCIETY SYMPOSIUM PROCEEDINGS

**Prior Materials Research Symposium Proceedings available by contacting Materials Research Society**

**National and international programs**

Mater. Res. Soc. Symp. Proc. Vol. 1665 © 2014 Materials Research Society
DOI: 10.1557/opl.2014.622

# Treatment of Irradiated Graphite to Meet Acceptance Criteria for Waste Disposal: Problem and Solutions

Michael I. Ojovan[1] and Anthony J. Wickham[2]
[1] Waste Technology Section, Division of Nuclear Fuel Cycle and Waste Technology, Department of Nuclear Energy, International Atomic Energy Agency, PO Box 100, Wagramerstraße 5, Vienna, A-1400 Austria
[2] Nuclear Technology Consultancy, PO Box 50, Builth Wells, LD2 3XA, UK, and School of Mechanical, Aerospace and Civil engineering, The University of Manchester, Manchester M13 9PL, UK

## ABSTRACT

An overview is given of an International Atomic Energy Agency Coordinated Research Project (CRP) on the treatment of irradiated graphite (i-graphite) to meet acceptance criteria for waste disposal. Graphite is a unique radioactive waste stream, with some quarter-million metric tons worldwide eventually needing to be disposed of. The CRP has involved 24 organizations from 10 Member States. Innovative and conventional methods for i-graphite characterization, retrieval, treatment and conditioning technologies have been explored in the course of this work, and offer a range of options for competent authorities in individual Member States to deploy according to local requirements and regulatory conditions.

## INTRODUCTION

Graphite is a porous, chemically inert material, highly conductive and resistant to corrosion, in general retaining its properties after exposure to an intense radiation field and at high temperatures. It does, however, undergo structural changes as a result of exposure to fast neutrons, resulting in dimensional change of components and significant changes in their mechanical and physical properties. In addition, certain impurities, together with the 1.1% of $^{13}$C naturally present in the graphite, become activated through interactions with slow neutrons and thus present a significant radiation hazard post-exposure which must be accommodated in subsequent dismantling and disposal. Finally, in reactors where the graphite has been exposed to an oxidising coolant, some degree of oxidation of the material induced by the ionizing radiation field will have taken place, potentially affecting its strength.

Graphite is used in reactors as a neutron moderator and reflector, a structural material, and a fuel-element matrix material. It has been deployed in about 250 uranium- (or $UO_2$)-graphite reactors such as the United Kingdom Magnox and Advanced Gas-Cooled Reactors (AGR), the French UNGG, a small number of high-temperature reactors (HTRs), the Soviet-era RBMKs, and in numerous 'production' reactors and materials-testing reactors. Most of those reactors are now quite old, with many already shutdown.

The ability to dismantle and remove graphite stacks has already been demonstrated in a small number of different reactor designs *e.g.* at Fort St. Vrain (prototype HTR) in the USA, the air-cooled Brookhaven research reactor, also in the USA, and the GLEEP research reactor and Windscale prototype AGR in the UK. The resulting irradiated graphite waste (often referred to within the industry as *i-graphite*) then awaits acceptable treatment and disposal solutions whilst residing in temporary storage facilities, except for material from GLEEP which was subject to very low irradiation and could be disposed of by an industrial process.

The global inventory of i-graphite is currently about 260,000 metric tons (Fig. 1).

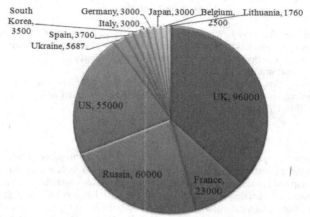

**Figure 1.** An assessment of the world inventory of irradiated graphite waste (metric tons).

The International Atomic Energy Agency (IAEA) has paid a great deal of attention to the problems arising from the need to dispose of *i*-graphite competently and safely. A series of IAEA technical meetings developing this topic area has been held *e.g.* in Bath, UK on "Graphite Moderator Lifecycle Technologies" in 1995 [1], in Manchester, UK on "Nuclear Graphite Waste Management" in 1999 [2], and again in Manchester, UK on "Progress in Radioactive Waste Management" in 2007 [3]. A compendium of information on the issue of *i*-graphite has also been published by IAEA [4].

The European Union (EU), then the CEC, sponsored an early (1984) study [5] on the "Assessment of Management Modes for Graphite from Reactor Decommissioning", using the UK Magnox reactors as an example. Because this task took place before a number of landmark international treaties and agreements concerning radioactive waste disposal, the study evaluated options such as sea disposal which are no longer available options. It is interesting that such a solution received rather favorable support, although the arguments were as much based upon economics as on the technical issues involved. Much more recently (2008), the EU initiated a major project under the 7th Framework initiative: CARBOWASTE "Treatment and Disposal of Irradiated Graphite and other Carbonaceous Waste" with 29 partners from 10 EU countries involved with a budget totalling about 11 million Euros including 6 million Euros EU-funding [6]. This study has considered not only *i*-graphite from existing power plants but also the regular carbonaceous waste arisings from future HTRs where the fuel itself consists of more than 95% carbonaceous material and thus forms the major consideration over time in terms of carbonaceous waste arisings. The final report of the CARBOWASTE study is expected to be published in 2014, but many of the experimental results have already been presented in conference and industry meetings.

A clear direction on the disposal of all radioactive wastes is now accepted as a necessity to underwrite the case for further development of nuclear power operations world-wide. In a 'greener' world, base-load and back-up generation for renewals clearly marks out a role for nuclear power, and it is urgently necessary to reverse the decline in the contribution of nuclear power generation (currently 13% overall world-wide) [7]. Thus, work on all aspects of radioactive waste is viewed by IAEA as very important and timely.

The IAEA has initiated the Coordinated Research Project (CRP) "Treatment of Irradiated Graphite to meet Acceptance Criteria for Waste Disposal"[1] in 2010, aiming to complement the CARBOWASTE study and to provide a platform for a number of the participants who are common to both projects to continue their studies following the end of CARBOWASTE funding. In addition, the CRP involves other Member States and organizations which were not part of CARBOWASTE – notably research groups from Russia, Ukraine and China. The CRP encompassed such issues as alternative techniques for removal of the graphite from the pressure vessels or containments, direct chemical or physical treatments, including pre-treatment to reduce the radio-isotope content and to facilitate the economics and radiological safety of the following process operations, and treatment of the products of innovative processes to improve radiological safety or for economic improvement (such as separation and recycling of useful isotopes for the nuclear and/or medical industries). The novel issues have also included evaluation of the possibilities for disposal of $^{14}CO_2$ produced by oxidation of reactor graphite by dilution with $^{14}C$-depleted $CO_2$ from the burning of fossil fuels prior to sequestering in exhausted gas fields and oil shales [8].

The IAEA CRP thus aims to add value to existing activities in various Member States, and has the potential to offer substantial savings in time and cost for the disposal of irradiated graphite. The added value included:

- Expertise and experimental facilities from additional Member States who were ineligible or otherwise not part of CARBOWASTE;
- Support to the less-well funded work packages of the CARBOWASTE project and facilitate the links between them;
- Additional partners and consultancy expertise;
- Additional treatment and handling options for investigation;
- Focusing on practical demonstration of potentially-useful technologies.

The overall objective of the CRP is to advise Member States of the various options which are being researched, to enable them to make an informed decision on the correct policy for their situation. It is not the intention to recommend specific methodologies on a global basis, since very different conditions and constraints apply in different Member States. As examples, ANDRA in France is obligated by law to find a solution to the disposal of graphite wastes by the end of 2015: in the UK, by contrast, the policy is to leave reactors intact in 'safe storage' for periods which may be as long as 130 years before eventual dismantling. Germany has severe constraints on the quantities of $^{14}C$ which may be disposed of in the identified repository (KONRAD) whilst other Member State authorities view the presence of $^{36}Cl$ to be a much more significant issue compared with $^{14}C$. The CRP will take due note of these differences and constraints when framing the text of its final TECDOC report.

---

[1] Project T21026

## CRP RESEARCH TOPICS

The CRP was launched by IAEA purposely to investigate treatment options available (or under consideration) in Member States. Participants facilitate the exchange of information and technological experiences on new developments in the area, and seek to identify innovative technologies which might be applied in conformity with modern safety and economic requirements. The research program aims to identify and evaluate, by practical means if possible, relevant handling and treatment technologies which may be of assistance in Member States' graphite-disposal strategies, complementary to the investigations of CARBOWASTE. The CRP has involved 24 organizations from 10 Member States whose agreed research-program titles are given in Table I.

**Table I.** Countries, organizations, researchers and research topics of IAEA CRP.

| Country, organization, and researchers involved | Research focus* | | | |
|---|---|---|---|---|
| | CH | PR | IM | DI |
| **China.** Tsinghua University, INET, Li Junfeng | | ▓ | | |
| **France.** IPNL: Christine Lamouroux (CEA), Gerard Laurent (EdF CIDEN), Laurence Petit (ANDRA). | ▓ | | | ▓ |
| **Germany.** FZJ, Werner von Lensa | | | | ▓ |
| **Germany.** FNAG, Johannes Fachinger | | | ▓ | |
| **Lithuania.** INPP. Alexander Oryŝaka | ▓ | | | |
| **Lithuania.** LEI. Ernestas Narkunas, Povilas Poskas | | | | ▓ |
| **Russia.** VNIINM, Vladimir Kascheev, FGUP RADON, Olga Karlina | | ▓ | ▓ | |
| **Spain.** ENRESA, Jose Luis Leganes Nieto | ▓ | | | |
| **Switzerland.** PSI, Hans F. Beer | ▓ | | | |
| **Ukraine.** IEG, Boris Zlobenko | ▓ | | | |
| **United Kingdom.** NDA, Simon Norris | ▓ | | | ▓ |
| **United Kingdom.** The University of Manchester, Abbie Jones, Tony Wickham | ▓ | | | ▓ |
| **United Kingdom.** University of Sheffield, Russell Hand | | | ▓ | |
| **United Kingdom.** Bradtec, David Bradbury, Hyder/Bradtec, Jon Goodwin, Studsvik UK, Maria Lindberg, Costain, T. Tomlinson, L'Arbresle Ingeniere (**France**), Laurent Rahmani | | ▓ | ▓ | ▓ |
| **United Kingdom.** NNL. Martin Metcalfe, Anthony Banford | ▓ | | | |
| **United States.** Idaho State University, Idaho National Laboratory, Mary-Lou Dunzik-Gougar | | | | |

*Areas of interest are labelled as follows: CH – characterization, PR – processing, IM – immobilization. DI – disposal.

The management of i-graphite waste is a complex task dealing with many different aspects of radioactive waste management starting with waste generation (on dismantling of nuclear facilities), followed by predisposal operations such as characterization, specialist chemical or physical treatments, immobilization, conditioning and storage and ending with disposal. It is therefore not easy to separate and focus on just one aspect in any comprehensive R&D program. Nevertheless the main focus of interest in one or another program can most often be identified as shown in last columns of Table I, which conventionally divides areas of interest as characterization (CH), processing (PR), immobilization (IM) and disposal (DI). A schematic of the CRP work and the interactions between various groups and topics is shown in Fig. 2.

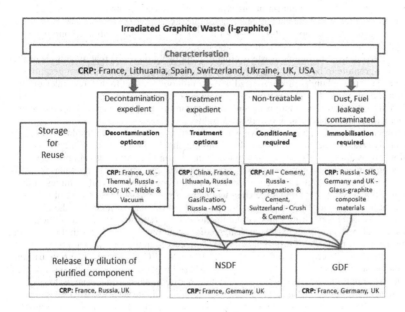

**Figure 2.** Schematic of CRP contributions toward irradiated graphite management.
(*N.b.* 'NSDF': Near-Surface Disposal Facility; 'GDF': Geological Disposal Facility)

A dedicated network, 'IMMONET' has been recently launched at IAEA [9] aiming to share important results among specialists. It contains amongst other sub-sites a dedicated web-site for this i-graphite CRP [10]. Although this site is not free-accessible and requires an IAEA registration, it facilitates not only sharing the materials amongst appropriate specialists but also the ensuing comments and discussions.

# CHARACTERIZATION

From Fig. 2 it follows that as a result of characterization the i-graphite can be conventionally divided into 4 streams:

(i)     Fuel-contaminated graphite typically classified as high-level waste (HLW). This i-graphite will obviously need a proper immobilization before acceptable for any safe storage or disposal routes;

(ii)     Not-expedient for treatment i-graphite due to some technical and/or non-technical reasons. This i-graphite will need nevertheless a proper conditioning to be accepted for safe storage and disposal;

(iii)     Treatment-expedient i-graphite where such a process might show a benefit in reduction of radio-isotope content and a consequent cost saving;

(iv)     Decontamination-expedient i-graphite. Suitable decontamination technologies need then to be used for this waste stream.

Proper characterization of i-graphite waste is crucial in appreciation of the real magnitude of the problem with i-graphite to be faced. It is very clear that, because numerous different graphite types have been deployed under different irradiation conditions, subject to different rates of the transfer of materials around reactor circuits which can become trapped in the graphite and then activated by slow neutrons, that there is no single definition of i-graphite and each source needs careful characterization and evaluation in its own right.

The main concerns expressed by radioactive waste authorities in dealing with irradiated graphite waste generally relate to the content of long-lived radionuclides such as $^{14}C$ and $^{36}Cl$ although other radioisotopes, particularly arising in impurities in the graphite and located in discrete positions or on pore surfaces, may also present specific issues. In some cases the material may be contaminated with nuclear fuel debris e.g. containing both fission products and actinides. It has been revealed during the program that the chemical form of specific isotopes is critical in understanding their behavior both during irradiation and post-irradiation, including their potential release mechanisms from failed containments in deep repositories over geological timescales. It is also clear that there remain disagreements and unresolved issues in this area, and also in the relative importance of different isotopes as perceived in the various Member States. The radionuclides of concern are typically $^{3}H$, $^{14}C$, $^{36}Cl$, $^{60}Co$, $^{137}Cs$ and $^{152}Eu$. The first, tritium, is important because of its ready transfer to any water-based organism, although its short half-life means that it is important only during the 'operational' period of a repository – i.e. before closure. The next two long-lived isotopes in the list, $^{14}C$ and $^{36}Cl$, attract very different levels of concern in different Member States, both for reasons of geology and the differing restrictions imposed on disposal sites and also because of preconceptions about their behavior in the geosphere and biosphere.

The remaining isotopes, gamma emitters of relatively short half-lives, are most problematical during the dismantling phase and will dictate the extent of shielding and personnel protection necessary. Thus, a very high $^{60}Co$ content might lead to considerations of alternative removal strategies compared with physical removal of intact blocks of the *i*-graphite.

Specifically, characterization data are required to:

- Classify the irradiated graphite waste for one or another type of disposal *e.g.* deep geological, intermediate depth or near surface disposal;
- Identify most appropriate processing routes aiming to treat the irradiated graphite to ensure compliance with waste acceptance criteria of disposal facilities;

- Create a source-term database for evaluation of performance of materials disposed of and behavior of radionuclides in a future disposal environment.

Characterization of irradiated graphite waste includes identification of its both radiological (mainly the content of radionuclides) and physico-chemical parameters including structural data and content of impurities (Table II).

**Table II.** Characteristics important for i-graphite management.

| Feature | Composition | Structural | Behavior | History |
|---|---|---|---|---|
| Components | Elemental, isotopic, chemical speciation/phase | Macro/bulk, Micro/bulk, Atomistic | Leaching, Gas-release, Wigner energy, Mass change (operational) | Maker, origin, irradiation schedule, times, accidences, Wigner energy actions, Storage. |
| Where investigated | All CRP members | UK, USA, Germany | All CRP members | All CRP members |

As discussed earlier, not only the structure but also physical and chemical properties of graphite change on irradiation. CRP researchers have confirmed that the properties of the *i*-graphite are different for each reactor type and even vary within a specific reactor according to the position of the component, presenting a complex function of histories of manufacture, construction operation, and retrieval and storage conditions. CRP characterization studies confirmed that the irradiated graphite is an inhomogeneous material which can be described as an amorphous-crystalline porous composite containing non-homogeneously distributed radionuclides (spot-type contamination). Some important CRP findings on characterization are:

- Data on characterization methodology, scaling factors identified and speciation and location of $^3H$ and $^{36}Cl$ (IPNL on behalf of EdF, CEA, ANDRA, France, see also [11]);
- Formation and behavior of $^{14}C$ within graphite structures, including investigations of production routes $^{13}C(n,\gamma)^{14}C$ versus $^{14}N(n,p)^{14}C$ using N-loaded precursors and sophisticated analysis of resulting surface groups and structures with a new phenomenon of nitrogen-'fixing' on graphite-pore surfaces identified (Idaho State University, INL, US);
- Data on structural changes in irradiated graphite including modelling of radionuclide generation and radiation damage (UK, Germany);

**PROCESSING**

Processing options considered within CRP range from decontamination aiming to remove radioactive contaminants from the irradiated graphite to complete burning of graphite followed by trapping of radionuclides from the off gases produced (Table III). The most difficult task on purifying off gases from incineration of graphite is the removal of $^{14}C$ present in off gases as $CO_2$. The main channels for $^{14}C$ accumulation in uranium-graphite reactors are: $^{14}N(n, p)^{14}C$ and $^{13}C(n, \gamma)^{14}C$ although the contribution of each reaction can vary. In some types of reactors, where air or carbon dioxide was used as a scavenge gas, the reaction $^{17}O(n, \alpha)^{14}C$ could be important.

Table III. Processing options considered.

| Option | Nibble-and-vacuum | Decontamination (including calcination) | Incineration* | Molten Salt Oxidation |
|---|---|---|---|---|
| Where most effective | Effective for dismantling and preparation for treatment | To reduce the volume, for declassification and release of a part of graphite | For maximum volume reduction | For volume reduction |
| Where considered | UK | France, UK, Russia | France, UK, Russia | Russia |

*Graphite does not effectively burn until a temperature in excess of 3300°C is attained in air. At lower temperatures there is an oxidation reaction to produce carbon dioxide admixed with quantities of carbon monoxide. The reaction becomes significant at around 400°C where it is controlled by the chemical kinetics and occurs throughout the pore structure: at higher temperatures issues of diffusion-control and ultimately mass-transport limitation of reagent supply become significant. For an incineration process to work, a system such as a fluidized bed is needed, with an additional oxygen supply to initiate the process [12].

Important data obtained by CRP participants to date include:

- Proofs on acceptable efficiency of thermal treatments both for mobilization of isotopes and conversion of graphite by oxidation and steam reforming *e.g.* results of UK (GLEEP graphite and others), Russia (RBMK graphite), Germany (for different atmospheres), France.
- Complex approach on *i*-graphite management starting with dismantling (nibble-vacuum tool), passing decontamination-gasification and ending with disposal (UK);
- Molten salt oxidation (MSO) investigated in Russia both for decontamination and full oxidation of i-graphite [13]
- Investigation of innovative chemical routes for decontamination of i-graphite (Spain, Russia, UK, US).

## CONDITIONING

The CRP researches have considered following conditioning and immobilization routes for i-graphite waste:

- Grouting of i-graphite blocks (all CRP participants);
- Epoxy resin impregnation of i-graphite blocks to decrease its porosity[2] (Russia);
- Crushing of i-graphite followed by mixing with cement grout (Switzerland)
- Use of geopolymers for i-graphite encapsulation (Germany);
- Self-propagating high temperature synthesis for fuel-contaminated i-graphite (SHS, Russia);
- Glass-graphite composite materials obtained both via sintering and melting routes (Germany and UK).

Cementation remains the favored technical option considered everywhere at present. However cements can be used simply to grout the i-graphite blocks or, as it has been proved in Switzerland, crushing the graphite waste and using it in combination with cements to immobilize

---

[2] Russian researchers in this area have a developed a specific formula known as 'F-Conserving Agent™', deployed to prevent release of contamination from certain damaged reactor cores such as Beloyarskaka NPP.

other solid radioactive waste in containers. This method opens the opportunity to use the waste graphite as a waste immobilizing medium.

## DISPOSAL

Depending of radionuclide content the i-graphite can be disposed either in a near surface disposal facility or in a deep geological disposal facility. The perceived primary need for encapsulation and deep burial of the graphite is the presence of long-lived radionuclides, principally $^{36}Cl$ (half-life of 301,000 years) and $^{14}C$ (half-life of 5,730 years). France has initially devoted attention to the design of a separate repository specifically intended for i-graphite together with radium-bearing waste. Finding a suitable location for this facility has, however, been unsuccessful, and an alternative route of thermal processing of *i*-graphite with conditioning of all resulting waste streams such as ash residues followed by an appropriate disposal route, is now under investigation [14]. Disposal options considered within CRP are outlined in Table IV.

**Table IV.** Disposal options for *i*-graphite.

| Option | Release by dilution with natural $CO_2$* | Geological disposal | Near-surface disposal | Deep underground injection |
|---|---|---|---|---|
| When applied | On graphite oxidation. At low toxicity e.g. low level of activity and low content of long-lived radionuclides. | At high toxicity e.g. high level of activity and high content of long-lived radionuclides | At low toxicity e.g. low level of activity and low content of long-lived radionuclides | By dilution with natural carbonaceous media. |
| Where investigated | France, Russia, UK | All CRP members | All CRP members | France, UK |

*This route can be used for $CO_2$ emitted during oxidation of i-graphite, whereas the remaining ash residue and other off gas contaminants are captured, conditioned and then disposed of via a suitable route.

## CONCLUSIONS

Dismantling of old reactors and the management of radioactive graphite waste are becoming an increasingly important issue for a number of IAEA Member States. Exchange of information and research co-operation in resolving identical problems between different institutions within Member States contributes towards improving waste management practices, their efficiency, and general safety. The execution of the CRP is promoting the exchange of advanced information on the on-going research and development activities and facilitates access to the practical results of their application for treatment and conditioning of specific graphite waste types. More details on i-graphite management will be provided in forthcoming IAEA publication (TECDOC) that is tentatively scheduled for 2014/15.

**ACKNOWLEDGMENTS**

The authors acknowledge the efforts of all the CRP researchers (Table I) who contributed to the success of the program. AJW acknowledges financial support from the UK Nuclear Decommissioning Authority.

**REFERENCES**

1. IAEA, *Graphite Moderator Lifecycle Behaviour*, Proceedings of a Specialists Meeting held in Bath, United Kingdom, 24-27 Sept 1995; IAEA-TECDOC-901, Vienna (1996).
2. IAEA, *Nuclear Graphite Waste Management: Technical Committee Meeting on Nuclear Graphite Waste Management*, Manchester UK, 18-20 October 1999, IAEA CD-ROM 01-00120, Vienna (2001).
3. IAEA (2010). *Progress in Radioactive Graphite Waste Management.* IAEA-TECDOC-1647, IAEA, Vienna (2010).
4. IAEA, *Characterization, Treatment and Conditioning of Radioactive Graphite from Decommissioning of Nuclear Reactors.* IAEA-TECDOC-1521, IAEA, Vienna (2006).
5. I.F. White, G.M. Smith, L.J. Saunders, C.J. Kaye, T.J. Martin, G.H. Clarke and M.W. Wakerley, *Assessment of Management Modes for Graphite from Reactor Decommissioning,* Commission of the European Communities, EUR 9232, Brussels, (1984).
6. A. Banford, H. Eccles, M. Graves, W. von Lensa, S. Norris. CARBOWASTE – An Integrated Approach to Irradiated Graphite. *Nuclear Future,* **4**, 268-270 (2008).
7. IAEA. (2013). *Nuclear Technology Review.* IAEA, Vienna, 170 p. (2013).
8. V.P. Rublevskiy. Commercial nuclear reactors and $^{14}$C. *Atomic Energy,* **113**, 143-147 (2012).
9. IAEA IMMONET SharePoint Site. http://nucleus.iaea.org/sites/nefw-projects/IMMONET/SitePages/Home.aspx (2013).
10. IAEA Irradiated Graphite Treatment CRP Site. http://nucleus.iaea.org/sites/nefw-projects/IMMONET/graphite-crp/SitePages/Home.aspx (2013).
11. B. Poncet, L. Petit. Method to assess the radionuclide inventory of irradiated graphite waste from gas-cooled reactors. *J. Radioanal. Nucl. Chem.,* **296**, 3, 15 p., DOI: 10.1007/s10967-013-2519-6 (2013).
12. J.-J. Guiroy, Graphite Waste Incineration in a Fluidised Bed, *in Ref [1] above,* 193-203 (1996).
13. O. Karlina, M. Ojovan, G. Pavlova, V. Klimov. Thermodynamic modelling and experimental tests of irradiated graphite molten salt decontamination. *Mater. Res. Soc. Symp. Proc.* **1518**, 6 p., DOI: http://dx.doi.org/10.1557/opl.2013.72 (2013).
14. Autorité de Sûreté Nucléaire, Ministère de l'Écologie du Développement durable et de l'Énergie, Plan Nacional de Gestion des Matières et des Déchets Radioactifs 2013 – 2015, Paris, April 2013.

# Performance assessment/geological disposal

Mater. Res. Soc. Symp. Proc. Vol. 1665 © 2014 Materials Research Society
DOI: 10.1557/opl.2014.623

# Projecting Risk into the Future:
## *Failure of a Geologic Repository and the Sinking of the Titanic*

Rodney C. Ewing
Department of Earth & Environmental Sciences
Center for International Security & Cooperation
Stanford University, Stanford, CA, 94305 U.S.A.

## ABSTRACT

This year marks the 101$^{st}$ anniversary of the sinking of the "unsinkable" RMS *Titanic*. On April 15, 1912, the *Titanic* struck an iceberg in the North Atlantic Ocean on its maiden voyage from Southampton, UK, to New York City. There was no single cause for the loss of the *Titanic*, rather the improbable combination of errors in human design and decision combined with unforeseeable circumstance lead to the loss of over 1,500 lives. The failure appears to have occurred over a range of spatial and temporal scales – from the atomic-scale process of embrittlement of iron rivets to global-scale fluctuations in climate and ocean currents. Regardless of the specific combination of causes, this failure in design and practice led to impressive improvements in both. Disaster and tragedy are harsh teachers, but critical to improvement and progress.

The important question for the nuclear waste management community is how do we learn and improve our waste management strategies in the absence of being able to fail. A geologic repository "operates" over a very distant time frame, and today's scientists and engineers will never have the benefit of studying a failed system. In place of a failure that is followed by improvement and progress, we can only offer a general consensus on disposal strategies supported by a wide array of evidence and risk assessments. However, it may well be that consensus leads to complacency and compromise, both of which are harbingers of disaster. With this concern in mind, this is the time to review our fundamental approach, particularly the methodologies used in risk assessments that have us calculate risk out to one million years. The structure of standards and implementing regulations, as well as the standard-of-proof for compliance, should be reexamined in order to determine whether their requirements are scientifically possible or reasonable. The demonstration of compliance must not only be compelling, but must also be able to sustain scientific scrutiny and public inquiry. We should benefit from the sobering reality of how difficult it is to anticipate future failures even over a few decades. We should be humbled by the realization that for a geologic repository we are analyzing the performance, success *vs.* failure, over spatial and temporal scales that stretch over tens of kilometers and out to a hundreds of thousands of years.

# SINKING OF THE RMS *TITANIC*

On the night of April 15th in 1912, the "unsinkable" RMS *Titanic*, on its maiden voyage, struck an iceberg and within just a few hours sank in the cold waters of the North Atlantic. Of the estimated 2,224 people aboard, more than 1,500 lives were lost. This was one of the defining maritime disasters of the 20th Century. In 1985, the wreck of the RMS *Titanic* was located in two pieces at a depth of 3,800 meters, some 600 km from the coast of Newfoundland. The discovery of the remains of the *Titanic* once again captured the imagination of the public, and this tragic event became the subject of movies, a TV series, exhibits, and even opportunistic advertising campaigns.

Even though the loss of the *Titanic* took only a few short hours, the tragedy was years in the making – the result of many separate decisions, all of which culminated in the tremendous loss of life just 101 years ago. There have been many theories proposed for the cause of the loss of the *Titanic*, and no single cause can be identified. One of the first to be blamed was Captain Edward J. Smith. Despite the fact that he had over 40 years of experience at sea, he was criticized for traveling at a speed of 22 knots (~ 40 km/h), near top speed for the *Titanic*, in the dark of night, even though the *Titanic* had received numerous iceberg warnings from other ships. A nearby ship, the HMS *Carpathia*, a Cunard Liner also on her maiden voyage, had stopped that night due to the danger of icebergs. After some delay, the *Carpathia* helped in the rescue of passengers from the *Titanic*. Later, as the *Titanic* sank, indecision and confusion would lead to some lifeboats being launched, even though they were not full. A number of design issues have also been identified as critical to determining the fate of the *Titanic*. At that time the HMS *Titanic* was the largest ship in the world, perhaps too large to be turned quickly in the face of an oncoming iceberg. The large gash from the iceberg across the starboard side opened five of the sixteen compartments to flooding, but the ship was designed to stay afloat with at most four compartments breached. The bulkheads were not high enough, so that as the ship pitched under the weight of the flooding waters, additional compartments were flooded, and the ship sank quickly. Sadly, there were not enough lifeboats. Thomas Andrews the ship's designer had wanted bigger bulkheads to more effectively seal the compartments from one another -- and more lifeboats. The number of lifeboats had been decreased for aesthetic reasons, with the confidence that there would be time to ferry passengers to safety because the *Titanic* was unsinkable, or at least would sink slowly. Andrews was one of those lost on that night. Even the details of materials science may have been part of the cause for the sinking. The plates of the *Titanic* were riveted by a combination of steel and iron rivets. The iron rivets contained, in some cases, large amounts of slag, making them more brittle, particularly in the cold waters of the North Atlantic. Some have suggested that the points of failure in the hull corresponded to the points of transition from steel to iron rivets. Finally, weather conditions may have conspired to place more than the usual number of icebergs across the course of the *Titanic*. Thousands of kilometers away the Caribbean and the Gulf of Mexico were experiencing an unusually hot summer that contributed to a more intense Gulf Stream. This intensified the boundary between the Labrador Current and the Gulf Stream near Newfoundland, creating a barrier of icebergs along their interface. Thus, there were more icebergs in a smaller area than usual – in the path of the *Titanic*. This may have been amplified by the mild winter that caused more icebergs to calve off the west coast of Greenland.

There was no single cause for the sinking of the *Titanic*, rather the improbable combination of errors in human design and decision combined with unforeseeable circumstance.

Failure occurred over a range of spatial and temporal scales – from the atomic-scale process of embrittlement of iron rivets to global-scale processes of fluctuations in climate and ocean current strength. The story of such failures is not unusual. Henry Petroski has offered an analysis of many cases of failure in engineered systems [1, 2]. In my lifetime, I have witnessed such stories repeat themselves in the U.S. space program. In 1967, Ed White, Gus Grissom and Roger Chafee lost their lives to a fire in *Apollo 1*, while conducting tests *on the ground*. In 1986, seven crew members lost their lives in the space shuttle, *Challenger*, which broke up within a few minutes of launch due to a failure of an O-ring seal on the solid-fuel rocket booster. In 2003, seven more crew members lost their lives as the space shuttle, *Columbia,* disintegrated on reentry. In each of these three cases, the judgments of risk were erroneous, deemed to be acceptable, or in fact, ignored.

## IN THE AFTERMATH OF THE RMS *TITANIC*

Public inquires as to the cause of the sinking of the *Titanic* began within the year in both the United States and Great Britain. Captain Smith was faulted for failing to take action in the face of repeated warnings about the presence of icebergs. The regulations for the number of lifeboats and evacuation training were found to be inadequate. Still, there was no finding of negligence, as all involved had followed standard practice. It was noted that previous to the accident that British ships transported millions of passengers with the loss of only a few lives. Still, important changes were made in the wake of the loss of the *Titanic*, including the creation of the *International Convention for the Safety of Life at Sea* (1914), *Rules for Life Saving Appliances* created by the Board of Trade (1914) and the International Ice Patrol, which has operated continuously since 1913, except during WWII. Undoubtedly, the loss of the *Titanic* led to corrective actions in design and practice that have reduced the risk for millions of subsequent sea-going passengers. Modern ships now favor a double hull design over compartmentalization, and there are enough life vests and lifeboats for all passengers. The U.S. space program has gone through similar reviews following the major disasters of *Apollo 1, Challenger* and *Columbia.* Each review took several years and lead to extensive corrective actions. Failures in design and practice have always lead to improvements in both. In fact, the word "success" is derived from the Latin *succedere*, "to come after", that is to come after failure.

*Curiously, failure is critical to success.* The challenging question for members of the nuclear waste management community is how do we compensate for the lack of experience, that is failure, in our design and evaluation of geologic repositories for spent nuclear fuel and highly-radioactive waste? After all, there will be very few geologic repositories designed and built, perhaps one in each country with nuclear power. Learning by real time experience is not possible. The success of the repository design can only be known over the operational period of the geologic repository – that is over hundreds of thousands of years. None of us will witness the failure, and no corrective action will be possible. Even the concept of retrievability only extends for a few hundred years, far short of the actual operational period of the repository. In fact, we often confuse the success of the construction of the underground workings and the emplacement of the waste, which can take up to one hundred years, with the operational period of the repository, which will, in fact, have to function over hundreds of thousands of years. *In the absence of the opportunity to experience failure what is the basis for our confidence in the performance of a geologic repository over such long periods?*

One answer could be that we have the easier job. We are not designing, for the first time, the largest ship in the world or space-age rockets and shuttles. Our engineering challenge is to create an underground excavation and to emplace highly radioactive waste into these workings. Further, the failure will not be catastrophic. At worst, we may anticipate the slow leakage of radioactive material from a few breached canisters stretched over a very long time, and even these radionuclides will be sorbed onto surrounding rock and diluted during transport. With the passage of time, radioactive decay works to our advantage. This is the basic story that we use when we communicate with the public, supported by elaborate, ever changing, risk assessments. *I wonder if this is enough?*

I would argue that the spatial and temporal time scales are at least as challenging as those of designing ships and spacecraft. Certainly a typical assessment of a geologic repository will span the range from atomic-scale, e.g., the corrosion of waste packages and waste forms, such as spent nuclear fuel, to global-scale changes in climate and its effect on hydrologic and geochemical systems. The time scales for predicted or inferred performance are all the more remarkable in that we calculate dose, or risk, out to a million years. We calculate this risk even though we know that the human species has spread across the surface of Earth only during the past 200,000 years. The assessment of the performance of a geologic repository does not seem like "the easier job."

## SIMPLE LESSONS FOR RADIOACTIVE WASTE MANAGEMENT

The story of the *Titanic* provides some simple insights for the management and disposal of radioactive waste, particularly for the design and evaluation of a geologic repository.

*Details matter* – just as the atomic-scale properties of the rivets mattered to the *Titanic*, the atomic-scale properties of spent fuel or vitrified waste matter. Simply placing spent nuclear fuel or vitrified waste in corrosion resistant canisters is not enough. A prudent approach involves careful consideration of the major components of the system – waste form durability, waste package, overpack, and the hydrology and geochemistry of the site. Some part of the system will inevitably perform below expectation, so all parts of the system must be well understood.

*Design matters* – just as the design of the *Titanic* failed to anticipate the breach of multiple compartments, the design of a geologic repository has to anticipate the failure of multiple barriers. For decades, a major tenet of geologic disposal has been a series of multiple barriers that prevent or slow the release of radionuclides. The multi-barrier concept is used to reduce uncertainty in long-term performance assessments and provides confidence and robustness in the design. During the last decade, however, particularly in the United States, the emphasis has shifted away from the performance of individual barriers to a total system performance assessment of the repository over its entire operational period, out to a million years. The performance is not measured by fractional release, but rather as a dose to a person or population in the distant future. This calculation of risk adds another layer to the analysis that further obscures elements of design and performance of the repository. For the *Titanic*, this would have meant not focusing on the brittleness of the rivets, but rather viewing their performance in the context of the probability of hitting an iceberg and the number of lives lost. The safety analysis should highlight the underlying science, rather than obscure our understanding of individual barriers.

*Boundary conditions change and matter* – just as long-term weather patterns affected the distribution of icebergs, geologic disposal must anticipate changes in long-term conditions in climate, hydrology and geochemistry. The solution in both cases, the *Titanic* and a geologic repository, is to look for sites where changing conditions are less likely or of less consequence. If climate change is the issue, then there are advantages to sites in the unsaturated zone where fluctuations in water table do not affect the analysis of water in contact with the waste packages and waste forms. If waste forms are important, and they should be, then reducing conditions should be favored over oxidizing conditions.

*Design for and anticipate failure* – just as the failure to have enough lifeboats led to the loss of many lives, the failure to select sites carefully, in anticipation of the failure of engineered barriers or changing conditions, should remain a fundamental tenet of geologic disposal. Site selection is the most important decision in the development of a strategy for geologic disposal.

## WHAT IF THE TITANIC HAD NOT SUNK?

The more subtle lessons of the *Titanic* come from asking the simple question, "What if the *Titanic* had not sunk on its maiden voyage or, for that matter, any subsequent voyage?" This question was raised by Henry Petroski in his recent book, *To Forgive Design - Understanding Failure*. The *no failure* scenario is speculative, but perhaps, no more speculative that those used to evaluate a geologic repository. Thus, one may imagine that the unsinkable *Titanic* arrived at in New York and safely disembarked over 2,000 happy souls. The success of the *Titanic's* design would have been "proven" by the absence of failure. The successful design, compartments with lower bulkheads, would have been copied in future ships. The design may have been modified based on the safety record, perhaps thinner plates on the hull to save on cost. Due to the unsinkable design, fewer lifeboats would have been required on future ships. Due to the lack of the loss of life, the safety conventions would not have been passed, even if proposed, due to an expense that would have been viewed as unnecessary. Certainly, progress and creative action would have been slowed by the absence of failure.

One may go even further and say that failure is good because it: *i.*) provides the basis for corrective action; *ii.*) is essential to building confidence; *iii.*) provides the basis for public acceptance. In contrast, for geologic disposal, we will have little opportunity for corrective action, confidence is based on consensus, and public acceptance is essential but hard to win. The key point for the nuclear waste management community is that our definition of success is based on consensus. The statement of consensus has, in fact, become a primary belief by nearly all bodies involved in geologic disposal:

> "An international consensus has emerged that burial of high-level activity waste in a deep geologic repository is technically feasible and that such an approach can provide adequate protection to humans and the environment." (U.S. Nuclear Waste Technical Review Board, *Technical Advancements and Issues Associated with the Permanent Disposal of High-Activity Wastes, Lessons Learned from Yucca Mountain and Other Programs,* 2011)

> "Disposal in a carefully sited and designed geologic repository is recognized by most of the international technical community, including the National Research Council, as a long-term management option for high-level waste that provides a high degree of safety and security." (U.S. National Research Council, *One Step at a Time: The Staged Development of Geologic Repositories for High-Level Radioactive Waste,* 2003)

"Disposal of these wastes in engineered facilities, or repositories, located deep underground in suitable geological formations is being developed worldwide as a reference solution in order to protect humans and the environment both now and in the future. Engineered geological disposal is thus seen as the radioactive waste management end-point providing safety without the need for renewed human intervention." (Nuclear Energy Agency, *Moving Forward with the Geological Disposal of Radioactive Waste,* 2008)

The consensus has a strong basis and is well founded, as has been argued by the NEA [3]. The consensus is based on extensive experimental data sets collected for a wide range of geologic formations and engineered materials. Much of the science and engineering have been demonstrated in specially built underground laboratories. In many cases, the engineering and science are state-of-the-art and employ the most advanced modeling techniques. Additional experience is derived from studies of the disposal of other types of waste. Safety assessments are based on best practice.

In fact, I am part of this consensus, a member of the community of scientists and engineers who believe that geologic disposal will provide the necessary safety for humans and the environment. Still, my concern is that in an endeavor where we will not and cannot experience failure, complacency, compromise and even consensus have to be challenged. There are three areas that especially require this challenge today:

- Methods used in the safety assessments and the calculation of uncertainty.
- Rationale for methods used to complete safety assessments over hundreds of thousands of years.
- The standard-of-proof used to support the safety analysis over very long times.

These are the three areas that require the most research attention today. We have made tremendous progress in the analytics of our approach, but only limited progress in developing an approach that is useful and believable. Unless, we develop strategies for testing geologic repositories over long periods to the point of failure, and unless we develop strategies that anticipate failure, then scientific and public acceptance will continue to elude our efforts.

## EPILOGUE

I sometimes chastise myself for being so stubborn on some of the issues of geologic disposal. It is hard to believe that being critical is time well spent in the face of the need for geologic disposal and the consensus of an entire community of scientists and engineers. However, I recently read a quote from a younger and talented member of our nuclear waste community, "I'd say we already know 95 to 99 percent of what we can do with waste." It is exactly this attitude against which we need to guard.

## ACKNOWLEDGMENTS

This paper is not a scientific paper, but is written as an essay. These thoughts were inspired by discussions with and the writings of Professor Henry Petroski, the Aleksandar S. Vesic Professor of Civil Engineering and Professor of History at Duke University. Henry first raised the question of what would have happened if the *Titanic* had not sunk in his recent book, *To Forgive Design – Understanding Failure*. His question woke me from the slumber of consensus.

As a member of the Nuclear Waste Technical Review Board in the United States, I emphasize that these views are entirely my own, and in no way represent the views of the NWTRB.

## REFERENCES

1. Henry Petroski, *To Engineer is Human: The Role of Failure in Successful Design,* (First Vintage Books, 1992) 251 pp.
2. Henry Petroski, *To Forgive Design – Understanding Failure*, (Belknap and Harvard Press, 2012) 410 pp.
3. NEA/OECD, *Moving Forward with Geological Disposal of Radioactive Waste,* NEA No. 6433 (2008) 21 pp.

Mater. Res. Soc. Symp. Proc. Vol. 1665 © 2014 Materials Research Society
DOI: 10.1557/opl.2014.624

# Evaluation of the long-term behavior of potential plutonium waste forms in a geological repository

Guido Deissmann[1,2], Stefan Neumeier[1], Felix Brandt[1], Giuseppe Modolo[1] and Dirk Bosbach[1]
[1]Forschungszentrum Jülich GmbH, Institute of Energy and Climate Research IEK-6: Nuclear Waste Management and Reactor Safety, 52425 Jülich, Germany
[2]Brenk Systemplanung GmbH, Heider-Hof-Weg 23, 52080 Aachen, Germany

## ABSTRACT

Various candidate waste matrices such as nuclear waste glasses, ceramic waste forms and low-specification "storage" MOX have been considered within the current UK geological disposal program for the immobilization of separated civilian plutonium, in the case this material is declared as waste. A review and evaluation of the long-term performance of potential plutonium waste forms in a deep geological repository showed that (i) the current knowledge base on the behavior and durability of plutonium waste forms under post-closure conditions is relatively limited compared to HLW-glasses from reprocessing and spent nuclear fuels, and (ii) the relevant processes and factors that govern plutonium waste form corrosion, radionuclide release and total systems behavior in the repository environment are not yet fully understood in detail on a molecular level. Bounding values for the corrosion rates of potential plutonium waste forms under repository conditions were derived from available experimental data and analogue evidence, taking into account that the current UK disposal program is in a generic stage, i.e. no preferred host rock type or disposal concept has yet been selected. The derived expected corrosion rates for potential plutonium waste forms under conditions relevant for a UK geological disposal facility are in the range of $10^{-4}$ to $10^{-2}$ g m$^{-2}$ d$^{-1}$ and $10^{-5}$ to $10^{-4}$ g m$^{-2}$ d$^{-1}$ for borosilicate glasses, and generic ceramic waste forms, respectively, and ~$5 \cdot 10^{-6}$ g m$^{-2}$ d$^{-1}$ for storage MOX. More realistic assessments of the long-term behavior of the waste forms under post-closure conditions would require additional systematic studies regarding the corrosion and leaching behavior under more realistic post-closure conditions, to explore the safety margins of the various potential waste forms and to build confidence in long-term safety assessments for geological disposal.

## INTRODUCTION

During the operation of nuclear reactors, about 1% of the uranium contained within the fuel is converted into plutonium and minor actinides, such as neptunium, americium, and curium, by capture of neutrons. After unloading the spent fuel from a reactor, the plutonium can be recovered during reprocessing. At present, the global inventory of separated plutonium from civilian reprocessing programs is estimated at about 250 t, with a large stockpile accumulated in the UK (~92 t), which is stored mainly as $PuO_2$ [1, 2]. The current preferred policy in the UK regarding the long-term management of separated civil plutonium is its reuse as MOX fuel, although a fraction of the UK plutonium inventory (i.e. some metric tons) is likely to be destined for geological disposal and consideration of disposal options will continue [3]. Currently, a facility is constructed in the UK to immobilize some 50 to 250 kg of impure separated plutonium residues into a ceramic waste form produced by hot isostatic pressing of a mixture of powdered plutonium oxide with calcium, zirconium and titanium oxide powders for 8 to 9 hours [4].

The understanding of the corrosion and leaching behavior of radioactive wastes under disposal conditions forms an essential prerequisite for long-term safety assessments of deep geological repositories. Throughout the last decades, a number of publications have reviewed and compared potential waste forms for plutonium disposal, covering topics such as waste loadings, waste form processing and manufacture, waste form stability and mechanical behavior, and waste form durability [e.g. 5-10]. However, the performance of the waste forms under post-closure conditions in a geological repository has not been dealt with in detail in previous studies and reviews, and was considered as a knowledge gap by relevant UK institutions. Therefore, in the context of the current UK program for the implementation of geological disposal, we performed a review on the long-term performance of potential plutonium waste forms in the repository environment on behalf of the Radioactive Waste Management Directorate of the UK Nuclear Decommissioning Authority (NDA RWMD). Focal points of this study included the long-term durability of the waste forms and their corrosion behavior under disposal conditions, to improve the knowledge base about waste form performance in a geological repository and to inform on future management strategies with respect to plutonium disposal. The key findings of the detailed review, based on an extensive literature survey (cf. [11] and references therein) and relevant to plutonium waste form performance in general, are summarized in the following sections.

The rather complex mechanisms of aqueous corrosion and radionuclide release differ between waste form types such as glasses or ceramic matrices, and may vary over time with evolving near-field conditions in a geological repository. To support a first indicative safety assessment, a pragmatic approach was developed to derive corrosion rates for generic plutonium waste form types for a range of long-term post-closure conditions relevant for a geological repository in the UK [12]. This approach based on scientific evidence and available experimental data was taken from a performance assessment perspective to provide the waste management organization with an impression on the long-term durability of and the radionuclide release from different waste forms for various post-closure scenarios. For details regarding the collated knowledge base on plutonium waste form behavior in the repository environment, the approaches and the reasoning behind the derivation of corrosion rates, and the compiled data base on waste form corrosion experiments, the reader is referred to [11] and [12].

## WASTE FORMS FOR PLUTONIUM DISPOSAL

A vast variety of potential waste forms for the immobilization of plutonium and the minor actinides have been proposed and explored throughout the last decades (cf. Table I). The focus of these investigations was mainly placed on various types of nuclear waste glasses (e.g. [6, 9, 13-15]) and numerous polyphase and single-phase crystalline ceramic matrices (e.g. [8, 16-20]). Alternative immobilization routes, such as the storage MOX concept (i.e. disposal of unirradiated MOX fuel rods [21, 22]) or the usage of polymer or cement encapsulants for plutonium immobilization have received considerably less attention.

Selection criteria for suitable plutonium waste forms comprise the achievable waste loading, the capability to incorporate neutron absorbers for criticality control, and especially the long-term durability against aqueous corrosion and the resistance against deleterious effects of the intense self-irradiation on physical and chemical properties (cf. Table II).

**Table I.** Potential waste forms for plutonium disposal.

| Waste form type | Examples |
|---|---|
| Glass waste forms | Borosilicate glasses:<br>- lanthanide borosilicate glass<br>- lead borosilicate glass<br>- calcium borosilicate glass<br>Phosphate glasses:<br>- iron phosphate glass<br>- aluminum phosphate glass<br>Alkali tin silicate glass |
| Ceramic waste forms | Polyphase ceramics:<br>- SYNROC (zirconolite, perovskite, hollandite, ... )<br>Single-phase ceramics:<br>- pyrochlore $A_2B_2O_7$ (e.g. $Gd_2Ti_2O_7$, $La_2Zr_2O_7$)<br>- zirconolite $CaZrTi_2O_7$<br>- monazite (REE, Th)$PO_4$<br>- zircon $ZrSiO_4$ |
| Storage MOX | MOX fuel not destined for reactor usage (fabricated by established technology with reduced technological specification) |
| Encapsulants | Cements<br>Polymers |

**Table II.** Important issues for plutonium waste forms.

| General issues | Issues concerning long-term behavior |
|---|---|
| - waste loading<br>- criticality control<br>- fabrication route / technological maturity<br>- safeguards | - aqueous durability<br>- radiation stability<br>- thermal stability<br>- He build up / mechanical integrity |

## WASTE FORM BEHAVIOR IN THE REPOSITORY ENVIRONMENT

The potential waste matrices considered by the NDA in the context of a potential future geological disposal of (parts) of the separated stocks of civilian plutonium in the UK, and which are discussed in the following sections comprise glasses and ceramic waste forms as well as storage MOX [23].

### Aqueous durability

In general, the knowledge base on the durability and leaching resistance of plutonium waste forms in aqueous environments is highly variable. Investigations performed during the last decade were focused mainly on (single-phase) ceramics and (lanthanide) borosilicate glasses, whereas information regarding the performance of waste forms such as storage MOX is generally sparse. To a large extent, the durability of plutonium waste forms and radionuclide leaching rates have been determined with (short-term) static test methods (e.g. MCC, PCT) often using deionised water under drifting pH-conditions. Only few data on waste form corrosion behavior have been obtained in long-term tests under dynamic conditions in flow-through or column tests. Moreover, in most investigations surrogates such as cerium, hafnium or neodymium were used instead of plutonium. Thus, experimental data on the durability of plutonium waste forms under repository-relevant conditions are generally limited or in many cases absent (cf. [24, 25]), since (i) most tests are closed-system tests, (ii) the experimental conditions employed (e.g. regarding solution chemistry and redox conditions) are rather different compared to the conditions in a geological repository in the long-term, and (iii) none of the surrogate elements can mimic all facets of the geochemical behavior of plutonium. In general, the corrosion/leaching rates of ceramic waste forms such as pyrochlore, zirconolite, or monazite, as well as storage MOX are significantly lower compared to nuclear waste glasses (cf. [19]).

The effect of internal alpha decay on the durability of plutonium waste forms has been raised as a concern especially for ceramic waste forms. Based on recent investigations of nuclear waste glasses it is generally thought that their chemical durability is not significantly impaired due to self-irradiation from incorporated actinides, due to the relatively small effects on the glass structure and volume, or on the amount of stored energy (cf. [9, 13, 19, 26]). In contrast, the response of the various ceramic waste forms to self-irradiation and the consequences on their leaching resistance are rather diverse (e.g. [17-19, 27]). Some waste matrices based on zircon or zirconolite can become completely amorphous ("metamict") due to internal alpha decay, whereas waste forms based on monazite and zirconia-based pyrochlores do not undergo a radiation-induced transformation to the amorphous state. For some ceramic matrices such as zircon an increase of the dissolution rates after radiation induced amorphization was observed in static tests (e.g. [27]). In contrast, some completely metamict zirconolites have been found to exhibit a high corrosion and leaching resistance (e.g. [18]).

In general, the relevant processes that govern the corrosion of and the radionuclide release from plutonium waste forms and the systems behavior under repository conditions (i.e. low temperature, near-neutral to alkaline pH, reducing conditions) are not yet fully understood in detail on a molecular level, taken into account the large number of factors that can affect the waste form behavior (nature of the matrix and its (crystallographic) structure, chemical composition within a specifically chosen system, waste loading, tolerance to internal radiation and mechanical stress, as well as – especially for ceramic waste forms - crystallite size, fabrication route, and the amount and nature of impurities). Compared to the efforts made especially during the last decade to understand the corrosion behavior of HLW-glasses from reprocessing and/or spent fuel under repository conditions, information with respect to the performance of alternative plutonium waste forms is rather limited.

## Derivation of corrosion rates

One important component of a safety assessment for a geological disposal facility for radioactive waste is to demonstrate an understanding of the corrosion behavior of the disposed waste matrices under repository conditions and the consequent release of radionuclides that can migrate into the near- and far-field. Therefore in the frame of this project "bounding values" for corrosion rates of generic candidate waste form types for plutonium under UK disposal conditions were derived for NDA RWMD that can be used in Post-Closure Safety Assessments. The derived rates were based on available experimental data and analogue evidence from other nuclear waste forms using geochemical reasoning. Although the release of plutonium from the waste matrices into the near-field may be controlled by plutonium solubility, waste form corrosion rates and associated radionuclide rates under disposal conditions can provide valuable insight into safety relevant near-field processes, for example, if colloidal transport of plutonium may be an issue, and can help to explore safety margins of waste forms that exhibit high corrosion resistance. Moreover, the potential for post-closure criticality events due to the accumulation of fissile material leached from the waste form or due to a preferential leaching of neutron poisons is also directly related to the corrosion behavior of the waste matrix under disposal conditions. In addition, waste form durability indicators can be essential for the evaluation of disturbed repository evolution scenarios (e.g. human intrusion, tectonic activity/fracturing). The generic candidate waste form types for plutonium considered in this study comprised nuclear waste glasses (i.e. borosilicate and phosphate glasses) and generic ceramic waste forms, as well as low-specification (storage) MOX.

The current UK disposal program is in a generic stage, i.e. no preferred host rock type or disposal concept has yet been selected. NDA RWMD has identified three generic rock types as potential hosts for a geological disposal facility and developed illustrative repository designs (cf. Table III). Due to the site-generic character of the assessments, a broad range of possible environmental conditions in the repository near-field, including effects of an alkaline plume potentially arising from a co-located cementitious LILW-repository module were considered.

**Table III.** Illustrative scenarios for geological disposal in the UK (after [28]).

| Generic host rock type | Description |
|---|---|
| Higher strength crystalline rocks | - potential host lithologies: granite, gneiss<br>- illustrative disposal concept: SKB KBS-3V<br>- EBS: copper canister, bentonite buffer |
| Lower strength sedimentary rocks | - potential host lithologies: clay, clay rocks, mudrocks<br>- illustrative disposal concept: NAGRA Opalinus Clay<br>- EBS: carbon steel canister, bentonite buffer |
| Evaporites | - potential host lithologies: bedded rock salt, anhydrite<br>- illustrative disposal concept: DBE drift concept<br>- EBS: steel canister, crushed salt backfill |

Different conceptual approaches for the assessment of the corrosion behavior were pursued for the evaluation of the long-term behavior in the repository environment, depending on waste form type and the availability of relevant data and information [12]. The information basis regarding waste form durability and leaching resistance was found to be rather diverse for the different generic matrices. Information on the long-term performance of plutonium-bearing nuclear waste glasses is rather limited to date. However, considerable knowledge exists from laboratory and in-situ studies about borosilicate-based glasses for the immobilization of reprocessing wastes and their corrosion behavior (explored and modelled, e.g., in the frame of EC funded research projects such as GLAMOR, GLASTAB, CORALUS, and NF-PRO) that was referred to for the evaluation of the durability of the generic plutonium waste glasses. Corrosion rates of the generic ceramic waste forms were derived based mainly on available experimental data for various single- and polyphase matrices. However, information with respect to the performance of ceramic waste forms in the repository environment is rather scarce, and a systematic approach to an understanding of the aqueous durability of the various ceramic matrices (regarding, e.g. the dissolution behavior as function of the crystallographic structure, chemical composition, lattice substitutions, radiation damage, etc.) is still lacking. Experimental investigations on the dissolution and long-term performance of storage MOX and/or calcined $PuO_2$ are also rather limited to date. Thus the assessments and the derivation of the corrosion rates of this waste form type were based in particular on experimental and modelling studies related to the matrix dissolution of spent nuclear fuels (i.e. UOX and MOX), and on the understanding of relevant processes affecting their long-term behavior in a repository, which has been significantly expanded throughout the last decades (e.g. in EC-funded programs such as SFS, NF-PRO, and MICADO). The corrosion rates derived in [12] for the different (generic) plutonium waste forms under conditions relevant for a UK geological disposal facility are summarized in Table IV. The derived distributions provide a 'non-conservative lower bound', a 'conservative upper bound' (i.e. a value not exceeded even under adverse conditions) and an 'expected range' (i.e. a best current estimate) for the corrosion rates under post-closure conditions.

**Table IV.** Derived corrosion rates for potential plutonium waste forms under conditions relevant for a UK geological disposal facility [12].

| Waste form | lower bound [g m$^{-2}$ d$^{-1}$] | expected range [g m$^{-2}$ d$^{-1}$] | upper bound [g m$^{-2}$ d$^{-1}$] |
|---|---|---|---|
| Borosilicate glass | $<10^{-4}$ | $10^{-4} \ldots 10^{-2}$ | 100 |
| Phosphate glass | $<10^{-5}$ | $10^{-5} \ldots 10^{-2}$ | 10 |
| Ceramic waste forms | $10^{-7}$ | $10^{-5} \ldots 10^{-4}$ | 0.5 |
| Storage MOX | $10^{-7}$ | $\sim 5 \cdot 10^{-6}$ | $10^{-2}$ |

## CONCLUSIONS

The review and evaluation of the long-term performance of potential plutonium waste forms in a deep geological repository showed that (i) the current knowledge base on the behavior and durability of plutonium waste forms under post-closure conditions is relatively limited compared to HLW-glasses from reprocessing and spent nuclear fuels, and (ii) the relevant processes and factors (incl. their permutations) that govern plutonium waste form corrosion, radionuclide release and total systems behavior in the repository environment are not yet fully understood in detail on a molecular level. Bounding values for the corrosion rates of potential plutonium waste forms under repository conditions can be derived from available experimental data and analogue evidence. However, more realistic assessments of waste form durability and radionuclide release behavior in the repository environment would require additional systematic studies regarding (i) the corrosion and leaching behavior of the candidate waste matrices under conditions more realistic for the repository environment, and (ii) the characterization of secondary phases formed during the corrosion processes, to explore the safety margins of the various potential waste forms and to increase the confidence in safety assessments. A large body of work on radiation damage especially on crystalline ceramic waste forms has been performed throughout the last decades, providing in-depth insight into damage mechanisms and radiation tolerance of different waste forms (e.g. [16-19, 26, 27]). However, the interplay between self-irradiation, radiation damage, radiation-induced phase transitions as well as annealing processes and the corrosion/leaching behavior under post-closure conditions requires some more attention to support long-term safety assessments for geological disposal of the candidate plutonium waste forms.

## ACKNOWLEDGMENTS

This work was performed for and funded by the UK Nuclear Decommissioning Authority – Radioactive Waste Management Directorate. The manuscript benefited from the comments of an anonymous reviewer.

# REFERENCES

1. International Panel on Fissile Materials, *Global fissile material report 2011*, (Princeton, NJ, 2012) 42 p.
2. Health & Safety Executive, *Annual figures for holdings of civil unirradiated plutonium*, www.hse.gov.uk/nuclear/safeguards/civilplut12.htm (2013).
3. Department of Energy & Climate Change, *Management of the UK's Plutonium Stocks*, (London, 2011) 33p.
4. T. Clements, E. Lyman and F. von Hippel, Arms Control Today **43**, July/August (2013).
5. W. Lutze and R.C. Ewing, in *Radioactive waste forms for the future*, edited by W. Lutze and R.C. Ewing (Elsevier, Berlin, 1988) p. 699.
6. I.W. Donald, B.L. Metcalfe and R.N.J. Taylor, *J. Mater. Sci.* **32**, 5851 (1997).
7. A. Macfarlane, Science & Global Security **7**, 271 (1998).
8. R.C. Ewing, *Proc. Nat. Acad. Sci.* **96**, 3432, (1999).
9. S. Stefanovsky, S.V. Yudintsev, R. Gieré and G.R. Lumpkin, *Geological Society of London Special Publications* **236**, 37 (2004).
10. I.W. Donald, *Waste immobilization in glass and ceramic based hosts: Radioactive, toxic and hazardous wastes* (Wiley, Chichester, 2010), 507 p.
11. G. Deissmann, S. Neumeier, G. Modolo and D. Bosbach, *Review of the durability of potential plutonium wasteforms under conditions relevant to geological disposal*, (Aachen, 2011) 85 p.
12. G. Deissmann, S. Neumeier, F. Brandt, G. Modolo and D. Bosbach, *Elicitation of dissolution rate data for potential wasteform types for plutonium* (Aachen, 2011) 98 p.
13. D.M. Wellman, J.P. Icenhower and W.J. Weber, *J. Nucl. Mater.* **340**, 149 (2005).
14. M.T. Harrison and C.R. Scales, *Mater. Res. Soc. Symp. Proc.* **1107**, 405 (2008).
15. M.T. Harrison, C.R. Scales, P.A. Bingham and R.J. Hand, *Mater. Res. Soc. Symp. Proc.* **985**, 0985-NN04-03 (2007).
16. R.C. Ewing, *Prog. Nucl. Energy* **49**, 635 (2007).
17. R.C. Ewing, *Min. Mag.* **75**, 2359 (2011).
18. G.R. Lumpkin, *Elements* **2**, 365 (2006).
19. W.J. Weber, A. Navrotsky, S. Stefanovsky, E.R. Vance and E. Vernaz, *Mater. Res. Soc. Bull.* **34**, 46 (2009).
20. B.E. Burakov, M.I. Ojovan and W.E. Lee, *Crystalline materials for actinide immobilisation* (Imperial College Press, London, 2011) 197 p.
21. J. Kang, F.N. von Hippel, A. Macfarlane and R. Nelson, *Science & Global Security* **10**, 85 (2002).
22. A.M. Macfarlane, *Prog. Nucl. Energy* **49**, 644 (2007).
23. Nuclear Decommissioning Authority, *NDA Plutonium topic strategy: Credible options technical analysis* (Doc No: SAF/081208/006.2, 2009) 142 p.
24. E.M. Pierce, B.P. McGrail, P.F. Martin, J. Marra, B.W. Arey and K.N. Geiszler, *Appl. Geochem.* **22**, 1841 (2007).
25. G. Deissmann, S. Neumeier, G. Modolo and D. Bosbach, *Min. Mag.* **76**, 2911 (2012).
26. W.J. Weber, R.C. Ewing, C.A Angell, G.W. Arnold, A.N. Cormack, J.M. Delaye, D.L. Griscom, L.W. Hobbs, A. Navrotsky, D.L. Price, A.M. Stoneham and M.C. Weinberg, *J. Mat. Res.* **12**, p. 1946 (1997).

27.  W.J. Weber and R.C. Ewing, in *Uranium – Cradle to grave*, edited by P.C. Burns and G.E. Sigmon (Mineralogical Association of Canada, Short Course Series **43**, 2013), p. 317.

28.  Nuclear Decommissioning Authority, *Geological disposal: Generic disposal facility designs* (Report NDA/RWMD/048, 2010) 129 p.

Mater. Res. Soc. Symp. Proc. Vol. 1665 © 2014 Materials Research Society
DOI: 10.1557/opl.2014.625

# Investigation of mineralogy, porosity and pore structure of Olkiluoto bedrock

Juuso Sammaljärvi[1], Antero Lindberg[2], Jussi Ikonen[1], Mikko Voutilainen[1], Marja Siitari-Kauppi[1], Lasse Koskinen[3]

[1]Laboratory of Radiochemistry, University of Helsinki, Finland.
[2]Geological Survey of Finland, Finland
[3]Posiva Oy, Finland

## ABSTRACT

Spent nuclear fuel from TVO's (Teollisuuden Voima Oy) and Fortum's nuclear power plants will be deposited deep in the crystalline bedrock in Olkiluoto, Western Finland. The bedrock needs to be well characterized to assess the risks inherent to the waste disposal at the site. If radionuclides (RN) are transported, it happens via water conducting fractures. Retardation may occur either by diffusion into stagnant pore water or by immobilization on mineral surfaces of the rock matrix. RN's retardation from flowing water is linked to parameters defining porosity and microscopic rock pore structure, such as pore size distribution, connectivity, tortuosity and constrictivity, and by the mineralogy and chemical nature of the minerals and charge of the pore surfaces.
In this work, centimeter scale rock cores from Olkiluoto were investigated. The work is part of the in situ project REPRO (Experiments to investigate Rock Matrix Retention Properties) where the diffusion and sorption of RN are studied experimentally. Porosity and pore structures were characterized with the PMMA autoradiography method and polarized microscopy, which was used also to ascertain the mineralogy of the samples.
The results show that the rock from the REPRO site has low porosity with a mean value of 0.5% and a range of 0.1-1.5%. Rock heterogeneity explains the variation of porosity values. Correlation between the porosity and the mineralogy was found. Areas of high porosity correspond to areas of altered minerals, such as cordierite, biotite and plagioclase, which cover spatially between 10 and 20% of the rock volume

## INTRODUCTION

Understanding the nature of the RN migration through geological formations is essential in any assessment of the confining properties of the barriers [1]. RN migration under natural long-term conditions is a complex process controlled by many parameters. Physical parameters of the rock matrix such as porosity, hydraulic conductivity and diffusivity are used to describe transport properties of materials. RN migration is dependent on the physical parameters and chemical nature and charge of the mineral surfaces of the rock.
The interface between aqueous solutions and mineral phases is defined by the relative distribution of solution-filled pores and minerals. Porosity in rocks in the widest sense is understood to mean the entire solution-filled void space between and within the individual mineral grains [2]. Porosity in crystalline rocks can be mainly divided into two types: primary, acquired during magmatic crystallisation, and secondary, acquired since crystallisation through rock alteration. Minerals in the rock are changed chemically with time, first by retrograde metamorphism and then by water–rock interactions, resulting alteration and weathering [2]. The study of the porosity of cm scale rock cores is the focus of this paper.

31

Porosities were studied with water gravimetry and the PMMA impregnation technique, and the results were compared. Pore structures were also characterized with the PMMA autoradiography method and polarized microscopy, which was used to ascertain the mineralogy of the samples.

## EXPERIMENTAL

### Samples

Suomen Malmi Oy (Smoy) drilled ten drill holes (ONK-PP 318-327) for the Posiva's Experiments to investigate Rock Matrix Retention Properties (REPRO) in ONKALO ( Geological repository for spent nuclear fuel) at Eurajoki [3]. The drill holes were used for geological characterization, hydrological and geophysical studies and instrumenting in research for retention of radionuclides in rock matrix. The lengths of the drill holes range from 4.90 to 21.65 metres. The drill holes are 56.5 mm in diameter. Figure 1 shows the location of two investigation niches. The REPRO niche is ONK-TKU-4219. The rock cores for porosity and mineralogy characterization were taken from the above mentioned drill holes. The bedrock in this area is granitic pegmatite and migmatitic gneiss. The samples presented in this paper represent these rock types. The first drill core to be investigated, ONK PP 318, was a granitic pegmatite and the other drill core, ONK-PP 323, was migmatitic gneiss.

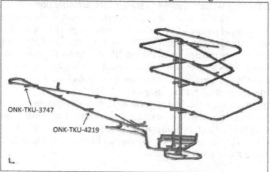

ONK-TKU-3747

ONK-TKU-4219

**Figure 1.** The location of the investigation niches at ONKALO [3]. Drill core samples used in this study were taken from niche ONK-TKU-4219.

### Water gravimetry

Water gravimetry measurements made at the Laboratory of Radiochemistry of the University of Helsinki (HYRL) were based on a technique found in the literature [4]. The samples were first dried for 5 days in an oven set to 105 ° C. Thereafter the samples were placed in a desiccator to prevent moisture condensation. The samples were then weighted and placed into plastic containers, and immersed in milliQ-water. The sample weight was then measured 2-5 times every work day for two weeks. The method is based on measuring the volume of water intruded into the pore space of the sample and comparing it to the total volume of the sample. The total volume of the sample was measured geometrically.

## [14]C PMMA impregnation technique

The polymethylmethacrylate (PMMA) method involves the impregnation of a rock sample with [14]C or [3]H labelled methylmethacrylate (MMA) in vacuum, polymerization, film and digital autoradiography, optical densitometry, and porosity calculation routines relying on digital image processing techniques [5-8]. The low molecular weight and low viscosity carrier monomer MMA, which can be fixed after impregnation by polymerization, provides direct information about the accessible pore space in rock. When applied to low porous rocks, autoradiography describes the spatial distribution of the porosity. The method integrates the microscopic scales to the core scale. The porosity distribution provided by the PMMA method is easily related to the mineralogy of the rock by digital image analysis. Total porosity is calculated by using 2D autoradiography of the sawn rock surfaces. Polymerisation of MMA was performed thermally as described in reference [9]. This work focuses on measuring the porosities and porosity patterns in different minerals of the samples. The results are presented as the porosity histograms and total porosities, which are compared to water gravimetry results.

## RESULTS

### Results of drill core ONK-PP 318

The mineralogy of the granitic pegmatite (GP) samples was measured by analyzing thin sections from the drill core ONK-PP 318. The mineralogy results show that the average composition of the GP in ONK-PP 318 is 47.6 % quartz, 31.5 % plagioclase, 16.5 % potassium feldspar, 3.4 % muscovite and 0.7% garnet. Quartz is the predominant component in this drill core with plagioclase and potassium feldspar as other major components. Muscovite and garnet are minor components. The grain size of main minerals is from a few mm to a centimeter.
The photograph of ONK-PP 318 15.62 m and the corresponding autoradiograph is shown in Figure 2. MMA has intruded into the rock matrix thoroughly; grain boundaries of medium and coarse mineral grains are filled by MMA. Intragranular porosity is found in plagioclase mineral grains. In addition there are plenty of intragranular fissures and micro fractures cutting the mineral grains. Total porosity of the sample was determined to be 0.3% by integrating the porosity values over the area of the image. The porosity distribution extends to about 2.5% but only a small portion of the porosity is in highly porous areas. There were notable empty fissures in this sample filled by MMA.

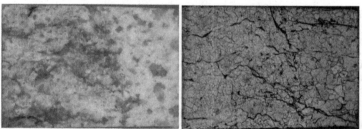

**Figure 2.** Photograph (left) of the ONK-PP 318 15.62 m and its corresponding autoradiograph (right). Darkest areas on the autoradiograph correspond to 2.5 % porosities. Dimensions of the sample: 4 cm *3 cm.

A thin section from the ONK-PP 318 15.62 m (Figure 3) shows the main minerals in GP sample and a micro fissure cutting the quartz minerals. GP is coarse grained (4-10 mm) Alteration is very weak, but marks of brittle deformation, such as broken grains and fractures cutting several minerals, are found. Grains of garnet in the thin sections are relics of assimilated gneiss. Figure 3 shows fractured quartz grains with fissures running in between. These fissures are likely filled with sericite. Plagioclase is present here in small grains.

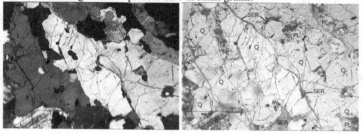

**Figure 3**. Photograph of thin section ONK-PP 318 15.62 m. Fractured quartz (Q) grains and some fissures which are filled by sericite (SER). Slightly altered plagioclase (PL). On left, crossed nicols.

### Results of drill core ONK-PP 323

The mineralogy of ONK-PP 323 was measured by taking thin sections from several parts of the core and averaging the results. The average composition of the migmatitic gneiss (MG) of the drill core ONK-PP 323 is 28.5 % biotite, 28.1 % quartz, 15.0 % potassium feldspar, 14.2 % plagioclase, 6.6 % cordierite, 4.5 % sillimanite, 2.0 % muscovite and circa 1 % other minerals. MG is a metamorphic, heterogeneous mixture of small grained mica gneiss and coarser leucosome veins of diameters varying from several millimetres up to ten centimetres. The rock samples are heterogeneous with areas of both high and low porosity.

ONK-PP 323 18.75m has clear orientation as seen in Figure 4, which is shown in the photograph of the sample and the corresponding autoradiograph. The average grain size seems to be fairly small but still visible to naked eye. From this sample the significant foliation of the minerals' texture as well as the pore structure is observed. The total porosity of this sample calculated to be 0.4 % with notable tail until about 3.0%. The grain boundaries are filled with MMA and porosities up to 3% correspond with the cordierite and sillimanite mineral grains.

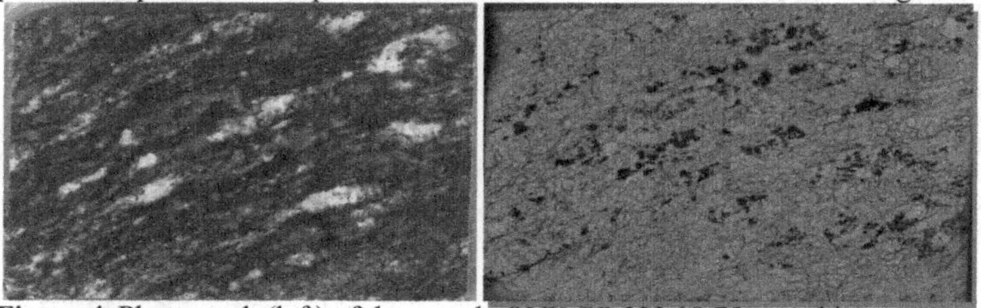

**Figure 4.** Photograph (left) of the sample ONK-PP 323 18.75 m and its corresponding autoradiograph (right) Darkest areas on the autoradiograph correspond to porosity of 5.5 %. Dimensions of the sample: 4 cm *3 cm.

Figure 5 shows a microscopic photograph of the sample 323 18.75m. The sample has large biotite grains amidst thin quartz and plagioclase veins. There are also large sillimanite veins and zircon flakes in the biotite grains.

**Figure 5.** Photograph of thin section ONK-PP 323 18.75 m . Clearly foliated mica gneiss (biotite-cordierite gneiss). Sillimanite (SIL) is coarser than usually. Biotite (BT) contains abundantly zircon (ZI) grains with dark rims. Other minerals are plagioclase (PL) and quartz (Q). Crossed nicols on left.

## DISCUSSION AND CONCLUSION

The porosity, pore structure and mineralogy of Olkiluoto drill cores were studied. The results show that the bedrock in the area has overall low porosity. The rock is heterogeneous with both areas of low porosity and high porosity. The heterogeneity of the rock samples is highlighted in the porosity curves in Figure 6. Both of the samples representing the two drill cores have fairly similar mean porosity values but the porosity distribution is wide in both cases. The width of the distributions is different in these two rock cores highlighting their different composition.

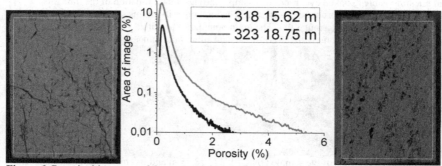

**Figure 6.** Porosity histogram of the samples ONK-PP318 15.62 m and ONK-PP323 (18.75 m) and their respective areas of interest highlighted. Total porosity of ONK-PP 318 is 0.35 % and the total porosity of ONK-PP323 is 0.40 %.

The areas of high porosity were found to mostly correspond to areas with altered minerals and areas of low porosity tend to correspond with unaltered minerals such as quartz. The results from autoradiography were compared to results obtained via water gravimetry. This comparison is presented in Figure 7.

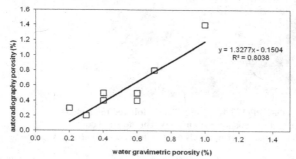

**Figure 7.** Comparison of autoradioraphy porosities and water gravimetry porosities.

From Figure 7 it can be seen that there is fair agreement between the two porosity measurement techniques. The main source of difference likely results from the error in the estimation of the sample volume in water gravimetry as well as from the uncertainties in PMMA autoradiographic technique [5,6]. These results will be used when interpreting the in situ RN diffusion and sorption experiments. In addition the REPRO project aims to compare the conditions between in situ and laboratory; for example the effect of stress relaxation which might open the pore space and increase the conductive diffusivity of the rock.

# REFERENCES

1. Möri, A., et al., 2003. The Nagra-JNC in situ study of safety relevant radionuclide retardation in fractured crystalline rock. IV: the in situ study of matrix porosity in the vicinity of a water conducting fracture. Technical Report 00–08. Nagra, Wettingen/ Switzerland

2. Norton, D., Knapp, R., 1977. Transport phenomena in hydrothermal systems: the nature of porosity. American Journal of Science 277, 913–936.

3. Toropainen, V., 2012 Core Drilling of REPRO Drillholes in ONKALO at Olkiluoto 2010-2011, May 2012 Posiva Working Report, 78

4. International Society for Rock Mechanics, Commission on standardization of laboratory and field tests,. International Journal of Rock Mechanics, Mining Science & Geomechanical Abstracts (1979) 16, 141–156

5. Hellmuth, K-H., Siitari-Kauppi, M. and Lindberg, A., (1993) Journal of Contaminant Hydrology (1993),13, 403-418

6. Hellmuth, K.-H., Lukkarinen, S. and Siitari-Kauppi, M., Isotopenpraxis. Isotopes in Environmental and Health Studies (1994), 30, 47-60

7. Hellmuth, K.-H. and Siitari-Kauppi, M., Klobes, P., Meyer, K., Goebbels, J. Applications, Phys. Chem. Earth A (1999), Vol. 24, No. 7, 569-573

8. Siitari-Kauppi, M., Flitsiyan, E.S., Klobes, P., Meyer, K. and Hellmuth, (1998). Progress in Physical Rock Matrix Characterization: Structure of the Pore Space. In: I.G. McKinley, C. McCombie (edits.), Scientific Basis for Nuclear Waste Management XXI, Mat. Res. Soc. Symp. Proc. 506, 671-678

9. Sammaljärvi, J., Jokelainen, L., Ikonen, I. and Siitari-Kauppi, M., Engineering Geology (2012) 135-136, 52-59

Mater. Res. Soc. Symp. Proc. Vol. 1665 © 2014 Materials Research Society
DOI: 10.1557/opl.2014.626

# "Relative Rates Method" for Evaluating the Effect of Potential Geological Environmental Change due to Uplift/Erosion to Radionuclide Migration of High-level Radioactive Waste

Takeshi Ebashi[1], Makoto Kawamura[1*], Manabu Inagaki[1**], Shigeru Koo[2], Masahiro Shibata[1], Toru Itazu[1***], Kunihiko Nakajima[2], Kaname Miyahara[1] and Michael J Apted[3]
[1] Japan Atomic Energy Agency, 4-49, Muramatsu, Tokai-mura, Ibaraki, 319-1184, Japan.
[2] NESI Inc., 4-33, Muramatsu, Tokai-mura, Naka-gun, Ibaraki, 319-1112, Japan.
[3] INTERA Inc., 3900, S. Wadsworth, Blvd., Suite 555, Denver, CO 80235, USA.
Current affiliation: *Mitsubishi Materials Techno Corporation, Saitama, Japan. **Nuclear Waste Management Organization of Japan, Tokyo, Japan. *** Visible Information Center, Inc. Ibaraki, Japan

## ABSTRACT

We have developed a "Relative Rates Method" to make bounding calculations regarding radionuclide migration due to uplift/erosion ("exhumation") of a HLW repository. Results show that this method can apply to a wide range of different uplift rates and erosion rates. In addition, for the long time period, it was shown that the relative difference of uplift rate / erosion rate and potential hydraulic change arising from extreme uplift/erosion could affect radionuclide release and migration, thus uplift/erosion concerns should be fed back to site selection. Our method provides a credible and defensible basis for analysis and interpretation of possible uplift/erosion impacts for future volunteer sites.

## INTRODUCTION

In Japan, the likelihood of uplift/erosion ("exhumation") on repository performance and waste isolation can be greatly reduced or excluded by careful siting. For example, areas where there is clear literature evidence of uplift amounting to greater than 300m during the last one hundred thousand years will be precluded [1]. However, the inability to completely exclude the uplift/erosion scenario would require an analysis of the consequences of such uplift/erosion scenario for long time period. Uplift and erosion could gradually cause evolution of the environmental conditions for the engineered and natural barriers when the repository nears to surface with time. For example, the change of the depth and the geography could affect groundwater flow and chemistry. Therefore, uplift and erosion scenarios must be analyzed, as no assessment cut-off times have yet been defined in Japanese regulations.

In a previous conservative study [2], it was assumed that the depth of repository to the ground surface gradually decreases and then groundwater flow and chemical condition changes at the time when the repository reaches the depth 100m. Such analyses were based on the assumption that uplift rate would be equal to erosion rate. Based on knowledge obtained from future site characterization, it is possible that uplift rate would exceed erosion rate. On this issue, Kawamura et al. [3] proposed an original method to describe the potential geological environmental changes due to uplift and erosion in a qualitative manner for a hypothetical sedimentary rock [3]. However, although such research has demonstrated that it is useful for deriving bounding evolution of possible future, it remains unclear how radionuclide migration analyses should be conducted to provide quantitative impacts.

The purpose of this paper is to illustrate "Relative Rates Method" to evaluate the effect of potential geological environmental change due to uplift/erosion on the radionuclide migration. Concretely, based on a hypothetical sedimentary rock discussed by Kawamura et al. [3], quantitative values are set for impacts on thermal-hydrological-mechanical-chemical (THMC) conditions and parameters in radionuclide migration analyses.

## APPROACH

Many parts of Japan have been subject to uplift over the last million years. The average uplift rate is generally less than 1mm / y for areas outside mountainous regions [4]. As noted previously, there can be a relative difference between uplift rate and erosion rate (generally uplift rate > erosion rate). Uplift/erosion mode including such relative difference would gradually affect the evolution of surface topography and THMC conditions and these evolutions would affect the safety function/parameters of geological disposal system. Typical cases are when uplift rate exceeds erosion rate, leading to formation of hills and mountains of the uplifted block, with attendant changes in THMC conditions and parameters at the repository depth. This means that change of THMC conditions differs depending on the relative uplift/erosion rate and original repository position. Thus, it is necessary to define a range of calculation cases for situations in which the erosion rate differs from uplift rate.

The method based on Kawamura et al. [3] are particularly useful in helping to develop a suite pattern of 'sensitivity cases' that bound the possible THMC impacts arising from different generic uplift/erosion situations in qualitative manner. In this study we have developed the "Relative Rates Method" to make bounding radionuclide migration calculation cases for each of these sensitivity cases. The method includes the following steps:

> Step 1: Establish premises on relative uplift and erosion rates
> Step 2: Set initial THMC parameters
> Step 3: Define calculation cases based on relative difference between uplift rate and erosion rate
> Step 4: Set radionuclide migration parameters
> Step 5: Calculation and evaluation

Furthermore, erosion form near the earth's surface area is complex and then some fluvial erosion models have been developed in recent year [5]. In addition, in Japan, the consistency of legislation between geological disposal (deeper-than- 300m initial depth) and subsurface disposal (deeper-than- 50m initial depth) is being reviewed by the regulator. Thus, this study covers the range from an initial depth for a HLW repository in Japan (>300m) to a depth of 50m. Application of this procedure is illustrated in the case study of the next section.

## RESULT AND DISCUSSION

### Step 1: Establish premises

To illustrate our method, this study assumes a simple, layered sedimentary rock characterized by information that could be obtained from a previous literature survey and eventually supplemented by borehole investigations. Fig.1 (a) shows a schematic diagram of the target domain. Geological evidence supporting such target domain is discussed in detail by Kawamura et al. [3].

We assume the following regional features for the assumed sedimentary host rock: located in a coastal region of Japan far away from the volcano, inclusion of surface geography (several generic types such as hill, plain, coast, etc) features within tens of kilometers of the repository site in order to include topography aspects into the analysis, and absence of fault movements, folding during the uplift/erosion period. There are two types of erosion form - a "linear erosion" and "regional erosion", and then this study assumes the regional erosion because regional erosion can be treated as a cluster of linear erosion.

As an initial thermal condition, the thermal gradient is 2°C per 100m [4, 6, 7] and then temperature is 25°C (H-1, P-1, C-1 at -500m) / 17°C (H-2, P-2, C-2 at -100m) for surface 15°C as a benchmark. As a hydrological initial condition, there are two types of hydraulic regions (deeper than 100m and shallower than 100m depths). It is assumed that no big fault cuts through the repository. As a mechanical initial condition, due to diagenetic effect of ground temperature and consolidation, the mechanical strength and durability are assumed to be increasing with depth. As an initial chemical condition, groundwater of the freshwater type is assumed to be found beneath hill and plain topography. Reducing freshwater shallower than 100m is dominantly Na-Ca-HCO₃, same as the previous study's case [2]. Meanwhile, reducing freshwater deeper than 100m is dominantly Na- HCO₃. For a coastal site, the area shallower than 100m is dominantly reducing freshwater (Na-Ca-HCO₃/ Na- HCO), the area deeper than 100m is dominantly reducing saline water (Na-Cl). The mineral composition of the assumed sedimentary formation is assumed constant.

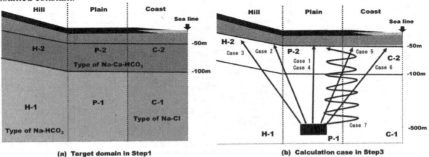

(a) Target domain in Step1    (b) Calculation case in Step3

**Figure 1.** Schematic diagram of the target domain

**Table I.** THMC parameters for each block

| Geological environment condition | | | Unit | Block H-1 | H-2 | P-1 | P-2 | C-1 | C-2 | Ref. |
|---|---|---|---|---|---|---|---|---|---|---|
| T | Rock temperature | | degree C | 25 | 17 | 25 | 17 | 25 | 17 | [4, 6, 7] |
| | Temperature gradient | | degree C/100m | 2 | 2 | 2 | 2 | 2 | 2 | |
| H | Hydraulic conductivity | | m/s | $1\times10^{-7}$ | $1\times10^{-5}$ | $1\times10^{-6}$ | $1\times10^{-6}$ | $1\times10^{-9}$ | $1\times10^{-7}$ | [4, 6] |
| | Hydraulic gradient | | - | 0.035 | 0.138 | 0.016 | 0.100 | 0.008 | 0.067 | |
| M | Porosity | | % | 35 | 35 | 35 | 35 | 35 | 35 | [6, 8] |
| | Rock density | | g/cm³ | 1.7 | 1.7 | 1.7 | 1.7 | 1.7 | 1.7 | |
| C | Water quality | | pH | - | 7.8~8.5 | 7.5~8.5 | 7.8~8.5 | 7.5~8.5 | 7.8~8.5 | 7.5~8.5 | [4, 6, 8, 9] |
| | | | Eh | mV | -340~-280 | -300~-200 | -340~-280 | -300~-200 | -340~-280 | -300~-200 | |
| | | Cation | [Na⁺] | | 2300 | 2300 | 2300 | 2300 | 6400 | 2300 | |
| | | | [K⁺] | | 20 | 20 | 20 | 20 | 70 | 20 | |
| | | | [Ca²⁺] | mg/l | 30 | 30 | 30 | 30 | 85 | 30 | |
| | | | [Mg²⁺] | | 20 | 20 | 20 | 20 | 200 | 20 | |
| | | Anion | [Cl⁻] | | 2100 | 2100 | 2100 | 2100 | 9700 | 2100 | |
| | | | [HCO₃⁻] | mg/l | 1600 | 1600 | 1600 | 1600 | 1100 | 1600 | |
| | | | [CO₃²⁻] | | 10 | 10 | 10 | 10 | 500 | 10 | |
| | | | [SO₄²⁻] | | 350 | 350 | 350 | 350 | 1500 | 350 | |
| | Mineral composition | Clay mineral | | % | 1~18 | 1~18 | 1~18 | 1~18 | 1~18 | 1~18 | |
| | | Mica | | | ~1 | ~1 | ~1 | ~1 | ~1 | ~1 | |
| | | Chlorite | | | n.d. | n.d. | n.d. | n.d. | n.d. | n.d. | |
| | | Calcite | | | 1~5 | ~1 | 1~5 | ~1 | 1~5 | ~1 | |
| | | Pyrite | | | 2~3 | 2~3 | 2~3 | 2~3 | 2~3 | 2~3 | |

## Step 2: Set Initial THMC parameters

For each block, quantitative THMC parameters (see Table. 1) are defined based on previous geological environmental investigations in Japan. Bedrock temperature and temperature gradient were set based on borehole investigation in regions of far from volcanoes [4, 6-7]. Hydraulic conductivity and hydraulic gradient of bedrock, as well as rock porosity, rock density and mineral composition of rocks, were representatively set based on results obtained from previous borehole investigations for Neogene and Quaternary sedimentary layers, for example, Soya (northern Hokkaido), Boso (middle Honshu) or Akita-Niigata (north-eastern Honshu) regions [4, 6, 8]. In addition, groundwater compositions were set based on the result obtained from previous borehole investigation for Neogene and Quaternary sedimentary layers [4, 6, 8], and investigation result in seabed coalfield and tunnel constructions in Japan [9].

**Table II.** Calculation cases and Switch time of block for each case

| Case No | Evolution of block | Uplift rate (mm/y) | Erosion rate (mm/y)*1 | Sea-level change | Evolution of the block | Distance from depature(m)*2 | Surface altitude(m)*3 | Depth(m)*4 | Switch Time (y) |
|---|---|---|---|---|---|---|---|---|---|
| 1 | Plain -> Plain (P-1 -> P-2) | 1 | 1 | No consideration | Arr.(Depth-50m) | 450 | 0 | -50 | 450,000 |
| | | | | | P-1 -> P-2 | 400 | 0 | -100 | 400,000 |
| | | | | | Dept.(P-1) | 0 | 0 | -500 | 0 |
| 2 | Plain -> Hill (P-1 -> H-2) | 1 | 0.8 | No consideration | Arr.(Depth-50m) | 563 | 113 | -50 | 562,500 |
| | | | | | P-1 -> H-2 | 500 | 100 | -100 | 500,000 |
| | | | | | Dept.(P-1) | 0 | 0 | -500 | 0 |
| 3 | Plain -> Hill (P-1 -> H-1 -> H-2) | 1 | 0.1 | No consideration | Arr.(Depth-50m) | 4,500 | 4,050 | -50 | 4,500,000 |
| | | | | | H-1 -> H-2 | 4,000 | 3,600 | -100 | 4,000,000 |
| | | | | | P-1 -> H-1 | 111 | 100 | -489 | 111,111 |
| | | | | | Dept.(P-1) | 0 | 0 | -500 | 0 |
| 4 | Plain -> Plain (P-1 -> P-2) | 0.1 | 0.1 | No consideration | Arr.(Depth-50m) | 450 | 0 | -50 | 4,500,000 |
| | | | | | P-1 -> P-2 | 400 | 0 | -100 | 4,000,000 |
| | | | | | Dept.(P-1) | 0 | 0 | -500 | 0 |
| 5 | Plain -> Coast (P-1 -> C-2) | 1 | 1 | Only consideration of Transgression 0.25mm/y | Arr.(Depth-50m) | 450 | -50 | -50 | 450,000 |
| | | | | | C-1 -> C-2 | 400 | -100 | -100 | 400,000 |
| | | | | | Dept.(P-1) | 0 | 0 | -500 | 0 |
| 6 | Plain -> Coast (P-1 -> C-1 -> C-2) | 1 | 1 | Only consideration of Transgression 0.27mm/y | Arr.(Depth-50m) | 450 | -122 | -50 | 450,000 |
| | | | | | C-1 -> C-2 | 400 | -108 | -100 | 400,000 |
| | | | | | P-1 -> C-1 | 370 | -100 | -130 | 370,370 |
| | | | | | Dept.(P-1) | 0 | 0 | -500 | 0 |
| 7 | Plain -> Coast (P-1 -> C-1 -> C-2) | 1 | 1 | Transgression: 80,000 y Regression: 20,000 y | Arr.(Depth-50m) C-1 -> P-2 | 450 | 0 | -50 | 450,000 |
| | | | | | ··· | 400 | 0 | -100 | 400,000 |
| | | | | | C-1 -> P-1 | 100 | 0 | -400 | 100,000 |
| | | | | | P-1 -> C-1 | 80 | 0 | -420 | 80,000 |
| | | | | | Dept.(P-1) | 0 | 0 | -500 | 0 |

*1 Depth of repository reduces due to the erosion rate.
*2 The value for only effect of the uplift
*3 The value for the difference between uplift rate and erosion rate
*4 The value for current surface altitude as a benchmark

## Step 3: Define calculation cases based on relative difference between uplift rate and erosion rate

Calculation cases are defined based on the relative difference in uplift and erosion rates (Table.2). In addition, common to all analysis cases, the initial value of the repository depth is -500m. According to previous study [4], the amount of uplift rate estimated for the next 100,000 years is less than 100m – the range roughly from 1mm/y to 0.1mm/y in many areas, except in some high mountains and the tips of peninsulas, in Japan [4, 10]. Thus, this study has some cases in point, namely, there are the case of starting point for uplift rate 1mm/y and 0.1mm/y (Table.2 and Fig.1 (b)). Case 1 and Case 4 assume that uplift rate is equal to erosion rate and such rates are 1mm/y and 0.1mm/y. The depth of the uplifting repository simply reduces due to uplift rate (= erosion rate). Case 2 and Case 3 assume that uplift rate is greater than erosion rate. In these cases, altitudes will become increasingly high for high uplift rate. Therefore, uplift rate was set at the 1mm/y maximum. Here, we define surface environment would be changed in the Hill from the Plain, when the altitude is higher 100m from the initial value (not necessarily 0m) by uplift. To illustrate sensitivity, 0.8mm/y and 0.1 mm/y are assumed as erosion rates for Case 2 and Case 3. Additionally, Case 5 and Case 6 in particular focus on transgression of the sea-level change caused by climate changes, for example, for consideration of impact on the decline in land coast region in Fig. 1. In these cases, altitudes above sea level will become progressively lower for transgression rate. Here, we define when the altitude above sea level is lower 100m from the initial value by transgression, surface environment would be changed in the Plain from the Coast, in contrast to the Case 2 and Case 3. This sensitivity study assumes that uplift rate and erosion rate are both 1mm/y, but transgression rates are set up 0.25mm/y (Case 5) and 0.27mm/y (Case 6). Moreover, Case 7 focus on the sea-level change by cyclic climate changes – glacial (regression) and interglacial (transgression) period. In this case, uplift rate and erosion rate are same as Case 5 and Case 6. However, transgression and regression period are set up 80,000 and 20,000 years, that is 100,000 years in one climatic cycle. Surface environment instantly changes from Plain to Coast at the end of transgression, in contrast, it instantly changes to reverse at the end of regression. However, these switching are assumed to be continuously.

## Step 4: Set radionuclide migration parameters

In this study, the previously established one-dimensional porous medium model for the engineered barrier system and rock are used [2, 11]. Also, the governing equations and initial/boundary conditions are based on the previous study [2]. Based on such models, the effect of uplift/erosion is treated as change of parameters. Cs-135 was selected as an illustrative, long-lived nuclide, taken into account as a dominant radionuclide in a previous study [2]. Table 3 shows the result of radionuclide migration parameters for each block. Based on the THMC parameters from Table.1, radionuclide migration parameters for each block were defined as follows

- Distribution coefficient (Kd) and effective diffusion coefficient (De) are selected by using the sorption/diffusion database [12] considering the quality of data and similarity of experimental conditions with the given conditions.
- Glass dissolution rate and De are considered as temperature dependent. The glass dissolution rate and the De in buffer material are set for 40°C for the initial 10,000 years, considering the results of thermal analysis of repository [13], and the temperature calculated from thermal gradient are used for those parameters after 10,000 years and De in rock.

Meanwhile, previous study [2] has shown that glass dissolution rate and De tend to be low sensitive to peak of release rate. However, this study is intended to illustrate the procedure to handle the propagation of THMC conditions to the radionuclide migration parameters, therefore, time-changes are set even for the low sensitivity parameters. An assessment point 100m away from the edge of the repository is set. As noted previously, this study does not assume the region of a depth shallower than 50m. Thus, biosphere is not included and then the unit is the flux (Bq /y per waste package).

**Table III.** Radionuclide migration parameters for each block

| Nuclide migration parameters | Unit | Hill | | Plain | | Coast | |
|---|---|---|---|---|---|---|---|
| | | H-1 | H-2 | P-1 | P-2 | C-1 | C-2 |
| Near field | | | | | | | |
| Kd for buffer (Cs) | m³/kg | 0.001 | 0.001 | 0.001 | 0.001 | 0.001 | 0.001 |
| Diffusion coefficient for buffer (Cs) | m²/s | $2.7 \times 10^{-10}$ | $2.3 \times 10^{-10}$ | $2.0 \times 10^{-10}$ | $2.5 \times 10^{-10}$ | $2.0 \times 10^{-10}$ | $2.3 \times 10^{-10}$ |
| Glass dissolution rate | g/m²/day | $8.7 \times 10^{-5}$ | $4.6 \times 10^{-5}$ | $8.7 \times 10^{-5}$ | $6.4 \times 10^{-5}$ | $8.7 \times 10^{-5}$ | $4.6 \times 10^{-5}$ |
| Far field | | | | | | | |
| Kd for rock (Cs) | m³/kg | 0.5 | 0.5 | 0.1 | 0.5 | 0.1 | 0.5 |
| Diffusion coefficient for rock | m²/s | $1.2 \times 10^{-11}$ | $1.1 \times 10^{-11}$ | $1.2 \times 10^{-11}$ | $1.1 \times 10^{-11}$ | $1.2 \times 10^{-11}$ | $9.8 \times 10^{-12}$ |
| Porosity for rock | % | 35 | 35 | 35 | 35 | 35 | 35 |
| Dry density for rock | kg/m³ | 1105 | 1105 | 1105 | 1105 | 1105 | 1105 |
| flow velocity | m/y | $3.2 \times 10^{-1}$ | $1.2 \times 10^{2}$ | $1.4 \times 10^{-2}$ | $9.0 \times 10^{0}$ | $7.2 \times 10^{-4}$ | $6.0 \times 10^{-1}$ |

·The equivalent value of zero concentration boudary is set to the groundwater flow rate through EDZ per waste package.
·Flow veolcity is set based on the relationship equation (hydraulic conductivity × hydraulic gradient / porosity)
·The glass dissolution rate and the De in buffer material are set for 40°C for the initial 10,000 years  ->$2.7 \times 10^{-4}$ g/m²/day and $2.7 \times 10^{-10}$ m²/s
·0.41 and 1.6 Mg/m³ are set to porosity and dry density of buffer.

## Step 5: Calculation and evaluation

Fig. 2 shows the results of six calculations for Cs-135 by using the GoldSim. The results presented in Fig. 2 show that the influence of future evolution of geological disposal conditions and relative difference of rate due to uplift and erosion can be an important factor for Cs-135 release rate from the host rock, which was the performance indicator in this study. This is because the release of assumed soluble Cs-135 is especially sensitive to the evolution of the flow velocity. These results are only intended to illustrate the methodology and assess sensitivity of assumptions (e.g., conceptual models, safety metric, assumed parameters, etc.). Note that repository uplift to near-surface, oxidizing conditions will also likely have a pronounced effect on release of long-lived, redox-sensitive radioelements, such as Se, Tc, Pu, Np and U.

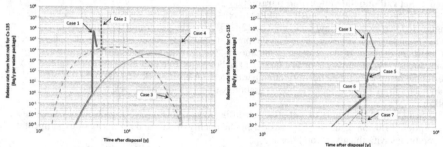

**Figure 2.** Illustration of calculation results for all Cases (Cs-135)

## CONCLUSIONS

We have developed a "Relative Rates Method" to make bounding calculations of impacts on radionuclide migration due to uplift/erosion ("exhumation"). Relative differences between uplift rate and erosion rate are examined in a quantitative manner. This methodology allows relative comparison of ranges of possible evolution in THMC conditions on radionuclide migration for different uplift/erosion rates. For the long time period, it was shown that the relative difference of uplift rate / erosion rate and potential hydraulic change arising from uplift/erosion can have a significant effect on migration of long-lived/non-solubility limited Cs-135 (left figure in Fig.2). Thus, uplift/ erosion should be strongly considered in any future site characterization and selection. In order to improve the methodology, more realistic treatment on (1) evolved condition of the engineered barriers and HLW at long time, (2) change from reducing to oxidizing conditions as the repository nears the surface, and (3) possible fault movement (4) impacts on other long-lived radioelements will also be needed.

## REFERENCES

1.    NUMO, NUMO-TR-04-04 (2004).
2.    JNC, JNC TN 1410 2000-004, (2000).
3.    M. Kawamura et al., *J. Japan Soc. Eng. Geol.*, Vol.51, pp.229-240, (2010) [in Japanese].
4.    JNC, JNC TN 1410 2000-002, (2000).
5.    K. Miyahara et al., *J. Nucl. Sci. Technol.*, Vol. 48(7), pp. 1069-1076, (2011).
6.    JNC, JNC TN1400 2005-014 (2005) [in Japanese].

7.  Y. Yano et al., Geothermal gradient map of Japan, (1999) [in Japanese]
8.  K. Ota et al., JAEA-Research 2010-068 (2011).
9.  JAEA, http://www.jaea.go.jp/04/tono/siryou/dbghs/dbghs.html [in Japanese].
10. Japan: Committee for Geosphere Stability Research, ISSN 2185-8548, 2011 [in Japanese].
11. S. Koo and M. Inagaki, JAEA-Data/Code 2010-006 (2010) [in Japanese].
12. Y. Tachi et al., JAEA-Data/Code 2010-031 (2011).
13. JNC, JNC TN 1410 2000-003, (2000).

Mater. Res. Soc. Symp. Proc. Vol. 1665 © 2014 Materials Research Society
DOI: 10.1557/opl.2014.627

## Construction of Microbial Kinetics Database for PA of HLW disposal

Hideki Yoshikawa[1], Tsuyoshi Ito[2], Kotaro Ise[1] and Yoshito Sasaki[1]
[1] Geological Isolation Research and Development Directorate, Japan Atomic Energy Agency
4-33, Muramatsu, Tokai-mura, Naka-gun, Ibaraki 319-1194, Japan.
[2] Visible Information Center Inc., Tokai-mura, Naka-gun, Ibaraki 319-1112, Japan

### ABSTRACT

To evaluate a change of chemical species of groundwater composition by the metabolism of microbes, which will be introduced to deep underground from the surface and be in a deep underground, is important for the discussion of the microbial effects on the performance assessment of the high-level radioactive waste repository. The purpose of this study is to develop of a microbial kinetics database to evaluate their activities in the deep underground environment.

Some microbial metabolism data were collected and constructed their kinetics database for aerobic, denitrifying, manganese reducing, iron reducing, sulfate reducing, methanogenic and acetogenic bacteria to evaluate above groundwater chemistry. About 1260 data were selected by literature survey for some journals and books published from 1960s and summarized in this microbial kinetics database. Some sensitivity analyses were performed for some parameter of metabolism of microbes.

### INTRODUCTION

The microbial effect on groundwater evolution present as an important issue in safety assessment of geological disposal of high-level radioactive waste (HLW). For example, some microbes consume dissolved oxygen, change chemical species from $NO_3^-$ to $N_2$, and reduce charge valence of iron or manganese from oxidizing to reducing condition in groundwater. The authors have been investigating the microbial growth rate of each microbe metabolism group in order to predict the quantitative evaluation of microbial activity affect on the groundwater component, and developing an evaluation method by using PHREEQC code to calculate the chemical composition in groundwater under the microbial effect. The microbial growth rate is represented generally by the Monod equation which is empiric formula. Value of the each parameter of the Monod equation varies due to differences microorganism type, growing environment, and the metabolic reaction. In addition, there is little information about the microbial activity present in the deep geological environment, such as their detail metabolism and death. Previous paper has shown that the microbial activity parameters related to the rate equation for microbe metabolisms and death from existing literature, and were collected and a microbial kinetic parameter database was constructed [1]. In this paper, moreover, a sensitivity analysis of each parameter and the comparison of their values for each parameter were performed for aerobes. A proposal to set the appropriate parameters to calculate the microbial activity are described in this paper for the potential use of the database for geological disposal.

# STRUCTURE OF MICROBIAL PARAMETER DATABASE

Some calculation example of microbial activity using PHREEQC was shown in the manual PHREEQC [2] and paper by Parkhurst and Appelo [3]. A model and a database were constructed to use the code in this study. The database has been made to develop an estimation method of biological effect on typical groundwater components using modified double Monod equation.

$$r_{bio} = \frac{dX}{dt} = -Yr_{sub} - \_X \qquad (1)$$

$$= Yk_{smax} \frac{C_s}{K_s + C_s} \frac{C_a}{K_a + C_a} X - \_X \qquad (2)$$

The parameter used in this equation is such as $r_{bio}$; growth rate ($s^{-1}$), $X$; biomass(cells/L), $Y$; yield coefficient (mg-$C_5H_7O_2N$/mg-electron donor(ED)), $k_{smax}$; maximum specific consumption rate of ED (mmol-ED/s), $K_s$; half saturation constant of ED (mg/L), $K_a$; half saturation constant of electron acceptor(EA) (mg/L), and $\beta$; decay constant ($s^{-1}$). These parameters for six major microbial groups which were classified depend on their effects on the redox potential (Eh) of the groundwater by their metabolism were collected. The six groups are 'aerobic,' 'denitrifying,' 'manganese reducing,' 'iron reducing,' 'sulfate reducing' and 'methanogenic'.

**Table I.** The half-reaction formula of each microbe group.

| microbes group | ED | half reaction |
|---|---|---|
| aerobic | $O_2$ | $O_2 + 4\,H^+ + 4\,e^- \rightarrow 2\,H_2O$ |
| denitrifying | $NO_3^-$ | $NO_3^- + 6\,H^+ + 5e^- \rightarrow 1/2N_2 + 3H_2O$ |
| manganese reducing | MnOOH | $MnOOH + 4\,H^+ + 2e^- \rightarrow Mn^{2+}\,2H_2O$ |
| iron reducing | $Fe(OH)_3$ | $Fe(OH)_3 + 3H^+ + e^- \rightarrow Fe^{2+} + 3H_2O$ |
| sulfate reducing | $SO_4^{2-}$ | $SO_4^{2-} + 9.5H^+ + 8e^- \rightarrow 0.5H_2S + 0.5HS^- + 4\,H_2O$ |
| methanogenic | $CO_2$ | $CO_2 + 8H^+ + 8e^- \rightarrow CH_4 + 2H_2O$ |

About 1260 data were obtained from biological papers and reports, which described clearly experimental condition, published during 45 years from 1967. This database contains a compilation of original data values as well as detailed information describing experimental procedure and calculation data which are used in some models to understand certain microbial activities [1].

# SENSITIVITY ANALYSIS

A sensitivity analysis was carried out for the parameter of the maximum specific consumption rate of ED ($k_{smax}$), the half saturation constant of ED ($K_s$), and that of EA ($K_a$) in this database. An initial input data of simple water chemical component were used to be able to easily incubate for each bacteria. The target of this analysis was aerobic bacteria with the lactate as an electron donor, considering an initial change of the repository environment from aerobic to anaerobic and existing low-molecular organic substrate in deep groundwater. Table II show the values set as a reference case.

**Table II.** Reference case parameter of a sensitivity analysis

| parameter | aerobobes |
|---|---|
| $k_{smax}$ : [mol of ED / mol of $C_5H_7O_2N$ s] | 3.4E-04 |
| $K_s$ : [mol/L] | lactate:3.4E-05 |
| $K_a$ : [mol/L] | $O_2$ 7.0E-06 |
| $\beta$ : [s$^{-1}$] | 4.3E-06 |
| $Y$ : [mol of $C_5H_7O_2N$ / mol of ED] | 0.39 |
| $X$ : [cells/mL] | 1.1E+03 |

## The maximum specific consumption rate of ED ($k_{smax}$)

Because of no $k_{smax}$ data of aerobic bacteria using lactate in their metabolism present in this database, the $k_{smax}$ was calculated from data of the maximum specific growth rate of aerobic microbes using lactate. The results of sensitivity analysis show that difference of end of consumption period is only about three days with the same amount of lactate comparing the minimum and maximum values. It is considered to be suitable to set a low $k_{smax}$ because the concentration of organic substance, which is easy to use in the disposal environment by the microbes, will be low in the deep groundwater. However, the medium value has set in this study, as difference of a few days is negligible in the long-term evaluation of the disposal environment.

**Table III.** Analysis condition for the maximum specific consumption rates

| | term | the maximum specific growth rate [1/s] | the maximum specific consumption rates [mol of ED / (mol of $C_5H_7O_2N$ s)] |
|---|---|---|---|
| Run1 | minimum value | 4:2E-05 | 9.0E-06 |
| Run2 | medium value | 1.3E-04 | 3.4E-05 |
| Run3 | maximum value | 3.1E-04 | 1.7E-04 |

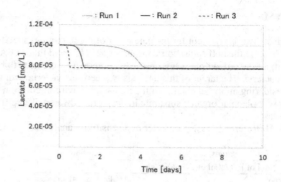

**Figure 1.** Results of sensitivity analyses for the maximum specific consumption rates

## The half saturation constant of ED ($K_s$)

Setting value of the half saturation constant of ED ($Ks$) is shown in Table IV. As this parameter will be effective in a low concentration of ED, further analyses are also performed in the case of a tenth part of the initial concentration of lactate respectively. The results of sensitivity analysis of $Ks$ were shown in following. The minimum, medium, and maximum value were set to 8.9E-06, 3.4E-05, and 1.7E-04 respectively. Figure 2 shows the half saturation constants of all organic substance for not only aerobic but also other microbes in the database for reference.

**Table IV.** Analysis condition for the half saturation constant of ED($Ks$)

|      | term          | $K_S$ [mol/L] | the amount of initial lactate[mol/L] |
|------|---------------|---------------|--------------------------------------|
| Run1 | minimum value | 8.9E-06       |                                      |
| Run2 | medium value  | 3.4E-05       | 1.0E-04                              |
| Run3 | maximum value | 1.7E-04       |                                      |
| Run4 | minimum value | 8.9E-06       |                                      |
| Run5 | medium value  | 3.4E-05       | 1.0E-05                              |
| Run6 | maximum value | 1.7E-04       |                                      |

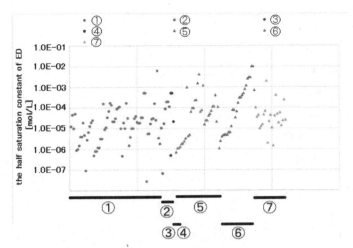

**Figure 2.** Half saturation constant of ED for all organic substances.
1:aerobic bacteria, 2:denitrifying bacteria, 3:manganese reducing bacteria,
4:iron reducing bacteria 5:sulfate reducing bacteria,
6:methanogenic bacteria, 7:other facultative anaerobe

The results of the analysis are shown in Figure 3. When a concentration of lactate was set at 1.0E-04 mol/L, there was a difference of only two days to consume the same amount of lactate. In the case of the lower concentration of lactate as 1E-05mol / L and maximum value of $Ks$, the lactate was not consumed in ten days by the reason of the microbial slower growth in the low amount of lactate. Therefore, when the electron acceptor concentration is low, $Ks$ is considered an important parameter in assessing for the long-term effect by the microbes. In the case of aerobic bacteria will becomes active, it is a non-conservative situation on the scenario of the geological disposal process for the oxygen consumption activity. Assuming the conservative situation, we recommend that the half saturation constant of ED ($Ks$) set between the median value 3.4E-05 and maximum value 1.7E-04 mol / L under the maximum amount of organic substances.

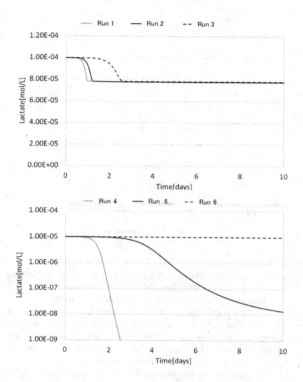

**Figure 3.** Results of sensitivity analyses for the half saturation constant of ED for lactate

On the other hand, it is found that the half saturation constant EA($K_a$) is also important and sensitive to the EA concentration from the sensitivity analysis in this investigation. Assuming the conservative situation of oxygen concentration change in the repository environment, there needs to be set up between the maximum and medium value as well as $K_s$.

## CONCLUSIONS

A microbial kinetics database for PA have been constructing and some sensitivity analysis were performed by this database. This database has 1260 data selected by literature survey and is available to calculate the microbial effect on the groundwater component by using a geochemical code PHREEQC.

It was found that the maximum specific consumption rate of ED($K_{smax}$) was not effective in a range of the existing data in this database from the results of the sensitivity analysis. When each amounts of the ED and EA is low, these half-saturation constant($K_s$ and $K_a$)will be a large effect on not only their ED and EA concentrations but also on consumption rate of organic substance and build-up rate of microbes.

In general, the half saturation constant of ED and EA is in the order of $10^{-5}$ (M) for the carbon source and that of from $10^{-5}$ to $10^{-6}$ (M) for the oxygen [4]. To set the larger half saturation constant would limit the microbial growth in the disposal environment which has low organic substance, however, it is conservative in considering the repository environment. Based on the results of this sensitivity analysis, the half saturation constant ED and EA are recommended to be set between the maximum and the medium value of this database like as in $10^{-4}$ to $10^{-5}$(M).

## ACKNOWLEDGMENTS

This database construction was carried out thanks to the support of Dr.Takahiro Asano (Chugai Technos Co. Ltd.) and Dr.Iku Miyasaka and Dr.Sakae Fukunaga(IHI Co., Ltd.). This study was performed as part of "Project for Assessment Methodology Development of Chemical Effects on Geological Disposal System" funded by the Ministry of Economy, Trade and Industry, Japan.

## REFERENCES

1. H.Yoshikawa, M.Inagaki, and I.Miyasaka, Mater.Res.Soc.Symp.Proc., 1193, pp.375-380, (2009)
2. D.L.Parkhurst, and C.A.J.Appelo, User's guide to PHREEQC (Version 2)--a computer program for speciation, batch-reaction, one-dimensional transport, and inverse geochemical calculations: *U.S. Geological Survey Water-Resources Investigations Report 99-4259.* (1999)
3. H.Prommer, and D.A.Barry, Modeling Bioremediation of Contaminated Groundwater In: *Bioremediation (edited by R.M. Atlas and J. Philp).* American Society for Microbiology, pp.108-138, (2005)
4. T.Yamane, in *Microbial Reaction Engineering $3^{rd}$ Ed.* (Japanese), (2002)

Mater. Res. Soc. Symp. Proc. Vol. 1665 © 2014 Materials Research Society
DOI: 10.1557/opl.2014.628

## Deposition Behavior of Supersaturated Silicic Acid in the Condition of Relatively High Ca or Na Concentration

Taiji Chida[1], Yuichi Niibori[1], Hayata Shinmura[1] and Hitoshi Mimura[1]
[1]Dept. of Quantum Science & Engineering, Graduate School of Engineering, Tohoku University, 6-6-01-2 Aramaki-aza-Aoba, Aoba-ku, Sendai 980-8579 JAPAN

### ABSTRACT

Around the radioactive waste repository, the pH of the groundwater greatly changes from 8 to 13 and the groundwater contains a relatively large quantity of calcium (Ca) and sodium (Na) ions due to cementitious materials used for the construction of the geological disposal system. Under such conditions, the deposition behavior of silicic acid is one of the key factors for the migration assessment of radionuclides. The deposition and precipitation of silicic acid with the change of pH and coexisting ions may contribute to the clogging in flow paths, which is expected as the retardation effect of radionuclides. Thus, this study focused on the deposition behavior of silicic acid under the condition of relatively high Ca or Na concentration.

In the experiments, $Na_2SiO_3$ solution (250 ml, 14 mM, pH>10, 298 K) was prepared in a polyethylene vessel containing amorphous silica powder (0.5 g) as the solid phase. Then, a buffer solution (to adjust to 8 in pH), $HNO_3$, and $Ca(NO_3)_2$ as Ca ions or NaCl as Na ions were sequentially added. Such a silicic acid solution becomes supersaturated, gradually forming colloidal silicic-acid and/or the deposit on the solid surface. In this study, the both concentrations of soluble and colloidal silicic-acid were monitored over a 40-day period. As a result, the deposition rate of silicic acid decreased with up to 5 mM in Ca ions. Besides, Na ions with up to 0.1 M slightly increased the deposition rate. Under the conditions of $[Na^+]>0.1$ M or $[Ca^{2+}]>5$ mM, the supersaturated silicic acid immediately deposited. These suggest that Na or Ca ions strongly affect the deposition behavior of supersaturated silicic-acid, depending on the surface alteration of solid phase, the change of zeta potential and the decrease of water-activity due to the addition of electrolytes (coexisting ions).

### INTRODUCTION

Silicic acid around the geological disposal system for radioactive wastes undergoes dissolution, deposition and polymerization depending on the change of pH in the range from 8 to 13 which is caused by the cementitious materials used for the construction of the repository [1]. Some dynamic behaviors of silicic acid, such as polymerization and alteration of solid surface, may facilitate the migration rate of radionuclides from the repository, e.g., because of the chemical alteration of the flow path surfaces and the complex formation between the polymeric silicic acid and radionuclides [2, 3]. On the other hand, it is also suggested that the deposition and precipitation of silicic acid may clog main flow paths and the micro-cracks in rocks, and retard the migration of radionuclides [4].

In general, the chemical behaviors of silicic acid are greatly affected by not only the change in pH but also the coexisting ions in the groundwater. Particularly, the cementitious materials alter the groundwater around the repository to a high calcium (Ca) and sodium (Na) condition while changing the pH of the groundwater [5]. Furthermore, the effects of the saline groundwater including Na ions on the repository system should be considered. As a preliminary study about the effects of coexisting ions on the deposition of silicic acid, this study examined the deposition behavior of supersaturated silicic acid under the condition of relatively high concentration of Ca ions (to 0.01 M) or Na ions (to 1.0 M).

## EXPERIMENTAL

The main experimental procedures are the same as the authors' previous works [6]. Figure 1 is the overview of the experimental procedure. The influence of Ca ions or Na ions on the deposition behavior of silicic acid was traced over a 1,000-hours period. $Na_2SiO_3$ solution (water glass obtained from Wako Pure Chemical Industries, Ltd.) was diluted to a given concentration, that is, 14 mM as a silicic acid solution. The silicic acid solution (pH>10) was put into the polypropylene vessel and mechanically stirred with a polypropylene stirrer. The volume of the solution was set to 250 ml. The temperature was kept constant within 298±0.5 K. As a solid sample, pure amorphous silica (silicic acid, $SiO_2 \cdot nH_2O$) was purchased as powder with grains smaller than 100 mesh from Mallinckrodt Co, and a size fraction of 74-149 μm particle diameters was separated by sieving. Its specific surface area (BET, $N_2$ gas) was 322 $m^2/g$ and X-ray diffraction pattern of this sample showed amorphous feature.

Weighted amount of pure amorphous silica 0.5 g was poured into the vessel. Then, the pH of the solution was set to 8 with a $HNO_3$ solution and a buffer solution mixing MES (2-(N-morpholion) ethanesulfonic acid) and THAM (tris(hydroxymethyl)aminomethane). Promptly after adjusting pH, $Ca(NO_3)_2$ or NaCl solution was added into the silicic acid solution. The concentration of Ca ions was set to 0.1 mM, 0.5 mM, 1.0 mM, 5.0 mM or 10 mM. The concentration of Na ions was set to 0.028 M (caused by just $Na_2SiO_3$ solution with no NaCl solution addition), 0.1 M or 1.0 M.

Since the solubility of silicic acid in the range of pH>9 extremely exceeds that in the pH<9 [7], the silicic acid solution becomes supersaturated, and deposits on the solid surface or

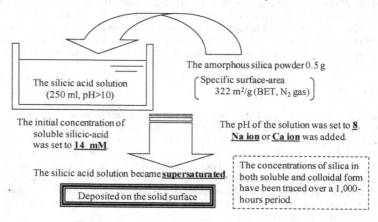

**Figure 1.** The experimental procedure.

changes into the colloidal form. In this study, after the aliquots had been filtered through the disposable membrane filter of pore size 0.45 μm, the concentration of soluble silicic-acid was determined by using silicomolybdenum-yellow method [8]. That is, the soluble silicic-acid (monomeric or oligomeric silicic-acid) was defined as silicic acid reacting molybdate reagent and coloring yellow, and colloidal silicic-acid was defined as silicic acid in liquid phase except for soluble silicic-acid [9]. The total concentration of silicic acid consisting of both the soluble form and the colloidal form in liquid phase was determined by ICP emission spectroscopy. The concentration of deposited silicic-acid was defined as "the initial concentration of soluble silicic-acid" minus "the total concentration of silicic acid in liquid phase".

As mentioned above, this study poured 0.5 g of amorphous silica in the vessel of 250 ml. Using BET's specific surface area, $a_B$ (m$^2$/g), we can estimate another specific surface area in the solution, $a$ (1/m). That is, its value is defined by $a=a_Bw/V$, where $w$ is the weighted amount of solid sample (g) and $V$ is the solution volume (m$^3$). Furthermore, when a simple parallel flat board is assumed as a model of fracture (flow-path) included in rock matrix, the specific surface area, $a$, is described also by $2/b$, where $b$ is fracture aperture (m). Since the value of $a$ was estimated at $6.44\times10^5$ 1/m in the experimental condition of this study, its value corresponds to around 30 μm in $b$. In other words, this experimental condition considers a specific surface area estimated in flow-paths consisting of fractures around the repository.

**RESULTS AND DISCUSSION**

Figure 2 shows the changes in the fractional contributions of soluble silicic-acid, colloidal silicic-acid and deposited silicic-acid, where the initial concentration of Ca was 5.0 mM. Each fractional contribution, $f$, is defined as the amount fraction of soluble silicic-acid, colloidal silicic-acid or deposited silicic-acid to the initial amount of soluble silicic-acid. As shown in Figure 2, the fraction of colloidal silicic-acid decreases, and the fraction of deposited silicic-acid gradually increases with time. On the other hand, the fraction of soluble silicic-acid decreases immediately after the experiment startup, and becomes nearly constant. In such a metastable state, the concentration of soluble silicic-acid slightly exceeded its solubility at pH 8, i.e., 2.0 mM [6]. These tendencies were also observed under the Ca free condition in the solution.

Figure 3 and Figure 4 show the effects of Na ions or Ca ions on the fractional contribution of deposited silicic-acid. As shown in Figure 3, Na ions in the range of up to 0.1 M increased the deposition rate of silicic-acid, and the supersaturated silicic-acid immediately deposited under the condition of [Na$^+$]=1.0 M. On the other hand, in Figure 4, the silicic acid in liquid phase immediately deposited under the condition of [Ca$^{2+}$]=10 mM. In a result, the concentration of Na ions >0.1 M or Ca ions >5 mM suggests an electrolyte concentration required for forming the aggregation of colloidal species. However, such a critical electrolyte concentration under the condition with no solid phase exceeds [Na$^+$]=1.0 M or [Ca$^{2+}$]=10 mM [8]. Furthermore, Ca ions in the range of to 5 mM suppressed the deposition rate, showing that the silicic acid was stable in liquid phase in comparison with that under the conditions of Ca-free. Even if the groundwater is altered by cement pore water, the concentration of Ca ions does not exceed 10 mM. Besides, in saline groundwater, the concentration of Na ions does not exceed 1.0 M. Therefore, Figures 4 and 5 suggest that the colloidal species of silicic acid is not easily aggregated around the repository, even if in the presence of solid phase.

Figure 5 shows the concentration of Ca ions in the liquid phase. The concentrations of Ca ions were nearly constant through these experiments, and more than 90 % of the initial Ca ions

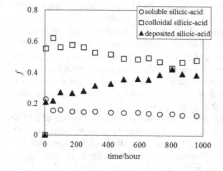

**Figure 2.** Fractional contributions of soluble silicic-acid, colloidal silicic-acid and deposited silicic-acid. The fractional contribution, $f$, is the fraction of each form to the initial concentration of soluble silicic-acid.

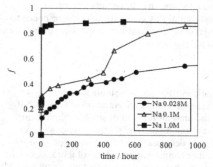

**Figure 3.** The dependencies of the deposition of silicic acid on Na concentration. The case of $[Na^+]=0.028$ M is quoted from the previous work [6].

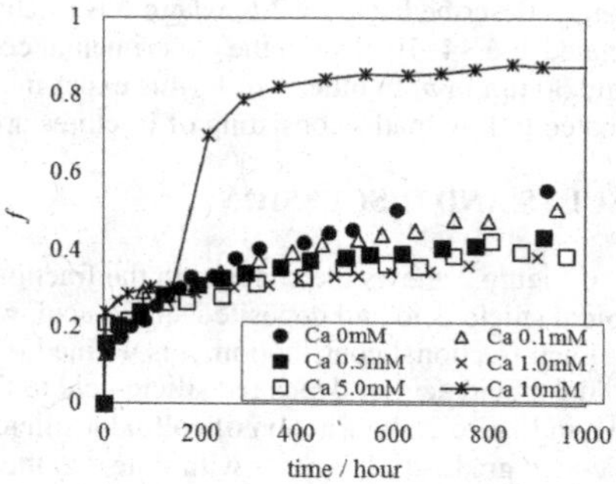

**Figure 4.** The dependencies of the deposition of silicic acid on Ca concentration. The cases of $[Ca^{2+}]=0$ mM, 0.1 mM, 0.5 mM and 1.0 mM are quoted from the previous work [6].

remained in liquid phase. Calcium-silicate-hydrate (CSH) is not much formed at 8 in pH [11]. Furthermore, the deposition form did not almost incorporate calcium ions. Additionally, the concentration of soluble silicic-acid immediately attained to the metastable state corresponding to the solubility of silicic acid as shown in Figure 3. These mean that the deposition of silicic acid consists of not CSH gels but the amorphous phase of silicic acid.

Generally, the addition of electrolytes to the supersaturated solution accelerates the aggregation and precipitation of colloidal species, and increases the deposition rate of supersaturated chemical species. That is, the larger the absolute values of zeta potential become, the more stable the colloidal species are because of the repelling force between colloidal particles. Inversely, the colloidal species easily aggregate when the absolute values of zeta potential get close to zero. Tamura et al. [10] reported that the zeta potential of the colloidal silicic-acid under

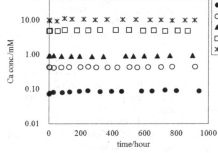

**Figure 5.** Ca concentration in the solution during the experiments. More than 90 % of the initial Ca ions remained in liquid phase.

Ca and solid-phase free condition was estimated at about -50 mV. (Here, the initial supersaturated concentration of silicic acid was set to 10 mM.) On the other hand, the zeta potential of colloidal silicic-acid under the condition of $[Ca^{2+}]$=10 mM (no solid phase) was around -20 mV. In other words, the zeta potential is in the range of -50 mV to -20 mV, depending on the concentration of Ca ions within the range of 0 mM to 10 mM. While the role of electrolytes becomes more complicated in the presence of solid phase, the zeta potential might be able to explain at least the results of Figure 4. However, we must also consider that the hydration of coexisting ions (electrolytes) decreases the activity of the water molecule in the solution. The electrolytes can contribute the aggregation of colloidal species, increasing the deposition rate of silicic acid.

Furthermore, the experimental results in the presence of Ca ions (Figure 4) showed that the deposition rate slightly decreased in the range of up to 5 mM in $[Ca^{2+}]$. Besides, in the case of $[Na^{+}]$=0.1 M (Figure 3), Na ions might contribute the growth/aggregation of colloidal silicic-acid with relatively long time-period in comparison with the case of $[Na+]$=1.0 M. These suggest that not only the change of zeta potential and the decrease of water-activity but also the altered surface of solid phase affect the deposition behavior of supersaturated silicic-acid including the colloidal form. For example, Chida et al. [6] reported that the deposition of supersaturated silicic acid apparently decreased the BET's specific surface area ($N_2$ gas) of the amorphous silica powders. As the surface area of the solid phase extremely decreases, the deposition rate may dramatically change, as shown in the case of 0.1 M in $[Na^{+}]$ (Figure 3).

For assessing the alterations of host rock and flow paths caused by cementitious materials, we need a more reliable model to clarify the behavior of the deposition of silicic-acid species by considering the colloidal form, the coexisting ions and the solid phase. Especially, the roles of Na and Ca ions considered around the repository ($[Na^{+}]$<1.0 M and $[Ca^{2+}]$<10 mM) would become complicated in the deposition behavior of supersaturated silicic-acid.

**CONCLUSIONS**

The influence of Na or Ca ions on the deposition behavior of silicic acid during 40 days was examined in the presence of the solid phase. The deposition rate of silicic acid decreased with up to 5 mM in Ca ions. Besides, Na ions with up to 0.1 M slightly increased the deposition rate. Furthermore, as Na or Ca ions exceed 1.0 M or 10 mM respectively, the supersaturated silicic-acid immediately deposited by the aggregation of colloidal species. These suggest that the specific surface area of solid phase, the change of zeta potential and the decrease of water-activity due to the addition of electrolytes (coexisting ions) affect the deposition behavior of supersaturated silicic-acid including the colloidal form.

Even if the groundwater is altered by cement pore water, the concentration of Ca ions does not exceed 10 mM. Besides, the concentration of Na ions in saline groundwater does not exceed 1.0 M. In such a condition, we must consider that the colloidal species of silicic acid is not easily aggregated around the repository. Furthermore, since the colloidal form of silicic acid can migrate along flow-paths around the repository and alters the surfaces of flow-paths [12], the dynamic behavior of colloidal silicic-acid is an important factor to estimate the spatial altered range. In order to estimate the alteration of host rock and the clogging effect due to the cement use for constructing the repository, this study suggested that a more reliable model to appropriately describe the kinetics of supersaturated silicic-acid species would be required to consider coexisting solid phase and electrolytes in groundwater.

## ACKNOWLEDGMENTS

This study was supported by Japan Society for the Promotion of Science, Grant-in-Aid for Scientific Research (A) No. 25249136.

## REFERENCES

1. Gaucher, E.C., Blanc, P., Matray, J. and Michau, N., *Appl. Geochem.*, **19**, 1505 (2004).
2. Kersting, A.B., Efurd, D.W., Finnegan, D.L., Rokop, D.J., Smith, D.K. and Thompson, J.L., *Nature*, **397**, 56 (1999).
3. Savage, D., *The Scientific and Regulatory Basis for the Geological Disposal of Radioactive Waste*, (John Wiley, New York, 1995).
4. Shinmura, H., Niibori, Y. and Mimura, H., *Proc. of WM 2013 Conference*, Paper No. 13270, (2013).
5. Atkionson, A., AERE-R 11777, (UKAEA, 1985).
6. Chida, T., Niibori, Y., Tochiyama, O., Mimura, H. and Tanaka, K., *Appl. Geochem.*, **22**, 2810 (2007).
7. Stumn, W. and Morgan, J.J., *Aquatic Chemistry*, 3rd ed., (John Wiley & Sons, New York, 1996).
8. Iler, R.K., *The Chemistry of Silica*, (John Wiley & Sons, New York, 1979).
9. Chida, T., Niibori, Y. Tochiyama, O., Mimura, H. and Tanaka, K. in *Dissolution Rate of Colloidal Silica in Highly Alkaline Solution*, edited by J.M. Hancher, S.S. Gasucoyne and L. Browning, (Mater. Res. Soc. Symp. Proc. **824**, San Francisco, CA, 2004), pp. 467-472.
10. Tamura, N., Niibori, Y. and Mimura, H., *Proc. of WM 2011 Conference*, Paper No. 11378, (2011).
11. Niibori, Y. et al., in *Scientific Basis for Nuclear Waste Management XXXVI*, edited by K. M. Fox, K. Idemitsu, C. Poinssot, N. Hyatt, K. Whittle (Mater. Res. Soc. Symp. Proc., **1518**, Boston, MA, 2012), pp. 255-260.
12. Niibori, Y. et al., *Journal of Power and Energy Systems*, **6**, 140 (2012).

# Radionuclides solubility, speciation, sorption and migration

Mater. Res. Soc. Symp. Proc. Vol. 1665 © 2014 Materials Research Society
DOI: 10.1557/opl.2014.629

# The DR-A in-situ diffusion experiment at Mont Terri: Effects of changing salinity on diffusion and retention properties.

Josep M. Soler[1], Olivier X. Leupin[2], Thomas Gimmi[3,4] and Luc R. Van Loon[3]

[1]IDAEA-CSIC, Barcelona, Catalonia, Spain.
[2]NAGRA, Wettingen, Switzerland.
[3]Paul Scherrer Institut, Villigen, Switzerland.
[4]University of Bern, Switzerland.

## ABSTRACT

In the new DR-A in-situ diffusion experiment at Mont Terri, a perturbation (replacement of the initial synthetic porewater in the borehole with a high-salinity solution) has been induced to study the effects on solute transport and retention, and more importantly, to test the predictive capability of reactive transport codes. Reactive transport modeling is being performed by different teams (IDAEA-CSIC, PSI, Univ. Bern, Univ. British Columbia, Lawrence Berkeley Natl. Lab.). Initial modeling results using the CrunchFlow code and focusing on $Cs^+$ behavior are reported here.

## INTRODUCTION

Clay formations are being considered internationally as potential host rocks for radioactive waste disposal due to their low permeability, diffusion-controlled solute transport and favorable retention properties. After a series of successful in-situ experiments in the Opalinus Clay (OPA) at the Mont Terri Underground Rock Laboratory, the new DR-A test, funded by Nagra (Switzerland), NWMO (Canada) and DOE (USA), has been designed to provide further insight into solute transport and retention, and more importantly, to test the predictive capability of reactive transport codes.

In previous experiments (DI, DI-A1, DI-A2, DR; [1-6]) the behavior of different tracers (water, cations and anions, conservative and sorbing) was studied using solutions at equilibrium with the host rock. Those results showed the value of the experimental setup (synthetic porewater in equilibrium with the rock; circulation of the synthetic porewater containing the tracers in a single borehole; closed circulation loop) and the importance of processes such as anion exclusion and sorption of cations.

The first stage of the DR-A test consisted of a conventional diffusion experiment using a synthetic version of the OPA porewater, including several tracers (HTO, $I^-$, $Br^-$, $Cs^+$, $^{85}Sr^{2+}$, $^{60}Co^{2+}$, $Eu^{3+}$). Reactive transport modeling is being performed by several teams (IDAEA-CSIC, PSI, Univ. Bern, Univ. British Columbia, Lawrence Berkeley Natl. Lab.). Predictive modeling is first performed based on existing knowledge (diffusion coefficients, accessible porosities, sorption parameters). Model calibration is performed later using the measured experimental data.

After 189 days, the solution in the borehole was replaced with a higher-salinity solution (0.50 M NaCl + 0.56 M KCl, approx.), with clear effects such as the back-diffusion of $Cs^+$ from the rock to the borehole due to desorption from the cation exchange complex in the rock. These results are now being modeled by the different teams. Initial modeling results using the CrunchFlow code and focusing on $Cs^+$ behavior are reported here.

## SETUP AND DIMENSIONS

Figure 1 shows the experimental setup (borehole circulation interval), and Table 1 shows the relevant parameters for the simulations.

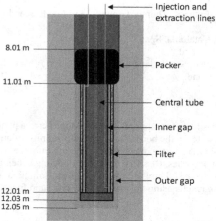

**Figure 1.** Schematic diagram showing the experimental setup (borehole circulation interval).

**Table I.** Relevant parameters for the simulations.

| | |
|---|---|
| Length of the injection interval | 104 cm |
| Volume of the circulation system | 11.2 L |
| Borehole diameter | 76 mm |
| Filter, outer diameter | 70 mm |
| Filter, inner diameter | 62 mm |
| Filter, porosity | 45% |
| Gap between filter and borehole wall | 3 mm |
| Central tube, outer diameter | 61 mm |
| Gap between central tube and filter | 0.5 mm |
| Porosity of Opalinus Clay | 15% |

## MODEL DESCRIPTION

Since transport in OPA is dominated by diffusion along the bedding planes (preferential fast pathways for diffusion; [7]) and the injection interval is long enough compared to expected transport distances, a 1D approach with symmetry with respect to the borehole axis is sufficient to model the results. However, the borehole in DR-A was drilled vertically, i.e. at an angle of about 60° with respect to bedding. In principle, this geometry does not allow a direct application of a 1D model. However, a modification of the borehole radius and capacity does allow the implementation (Fig. 2). The intersection of borehole and bedding is an ellipse. Using instead a circle with a perimeter equal to that of the ellipse, and compensating the slight increase in cross-section area (1%) by reducing the borehole capacity, identical results between 2D (elliptical borehole section) and 1D (circular borehole section) transport calculations are obtained.

Calculations have been performed using the CrunchFlow reactive transport code [8]. In these calculations, the code solves the equation of conservation of mass for the different components

$$\frac{\partial c_{tot}}{\partial t} = \nabla \cdot (D_e \nabla c) \tag{1}$$

where $c$ is concentration in solution (moles per volume of solution), $t$ is time, $D_e$ is the effective diffusion coefficient and $c_{tot}$ is the total concentration (moles per bulk volume), which is given by

$$c_{tot} = \phi c + \rho_d s = \phi c + \rho_s (1 - \phi) s \tag{2}$$

where $\phi$ is porosity, $\rho_d$ is the bulk dry density, $s$ is the concentration of sorbed component (moles per mass of solid), and $\rho_s$ is the solid density (2700 kg/m³). Single values for $D_e$ and $\phi$ are used for all species in solution.

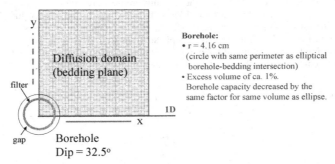

**Figure 2.** Schematic diagram showing the calculation domain. The boundaries of the domain are no-flux boundaries.

## CALCULATIONS

Table 2 shows the initial solution compositions ($t = 0$) and the new composition of the solution in the borehole at $t = 189$ d. These are measured compositions, together with charge balance and atmospheric equilibrium assumptions. A total of 40 species in solution were taken into account. Initially ($t = 0$) total carbonate was calculated by charge balance. At $t = 189$ d, Cl⁻ and HCO₃⁻ in the borehole were calculated by charge balance and equilibrium with the atmosphere, respectively. Sorption was calculated with a multisite cation exchange model (Table 3). This model produced a slightly better fit than the alternative one using log $K$ (II-Cs) = -3.2 [9].

Figure 3 shows modeling results for $Cs^+$ (concentrations in the borehole) together with the measured data. The model was fitted to the data up to $t = 189$ d (before replacement of solution in the borehole). Fitting required a well-mixed filter. The parameters for the calculation are listed below.

Borehole: $D_e = 3.3e{-}4$ m$^2$/s (well mixed)    Filter: $D_e = 3.0e{-}5$ m$^2$/s (well mixed)
Rock: $D_e = 1.2e{-}10$ m$^2$/s                      Gap: $D_e = 2.0e{-}9$ m$^2$/s (diff. coeff. in water; [10])

At $t = 189$ d the solution in the borehole was replaced with the high-salinity solution. Model results after 189 days correspond to a prediction based on the initial fit of the model. It can be observed that the increase in $Cs^+$ concentration in the borehole after replacement of the solution is reproduced by the model. $K^+$ displaces $Cs^+$ from the type-II exchange sites, which diffuses back to the borehole. $K^+$ also displaces $Ca^{2+}$ and $Mg^{2+}$ from the planar sites. However, the model overestimates $Cs^+$ concentrations immediately after the replacement.

Figure 4 shows model results and experimental data for the rest of major components in the model, which was calibrated with the $Cs^+$ data from the first stage of the experiment. It can be observed that the trends are successfully reproduced, especially for $Ca^{2+}$ and $Mg^{2+}$. Changes in $Na^+$ and $K^+$ are slightly slower than predicted, and the same happens with the anions $SO_4^{2-}$ and $Cl^-$. These differences may be caused by the different diffusion coefficients and accessible porosities for the different species. However, these results are only preliminary. In a next step, modeling will be attempted with the new CrunchFlowMC code, which explicitly includes species-specific diffusion coefficients and anion exclusion. Uncertainties will also be significantly reduced when the results of solid sample analyses from overcoring are available, which will allow simultaneous fitting of the model to (i) the evolution of the concentrations in the circulation system, and (ii) to the measured tracer distribution profiles in the rock.

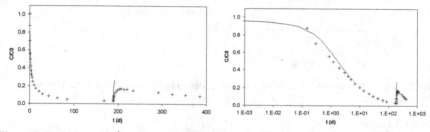

**Figure 3.** Evolution of $Cs^+$ concentration in the borehole. Line corresponds to model results and symbols to measured data. Borehole solution was replaced at $t = 189$ d.

## CONCLUSIONS

In the first stage of the experiment (before replacement of solution) $Cs^+$ was modeled using multicomponent diffusion and cation exchange. The filter seems to be well mixed (large diffusion coefficient), not acting as a diffusion barrier.

The model from the first stage was also used to predict the evolution of different components after replacement of the borehole solution with a high-salinity solution. The trends were successfully reproduced, especially for $Ca^{2+}$ and $Mg^{2+}$. The calculated $Cs^+$ concentration in the borehole right after replacement of solution is somewhat larger than observed. Changes in $Na^+$ and $K^+$ were slightly slower than predicted, and the same happened with the anions $SO_4^{2-}$ and $Cl^-$. The differences between the different tracers may be caused by the different diffusion coefficients and accessible porosities for the different species. These are preliminary results that may be refined when using with the new CrunchFlowMC code, which explicitly includes species-specific diffusion coefficients together with anion exclusion, and also when including the results from solid sample analyses after overcoring.

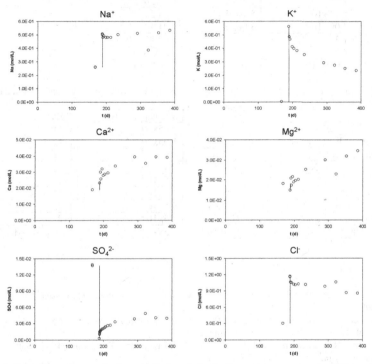

**Figure 4.** Evolution of concentrations in the borehole for the major components of the modeled system. Lines correspond to model results and symbols to measured data.

**Table II.** Initial solution compositions (total concentrations in mol/kg_$H_2O$).

| | Borehole – inner gap (t = 0) | Filter – outer gap – rock (t = 0) | Borehole – inner gap (t = 189 d) |
|---|---|---|---|
| T ($^{\circ}$C) | 18 | 18 | 18 |
| pH | 7.6 | 7.6 | 7.6 |
| $Na^+$ | $2.59\times10^{-1}$ | $2.59\times10^{-1}$ | $5.00\times10^{-1}$ |
| $K^+$ | $1.64\times10^{-3}$ | $1.64\times10^{-3}$ | $5.60\times10^{-1}$ |
| $Mg^{2+}$ | $1.80\times10^{-2}$ | $1.80\times10^{-2}$ | $1.47\times10^{-2}$ |
| $Ca^{2+}$ | $1.88\times10^{-2}$ | $1.88\times10^{-2}$ | $2.30\times10^{-2}$ |
| $Sr^{2+}$ | $5.10\times10^{-4}$ | $5.10\times10^{-4}$ | $4.54\times10^{-4}$ |
| $Cl^-$ | $3.00\times10^{-1}$ | $3.00\times10^{-1}$ | $1.14\times10^{-0}$ |
| $SO_4^{2-}$ | $1.37\times10^{-2}$ | $1.37\times10^{-2}$ | $2.37\times10^{-4}$ |
| $HCO_3^-$ | $8.19\times10^{-3}$ | $7.98\times10^{-3}$ | $5.63\times10^{-4}$ |
| $Cs^+$ (tracer) | $2.069\times10^{-4}$ | $2.00\times10^{-8}$ | $6.23\times10^{-6}$ |

**Table III.** Parameters of the multisite cation exchange model, with capacities based on a 21wt% illite in OPA [10].

| | log $K$ | Site capacity (eq/kg) | Reference |
|---|---|---|---|
| FES-Cs + $Na^+$ = FES-Na + $Cs^+$ | -7.0 | $1.05x10^{-4}$ | [11] |
| FES-K + $Na^+$ = FES-Na + $K^+$ | -2.4 | $1.05x10^{-4}$ | [11] |
| II-Cs + $Na^+$ = II-Na + $Cs^+$ | -3.6 | $8.4x10^{-3}$ | [11] |
| II-K + $Na^+$ = II-Na + $K^+$ | -2.1 | $8.4x10^{-3}$ | [11] |
| PS-Cs + $Na^+$ = PS-Na + $Cs^+$ | -1.6 | $9.5x10^{-2}$ | [11] |
| PS-K + $Na^+$ = PS-Na + $K^+$ | -1.1 | $9.5x10^{-2}$ | [11] |
| $PS_2$-Ca + 2 $Na^+$ = 2 PS-Na + $Ca^{2+}$ | -0.67 | $9.5x10^{-2}$ | [12] |
| $PS_2$-Mg + 2 $Na^+$ = 2 PS-Na + $Mg^{2+}$ | -0.59 | $9.5x10^{-2}$ | [12] |
| $PS_2$-Sr + 2 $Na^+$ = 2 PS-Na + $Sr^{2+}$ | -0.59 | $9.5x10^{-2}$ | Same as $PS_2$-Mg [13] |

**REFERENCES**

1. J.-M. Palut, Ph. Montarnal, A. Gautschi, E. Tevissen and E. Mouche, *J. Contam. Hydrol.* **61**, 203 (2003).
2. E. Tevissen and J.M. Soler, *Mont Terri Project. DI Experiment. Synthesis Report*, (Mont Terri Technical Report TR 2001-05, St-Ursanne, 2003), p. 55.
3. P. Wersin, L.R. Van Loon, J.M. Soler, A. Yllera, J. Eikenberg, Th. Gimmi, P. Hernán and J.-Y. Boisson, *Appl. Clay Sci.* **26**, 123 (2004).
4. L.R. Van Loon, P. Wersin, J.M. Soler, J. Eikenberg, Th. Gimmi, P. Hernán, S. Dewonck and J.-M. Matray, *Radiochim. Acta* **92**, 757 (2004).
5. P. Wersin, J.M. Soler, L. Van Loon, J. Eikenberg, B. Baeyens, D. Grolimund, T. Gimmi and S. Dewonck, *Appl. Geochem.* **23**, 678 (2008).
6. J.M. Soler, P. Wersin and O.X. Leupin, *Appl. Geochem.* **33**, 191 (2013).
7. L.R. Van Loon and J.M. Soler, *Diffusion of HTO, $^{36}Cl^-$, $^{125}I^-$ and $^{22}Na^+$ in Opalinus Clay: Effect of confining pressure, sample orientation, sample depth and temperature*, (Nagra Technical Report NTB-03-07, Wettingen, 2003), p. 119.

8.  C.I. Steefel, *CrunchFlow. Software for Modeling Multicomponent Reactive Flow and Transport. User's Manual*, (Lawrence Berkeley National Laboratory, Berkeley, 2009), p. 91.
9.  L.R. Van Loon, B. Baeyens and M.H. Bradbury, *Appl. Geochem.* **24**, 999 (2009).
10. Y.-H. Li and S. Gregory, *Geochim. Cosmochim. Acta* **38**, 703 (1974).
11. M.H. Bradbury and B. Baeyens, *J. Contam. Hydrol.* **42**, 141 (2000).
12. A. Jakob, W. Pfingsten and L. Van Loon, *Geochim. Cosmochim. Acta* **73**, 2441 (2009).
13. F.J. Pearson, D. Arcos, A. Bath, J.-Y. Boisson, A.M. Fernández, H.-E. Gäbler, E. Gaucher, A. Gautschi, L. Griffault, P. Hernán and H.N. Waber, *Mont Terri Project – Geochemistry of Water in the Opalinus Clay Formation at the Mont Terri Rock Laboratory*, (Reports of the FOWG, Geology Series, No. 5, Bern, 2003), p. 141.

Mater. Res. Soc. Symp. Proc. Vol. 1665 © 2014 Materials Research Society
DOI: 10.1557/opl.2014.630

# Incorporation of REE into Secondary Phase Studtite.

C. Palomo, N. Rodríguez, E. Iglesias, J. Nieto, J. Cobos and J. Quiñones
Centro de Investigaciones Energéticas, Medioambientales y Tecnológicas (CIEMAT),
Av. Complutense 40 28040 Madrid, Spain
javier.quinones@ciemat.es, carmen.pm22@gmail.com

## ABSTRACT

The formation of uranyl peroxide phases was identified as a corrosion product of spent fuel by Hanson et al [1]. The subsequent analysis of this phase showed that metastudtite retained [241]Am, [237]Np and [239]Pu [2]. In this study, the retention of radionuclide $Pu^{4+}$ and $An^{3+}$, released from the spent fuel matrix into studtite structure, has been evaluated by the precipitation of studtite from uranyl dissolution with variable concentrations of REE (Th, Nd, Sm and Eu). Three different precipitation conditions parameters were studied: media of synthesis, time of synthesis and REE concentration. Synthesized phases were characterized by XRD and the cell parameter was calculated. The REE incorporation was determined by ICP-MS analysis. The results showed that studtite could incorporate 63% of Th in solution during its precipitation. Changes in the "a" cell parameter were identified. The results suggest that studtite coprecipitated with REE could play a role as a limiting for the REE mobility.

## INTRODUCTION

The spent fuel (SF) is composed 95% of $UO_2$ and 5% of actinides (e.g., [239]Pu, [237]Np, [241]Am) and fission product (e.g., [137]Cs, [90]Sr, [99]Tc). The $UO_2$ is extremely insoluble under reducing conditions [3]. Alpha radiation field associated to SF is able to decompose water in radiolytic products. Those species generate oxidizing conditions in the near field of SF. $UO_2$ is not stable under this locally oxidizing environment [4] and peroxides coming from the irradiated water oxidize the U(IV) to the more soluble U(VI) oxidation state. Then, hexavalent uranium U(VI) could react with ligands in groundwater and precipitates as secondary phase[5, 1]. The precipitated secondary phases could be capable to retain several randionuclides as Burns et al in 1997 [6] theorized. In this way, Cs, Ba, Np and Pu were identified by radiochemical analysis of secondary phases [7, 2].

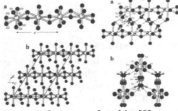

**Figure 1.** Structure of studtite.[9]

Studtite was characterized as corrosion product of SF by Hanson et al. Its structure was described by Walenta 1976 [8] and refined by Burns and Hughes (Figure 1) [9]. Possible incorporation of Np in the transformation of metaschoepite to studtite has been studied by Douglas et al [10].

## EXPERIMENTAL

### Materials - Synthesis of studtite

Solid-phase synthesis was carried out using the following reagents: uranyl nitrate hexahydrated 99%, Ln and Th dissolution of 1000 µg/l in nitric acid (1-5%wt) and $H_2O_2$ (30%).

Two sets of precipitation experiments were performed in order to test geological repository conditions: some of them in demineralized water and other in Grimsel Groundwater, .

The uranyl nitrate dissolution (0.1M) was prepared by mixing uranyl nitrate hexahydrated 99% Merk with ultrapure water or Grimsel groundwater from Switzerland in order to reproduce the synthesis media. A REE dissolution has been added before precipitation occurs, mixing with uranyl nitrate dissolution. Solid secondary phases have been obtained by addition of excess of $H_2O_2$ to each solution. Precipitation parameters were evaluated: different synthesis time, water composition and trace element concentration. Experimental conditions used in the synthesis are compiled in Table I.

**Table I.** Description of experimental conditions of synthesis

| Experiment | Media | Element | % mol Trace/ mol U | synthesis time (h) |
|---|---|---|---|---|
| SD-t | Demineralized | Th | 0.2 % | 3, 24, 72, 168 |
| SD-c | Deminerilized | Th | 0.2%, 2%, 20% | 72 |
| SD-tc | Deminerilized | Th | 2%, 4% | 3, 24 |
| SD -Ln | Deminerilized | Nd, Eu, Sm | 2% | 72 |
| SG-t | Grimsel | Th | 0.2 % | 3, 24, 72, 168 |
| SG-Ln | Grimsel | Nd, Eu, Sm | 2% | 72 |

**Experimental Techniques**

To determine the Th and Ln's content, synthesized phases were redissolved. An aliquot of solution was analyzed by ICP-MS using a Perkin Elmer ELAN6000.

All of the precipitated phases were characterized by X-ray Powder diffraction with a Philips XRD-X' Pert – MDP The International Centre for Diffraction Data Powder Diffraction File (ICDD-PDF) database of reference patterns was used to identify phases present in samples [11]. Also, examination with Microscopy electron diffraction was performed with a Hitachi – SV 6600.

The cell parameter was determined by CELREF v.3 software [12] applying the results of XRD.

**RESULTS AND DISCUSSION**

Precipitation was achieved in all synthesis conditions: formation of yellow powder of uranyl minerals were observed in all of the cases.

All samples precipitated characterized by XRD were identified as studtite; using the database (ICDD-98-016-7992) [11] (Figure 2). The diffraction patterns did not show significant differences between the samples precipitated in both media; neither was observed differences in the compounds with Th and Ln's. Formation of pure Th or Ln phases was not detected. The sample, precipitated in media 0.2 mol Th/mol U (Figure 2b), showed a better crystallinity level than the other samples tested.

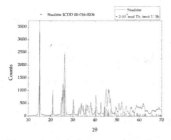
a)

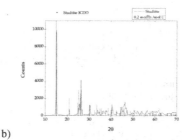
b)

**Figure 2.** Diffractograms of synthetized studtite with trace element.(lefty-axis) and ICDD reference pattern (right y-axis). (a) Studtite precipitated in ultrapure water with 0.002 Th/U rate mole,. (b) Studtite precipitated in ultrapure water with 0. 2 Th/U rate mole. Synthesis time: 72h.

Figure 3a shows the SEM image of studtite structure precipitated in Grimsel groundwater with Eu. Figure 3b shows studtite precipitated with Th (mole ratio Th/U of 0.2). The comparative analysis of samples did not show any important differences. The morphology was similar in all samples. A completely different structure to the other samples precipitated has been observed in the samples precipitated from a solution with 0.2 mol Th/mol U (Figure 2b). A

a)

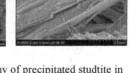
b)

**Figure 3.** a) Micrography of precipitated studtite in Grimsel groundwater with Eu, b) Studtite precipitated with Th 20%.

fibrous crystal formation characteristic of the studtite was detected [8, 9].

Table II shows the results of studtite powder analyzed by ICP-MS. First column corresponds to the ratio of mol of trace element (Tr) incorporated to studtite by mol of precipitated studtite. Assay SD-t showed Th incorporation was improved with time. The amount of precipitated studtite increases when the reaction time also increases, also raising the capability of Th incorporation into the uranyl phase, confirmed by the SG-t results. Looking at SD-c experiment, Th moles incorporated to studtite increased with Th percentage in the solution. Th is retained into studtite in the precipitation, so the presence of higher Th% in solution contributes to a higher Th included in studtite during precipitation process. In order to evaluate the parameter that will determine the Th incorporation, a SD-tc test has been set to compare to the SD-t and SD-c studies. Results showed that Th moles in studtite increased with the concentration of media, despite of the reaction time increasing. SD-Ln assay did not show important differences among elements. Ln retained into solid sample was two orders of magnitude lower than Th, these results demonstrate that the oxidation state would determine the incorporation in studtite.

The second column represents the efficiency ($\eta$) of incorporation. This parameter has been calculated as the ratio between the mol of Tr incorporated to studtite and initial Tr mol in the solution. The results showed that coprecipitation efficiency increased with time of synthesis. Distribution coefficient ($\lambda$) (column 3) is described by the equation:

$$\ln\frac{[T]}{[T]_0} = \lambda \cdot \ln\frac{[U]}{[U]_0}$$
<1>

where [T] and [U] are the concentration of Tr and U in solution, respectively, after studtite precipitation. $[T]_0$ and $[U]_0$ represent the initial concentration before studtite precipitation .

$\lambda$ results showed that trace ions was removed from the dissolution and incorporated to the solid while the studtite precipitated.

**Table II.** ICP-MS results obtained from the analyzed samples after dissolved studtite precipitated with Trace (Tr) element leached.

| Essay | Sample | mol trace element/ mol studtite precipitated | η | λ |
|-------|--------|---------------------------------------------|------|------|
| SD-t | SDTh-3 | $2.43 \cdot 10^{-04}$ | 0.09 | 0.1784 |
| | SDTh-24 | $9.27 \cdot 10^{-04}$ | 0.34 | 0.0692 |
| | SDTh-72 | $1.19 \cdot 10^{-03}$ | 0.15 | 0.4378 |
| | SDTh-168 | $1.62 \cdot 10^{-03}$ | 0.47 | 0.0686 |
| SD-c | SDTh | $3.16 \cdot 10^{-03}$ | 0.13 | 0.0235 |
| | SDTh-10 | $1.82 \cdot 10^{-01}$ | 0.63 | 0.2879 |
| SD-tc | SD2Th-24 | $7.93 \cdot 10^{-03}$ | 0.32 | 0.2702 |
| | SD2Th-72 | $6.86 \cdot 10^{-03}$ | 0.27 | 0.1058 |
| | SD4Th-24 | $8.21 \cdot 10^{-03}$ | 0.18 | 0.1156 |
| | SD4Th-72 | $2.06 \cdot 10^{-02}$ | 0.38 | 0.3315 |
| SG-t | SGTh-3 | $1.32 \cdot 10^{-03}$ | 0.36 | 0.0518 |
| | SGTh-24 | $8.34 \cdot 10^{-04}$ | 0.25 | 0.0833 |
| | SGTh-72 | $1.20 \cdot 10^{-03}$ | 0.37 | 0.0723 |
| | SGTh-168 | $1.90 \cdot 10^{-03}$ | 0.48 | 0.0933 |
| SD-Ln | SDNd | $9.29 \cdot 10^{-05}$ | 0.0046 | 0.0106 |
| | SDEu | $9.10 \cdot 10^{-05}$ | 0.0041 | 0.0014 |
| | SDSm | $4.19 \cdot 10^{-05}$ | 0.0013 | 0.0049 |
| SG-Ln | SGNd | $1.28 \cdot 10^{-04}$ | 0.0028 | 0.0012 |
| | SGEu | $7.85 \cdot 10^{-05}$ | 0.0020 | 0.0007 |
| | SGSm | $5.05 \cdot 10^{-05}$ | 0.0016 | 0.0006 |

The cell parameter a, b, c and β angle were calculated from the XRD results. In Figure 4 and 5 the value of parameter (a, b, c and β) as a function of mol trace/mol studtite can be observed. .

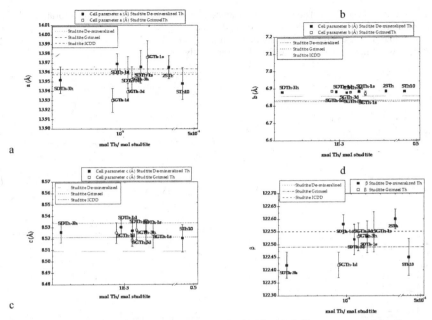

**Figure 4**. Cell parameter of studtite precipitated with Th. (a) Cell parameter a. (b) cell parameter b. (c) cell parameter c. (c) β angle

Figure 4 shows the cell parameters of the samples synthesized with Th. The cell parameter "a" (Figure 4a) was modified by the Th incorporation into studtite. An increase of this cell parameter for incorporation of $8 \cdot 10^{-4} - 10^{-3}$ mol Th/mol studtite was observed. However, it decreased with higher concentrations of retained Th. The "b" parameter kept constant for all samples independent on Th concentration incorporated. "c" parameter was slightly modified by the incorporation of Th, discovering that an increasing in Th composition could be related to "c" parameter reduction. The discrepancies in "c" could be associated to experimental errors in the "c" determination. The incorporation of Th modified the studtite "β" angle too. It keep nearly constant for $8 \cdot 10^{-4} - 10^{-3}$ mol Th/mol studtite, but this value decreases for higher Th incorporation. The synthesis media slightly modified the cell parameters.

As a summary, these results indicated that the incorporation of Th would not modify the uranyl chain (Figure 1), coming about Th did not replace U (VI) atoms in the studtite structure. The incorporation would take place in the interlayer of uranyl chains connected by hydrogen bonds (Figure 1) [9].The drop of cell parameters observed when the incorporation of Th > $10^{-2}$ mol Th/mol studtite could be explained by the retention of Th outside the crystal lattice (e.g dislocations [13])

Studying $Ln^{3+}$ incorporation as analogue of $Ac^{3+}$ (see Figure 5), the variations observed in the cell parameters showed that an increase of Ln concentration is associated to a reduction of "a" cell parameter (Figure 5a). The l "b" cell parameter was not modified by the incorporation of trace element in this structure as Th incorporation showed.

Summarizing, these results indicate that incorporation of trace element occurs in the interlayer between the uranyl chains, as the Th results firstly showed.

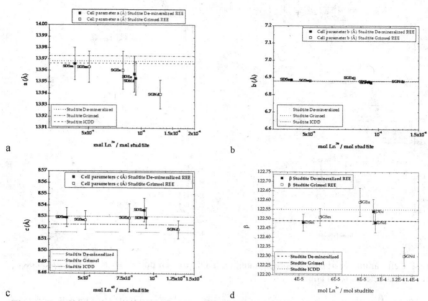

a

b

c

d

**Figure 5.** Cell parameter of studtite precipitated with Ln. (a) Cell parameter "a". (b) cell parameter "b". (c) cell parameter "c". (c) "β" angle

## CONCLUSIONS

The different synthesis conditions studied in this work allow us to conclude that the method of synthesis developed in this work it is as an adequate process to obtain the studtite pure and unique phase. Moreover, during the process, phases of trace elements have not been identified. The precipitation synthesis media did not alter the incorporation of radionuclide into the studtite structure. Besides, the time and composition in the media of synthesis improve the incorporation of radionuclides into the studtite structure. Solid phase characterization has confirmed that the incorporation of $Th^{4+}$ into the studtite is more favourable than that for the $Ln^{3+}$. On the other hand, the incorporation of trace element into studtite did not modify the cell. The REE tested in this work did not replace the $U^{6+}$ into the structure they could be retained between uranyl chains.

76

**ACKNOWLEDGMENTS**

This research was supported by the Spanish personnel research training program from Ministry of Economy and Competitiveness with a grant of National Plan of R+D+I 2009. We are grateful to L. Serrano, S. Duran for the ICP-Ms analysis, J.M. Cobo and L. Gutierrez

**REFERENCES**

1. Hanson, B., et al., Radioch. Acta 93 (2005) 159–168
2. McNamara, B., et al., Radioch. Acta 93 (2005) 169-175
3. Shoesmith, D. W., J. Nucl. Mater., 282 (2000) 1-30
4. Christensen, H. Nucl. Technology 131 (2000) 102-123
5. Wronkiewicz, D.J., et al., J. Nucl. Mater., 190 (1992) 101-106
6. Burns, P.C., et al., J. of Nucl. Mater., 245 (1997) 1-9
7. Buck, E.C., J. Nucl. Mater., 249 (1997) 70-76
8. Walenta, K. American Mineralogist, 59 (1974) 166-171, 1974
9. Burns, P.C., et al., American Mineralogist, 88(2003) 1165–1168
10. Douglas, M., et al., *Environmental Science & Technology 39 (*2005**), , 4117-4124.
11. JCPDS - ICDD -PDF-2 Release_2002. (International Centre for Diffraction Data, 2002).
12. Laugier, J. and Bernard, B. CELREF. (Laboratoire des Matériaux et du Génie Physique. Ecole Nationale Supérieure de Physique de Grenoble (INPG), 2003).
13. Curti, E. PSI-Bericht Nr. 97-10 (1997)

Mater. Res. Soc. Symp. Proc. Vol. 1665 © 2014 Materials Research Society
DOI: 10.1557/opl.2014.631

# Migration behavior of plutonium affected by ferrous ion in compacted bentonite by using electrochemical technique

Daisuke Akiyama[1], Kazuya Idemitsu[1], Yaohiro Inagaki[1], Tatsumi Arima[1], Kenji Konashi[2], Shinichi Koyama[3]

[1] Department of Applied Quantum Physics and Nuclear Engineering, Kyushu University, Fukuoka 819-0395, Japan

[2] Institute for Materials Research, Tohoku University, Oarai, Ibaraki 311-1313, Japan

[3] Japan Atomic Energy Agency, 4002 Narita, Oarai, Higashi-Ibaraki, Ibaraki 311-1393, Japan

## ABSTRACT

The migration behavior of plutonium is expected to be affected by the corrosion products of carbon steel in compacted bentonite at high-level waste repositories. Electrochemical experiments were carried out to simulate the reducing environment created by ferrous iron ions in equilibrium with anoxic corrosion products of iron. The concentration profiles of plutonium could be described by the convection -dispersion equation to obtain two migration parameters: apparent migration velocity $V_a$ and apparent dispersion coefficient $D_a$. The apparent migration velocity was evaluated within 1 nm/s and was found to be independent of the experiment duration and the dry density of bentonite in the interval 0.8-1.4 Mg/m$^3$. The apparent dispersion coefficient increased with the experiment duration at a dry density of 1.4 Mg/m$^3$. The results for other dry densities also showed the same trend. These findings indicate that plutonium migration likely starts after ferrous ions reach the plutonium, in other words, the reducing environment due to ferrous ions could change the chemical form of plutonium and/or the characteristics of compacted bentonite. The apparent diffusion coefficient was estimated to be around 0.5 to 2.2 μm$^2$/s and increased with decreasing the dry density of bentonite.

## INTRODUCTION

Bentonite buffers and carbon steel overpacks will constitute important parts of multi-barrier systems for geological disposal of high-level waste in Japan [1]. After closure of a high-level waste repository, the carbon steel overpack will produce anoxic iron corrosion products containing ferrous iron, which can migrate into bentonite. Iron corrosion products may alter bentonite and maintain a reducing environment, which may affect the migration behavior, especially for redox-sensitive elements such as plutonium. Plutonium has a very low diffusivity in compacted bentonite under an oxidizing environment. There have been a few studies on plutonium diffusion in compacted bentonite, for example, in a concrete-bentonite system with an experiment duration as long as 5 years [2]. However, there have been few reports on its diffusion behavior under a reducing environment. Therefore, the authors have developed and carried out electrochemical experiments with ferrous ions supplied to compacted bentonite by anodic corrosion of an iron coupon [3,4].

## EXPERIMENTAL

A typical Japanese sodium bentonite, Kunipia-F, was used in this experiment. It contains approximately 95 wt % of montmorillonite. The chemical formula of Kunipia-F is estimated to be $(Na_{0.3}Ca_{0.03}K_{0.004})(Al_{1.6}Mg_{0.3}Fe_{0.1})Si_4O_{10}(OH)_2$ [5].

Powdered bentonite was compacted into pellets with diameter of 10 mm and height of 10 mm and compacted to dry densities of 0.8 to 1.4 $Mg/m^3$. Each pellet was inserted into an acrylic resin column and saturated with an aqueous solution of 0.01 M NaCl for 30 days.

A carbon steel coupon with a concave section with diameter of 12 mm and depth of 1 mm was assembled with bentonite saturated by immersion in the 0.01 M NaCl solution in an apparatus for electromigration (Figures 1 and 2). Ten microliters of $^{238}Pu$ tracer solution of 0.5 to 1 kBq (3.5 to $7 \times 10^{-12}$ mol) was spiked on the interface between the bottom of a bentonite specimen inserted into an acrylic resin column and the top surface of a bentonite specimen inserted into the concave section of the carbon steel coupon. A Ag/AgCl reference electrode and a Pt foil counter electrode were immersed in 0.01 M NaCl solution in the upper part of the apparatus. The carbon steel coupon was connected to a potentiostat as a working electrode and was supplied with an electric potential of 0 mV versus the Ag/AgCl electrode at 25 °C. In all cases, the carbon steel coupon had a potential approximately 1 V higher than the counter electrode. After supplying electric potential for a certain period of time, the bentonite specimen was cut into slices of 0.3 to 2 mm in thickness. Each slice was immersed in a 1 M HCl solution to extract plutonium and iron, and the liquid phase was separated by centrifugation. Then, the supernatant was analyzed by using an

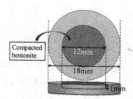

**Figure 1.** Carbon steel coupon with concave section

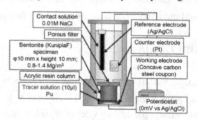

**Figure 2.** Experimental apparatus for electromigration.

alpha liquid scintillation counter (ORDELA) for plutonium and an atomic absorption spectrometer (AAS: Shimadzu; AA-6300) for iron.

## RESULTS AND DISCUSSION

### Profiles and determination of apparent migration velocity and apparent dispersion coefficient of plutonium and iron

The concentration profiles of iron are shown in Figures 3 and 4. Lower dry density of bentonite and longer experiment durations results in deeper iron migration. Iron as ferrous ion migrates via the interlayer of montmorillonite, where ion exchange occurs between ferrous and sodium ions.

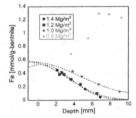

**Figure 3.** Iron profiles in bentonite at various dry densities of bentonite after 13 days.

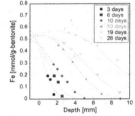

**Figure 4.** Iron profiles in bentonite at a dry density of bentonite of 1.4 Mg/m$^3$ after up to 26 days.

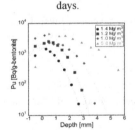

**Figure 5.** Plutonium profiles in bentonite at various dry densities of bentonite after 13 days.

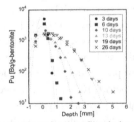

**Figure 6.** Plutonium profiles in bentonite at a dry density of 1.4 Mg/m$^3$ after various durations between 3 and 26 days.

The concentration profiles of plutonium in bentonite specimens are shown in Figures 5 and 6. Each profile of plutonium has a peak, which is slightly offset from the origin.

Electromigration can be described with the dispersion-convection equation [6]:

$$\frac{\partial C}{\partial t} = D_a \frac{\partial^2 C}{\partial x^2} - V_a \frac{\partial C}{\partial x}, \qquad \begin{array}{l} C(x,0) = M \\ C(\pm\infty, t > 0) = 0 \end{array} \qquad (1)$$

where $D_a$ is the apparent dispersion coefficient and $V_a$ is the apparent convection velocity, including the negligible electro-osmotic flow of water, $M$ is the total amount of plutonium in the plane source. The dispersion-convection Eq. (1) is analyzed numerically. The model fitting curves show good agreement with the profiles of plutonium obtained as shown in Figures 5 and 6.

## Apparent migration velocity, dispersion coefficient, and diffusion coefficient of plutonium

Figures 7 and 8 show that the apparent migration velocity obtained was within 1 nm/s, that was much smaller than that of iron [8]. On the other hand, the apparent dispersion coefficient was found to depend on the experiment duration and the dry density of bentonite. Figure 7 shows that the apparent dispersion coefficient increased with decreasing dry density, and the dispersion coefficient after 21 days is greater than that after 13 days. Figure 8 shows that in short-term tests of up to 6 days, the apparent dispersion coefficient was up to 0.1 $\mu m^2$/s at a dry density of 1.4 $Mg/m^3$, whereas in long-term tests of 10 to 26 days, it increased by 0.2 to 0.5 $\mu m^2$/s, depending on the experiment duration.

In general, the relationship between the apparent dispersion coefficient $D_a$ [$m^2$/s] and the apparent diffusion coefficient $D^*_a$ [$m^2$/s] is as follows:

$$D_a = D^*_a + \alpha V_a \qquad (2)$$

where $\alpha$ is the dispersion length [m] and $V_a$ is the apparent migration velocity [m/s]. This equation means that the dispersion coefficient is a sum of the diffusion coefficient and the mechanical dispersion coefficient. The dispersion coefficient was estimated to be approximately $10^{-13}$ $m^2$/s, and the mechanical dispersion coefficient was estimated to be approximately $10^{-14}$ $m^2$/s, which is the product of the dispersion length ($\sim 10^{-4}$ m, obtained in a previous study) and the migration velocity ($\sim 10^{-10}$ m/s). Therefore, the dispersion and diffusion coefficients of plutonium might be approximately equal.

The relation in which the diffusion coefficient increased with increasing the experiment duration might be attributable to plutonium migration starting when ferrous ions reached the plutonium and changed the chemical form of plutonium. This is why we assumed that plutonium migration started after 5 days after adding the plutonium spike and supplying an electric potential. Figure 9 shows that the diffusion coefficient of plutonium at a dry density of 1.4 $Mg/m^3$ was approximately constant at 0.6 $\mu m^2$/s. The diffusion coefficient for other densities was also estimated. In comparison with the diffusion coefficient reported in previous study, under reducing condition diffusion coefficient is much higher than under aerobic condition [7,8,9]. Our group has previously reported that the diffusion coefficient of plutonium does not exceed 0.1 μm²/s under reducing condition (Figure 10) [7,8]. This discrepancy might be due to the previous experiment being too short to obtain the true diffusion coefficient of plutonium and the diffusion coefficient increasing as the amount of ferrous ions increased. The diffusion coefficient obtained in this study is comparable with that in H-type bentonite which has low swelling capability and large pore space for diffusion pathway [10]. PH change in pore water by oxidation of ferrous ions could follow the reaction in Eq. (3,4) [11]. These reactions could change plutonium chemical form and pore structure of bentonite as H-type bentonite.

$$Pu(IV) + Fe(II) \leftrightarrow Pu(III) + Fe(III) \qquad (3)$$

$$Fe^{2+} + 3H_2O \leftrightarrow Fe(OH)_3(am) + 3H^+ + e^- \qquad (4)$$

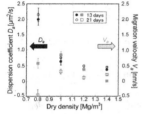

**Figure 7.** Dispersion coefficient and migration velocity of plutonium as a function of the dry density of bentonite.

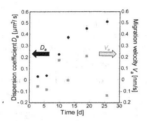

**Figure 8.** Dispersion coefficient and migration velocity of plutonium as a function of the experiment duration at a dry density of 1.4 Mg/m³.

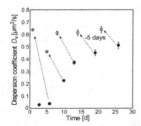

**Figure 9.** Dispersion coefficient of plutonium as a function of experiment duration, assuming that plutonium migration starts 5 days after spiking the solution on bentonite with a dry density of 1.4 Mg/m³.

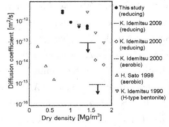

**Figure 10.** Comparison of diffusion coefficient values for plutonium obtained in this study and in previous studies.

## CONCLUSION

The apparent diffusion coefficient of plutonium was estimated to be $0.6 \ \mu m^2/s$ at a dry density of bentonite of $1.4 \ Mg/m^3$. This value is much larger than those reported in previous studies. In addition, the diffusion coefficient at other dry densities showed the same trend. This is likely due to the chemical form of plutonium changing, which may occur as a result of the reducing environment and decreasing pH in bentonite.

## ACKNOWLEDGEMENTS

This study was conducted at the Institute for Materials Research, Tohoku University. The authors thank M. Watanabe of the Institute for assistance in this study.

## REFERENCES

1. JNC, H12: *Project of Establish the Scientific and Technical Basis for HLW Disposal in JAPAN*, JNC, Tokai Japan (2000).
2. Albinsson, Y., K. Andersson, S. Börjesson, B. Allard, J. Contaminant Hydrology 12, 189 (1996).
3. K. Idemitsu, X. Xia, Y. Kikuchi, Y. Inagaki and T. Arima in *Scientific Basis for Nuclear Waste Management XXVIII*, edited by John M. Hanchar, Simcha Stroes-Gascoyne and Lauren Browning (Mater. Res. Soc. Proc. **824**, Pittsburgh, PA, 2004) pp.491-496.
4. K. Idemitsu, Y. Yamasaki, S. A. Nessa, Y. Inagaki, T. Arima, T. Mitsugashira, M. Hara, Y. Suzuki in Scientific Basis for Nuclear Waste Management XXX, edited by D.S. Dunn, C. Poinssot, B. Begg (Mater. Res. Soc. Proc. **985**, Pittsburgh, PA, 2007), NN11-7, pp.443-448.
5. H. Sato, T. Ashida, Y. Kohara, M. Yui, and N. Sasaki, J. Nucl. Sci. Tech. **29**, 873 (1992).
6. N. Maes, H. Moors, A. Dierckx et al., J. Contam. Hydrol., 36, 231–247 (1999).
7. K. Idemitsu, X. Xia, T. Ichishima, H. Furuya, Y. Inagaki, T. Arima et al., in *Scientific Basis for Nuclear Waste Management XXIII*, edited by Robert W. Smith, David W. Shoesmith (Mater. Res. Soc. Proc. **608**, Boston, MA, 2000) pp.261-266.
8. K. Idemitsu, H. Ikeuchi, S. A. Nessa, Y. Inagaki, T. Arima et al., in *Scientific Basis for Nuclear Waste Management XXXII*, edited by Neil C. Hyatt, David A. Pickett and Raul B. Rebak (Mater. Res. Soc. Proc. **1124**, Boston, MA, 2009) pp.283-288.
9. H. Sato, PNC TN8410 97-202 (1998)
10. K. Idemitsu, K. Ishiguro, Y. Yusa, N. Sasaki, N. Tsunoda, Engineer. Geol. 28, 455 (1990)
11. Rai D, Gorby Y, Fredrickson J, Moore D, Yui M. J. Solution Chem. **31** (2002) pp. 433-453.

Mater. Res. Soc. Symp. Proc. Vol. 1665 © 2014 Materials Research Society
DOI: 10.1557/opl.2014.632

## Modeling of an in-situ diffusion experiment in granite at the Grimsel Test Site

Josep M. Soler[1], Jiri Landa[2], Vaclava Havlova[2], Yukio Tachi[3], Takanori Ebina[3], Paul Sardini[4], Marja Siitari-Kauppi[5] and Andrew J. Martin[6]

[1]IDAEA-CSIC, Barcelona, Catalonia, Spain.
[2]UJV-Rez, Husinec-Rez, Czech Republic.
[3]JAEA, Tokai, Japan.
[4]University of Poitiers, France.
[5]University of Helsinki, Finland.
[6]NAGRA, Wettingen, Switzerland.

### ABSTRACT

Matrix diffusion is a key process for radionuclide retention in crystalline rocks. Within the LTD project (Long-Term Diffusion), an in-situ diffusion experiment in unaltered non-fractured granite was performed at the Grimsel Test Site (www.grimsel.com, Switzerland). The tracers included $^3$H as HTO, $^{22}$Na$^+$, $^{134}$Cs$^+$ and $^{131}$I$^-$ with stable I$^-$ as carrier.

The dataset (except for $^{131}$I$^-$ because of complete decay) was analyzed with different diffusion-sorption models by different teams (NAGRA / IDAEA-CSIC, UJV-Rez, JAEA, Univ. Poitiers) using different codes, with the goal of obtaining effective diffusion coefficients ($D_e$) and porosity ($\phi$) or rock capacity ($\alpha$) values. A Borehole Disturbed Zone (BDZ), which was observed in the rock profile data for $^{22}$Na$^+$ and $^{134}$Cs$^+$, had to be taken into account to fit the experimental observations. The extension of the BDZ (1-2 mm) was about the same magnitude as the mean grain size of the quartz and feldspar grains.

$D_e$ and $\alpha$ values for the different tracers in the BDZ are larger than the respective values in the bulk rock. Capacity factors in the bulk rock are largest for Cs$^+$ (strong sorption) and smallest for $^3$H (no sorption). However, $^3$H seems to display large $\alpha$ values in the BDZ. This phenomenon will be investigated in more detail in a second test starting in 2013.

### EXPERIMENTAL SETUP

Several tracers ($^3$H as HTO, $t_{1/2}$ = 12.33 a; $^{22}$Na$^+$, $t_{1/2}$ = 2.602 a; $^{134}$Cs$^+$, $t_{1/2}$ = 2.065 a; $^{131}$I$^-$, $t_{1/2}$ = 8.021 d, with stable I$^-$ as carrier) were continuously circulated through a packed-off borehole and the decrease in tracer concentrations in the liquid phase was monitored for a period of about 2.5 years. The experimental setup is shown in Fig. 1.1. The length and radius of the injection interval were 70 cm and 2.8 cm, respectively. The total volume of solution in the circulation system was 8 L [1].

At the end of the experiment, the borehole section was overcored and the tracer profiles in the rock analyzed. Transport distances in the rock were about 20 cm for $^3$H, 10 cm for $^{22}$Na$^+$ and 1 cm for $^{134}$Cs$^+$. Recovery of activities was not complete. Experimental results included the evolution of tracer concentrations (activities of $^3$H, $^{22}$Na$^+$, $^{134}$Cs$^+$) in the circulation system and tracer distribution profiles in the rock around the injection interval. $^3$H and $^{22}$Na$^+$ showed a similar decrease in activity in the circulation system (slightly larger drop for $^3$H). The drop in activity for $^{134}$Cs$^+$ is much more pronounced. Transport distances in the rock are about 20 cm for

$^3$H, 10 cm for $^{22}$Na$^+$ and 1 cm for $^{134}$Cs$^+$. I$^-$ diffusion was only evaluated from results of out-leaching tests performed on overcored rock blocks.

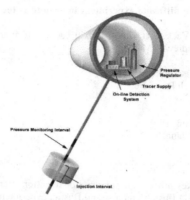

**Figure 1.** Schematic diagram showing the experimental setup.

## CALCULATIONS

### Modeling approach by NAGRA / IDAEA-CSIC

Modeling was performed using the CrunchFlow reactive transport code [2]. The 1D calculations (with symmetry around borehole axis) solved the equation of conservation of tracer mass, which can be written as

$$\frac{\partial c_{tot}}{\partial t} = \nabla \cdot \left( D_e \nabla c \right) \tag{1}$$

where $c$ is concentration in solution (mol/m$^3$), $t$ is time (s), $D_e = \phi D_p$ is the effective diffusion coefficient (m$^2$/s), $\phi$ is the porosity accessible for the tracer (-), $D_p$ is the pore diffusion coefficient (m$^2$/s) and $c_{tot}$ is the total concentration of tracer (mol/m$^3$bulk), which is given by

$$c_{tot} = \phi c + \rho_d s = \phi c + (1-\phi)\rho_s K_d c = (\phi + (1-\phi)\rho_s K_d)c = \alpha c \tag{2}$$

where $\rho_d$ is the bulk dry density (kg/m$^3$), $s$ is the concentration of sorbed tracer (mol/kg_solid), $\rho_s$ is the density of solids (2660 kg/m$^3$), $K_d$ is the distribution coefficient (m$^3$/kg) and $\alpha$ is the rock capacity factor (-). Only linear sorption ($s = K_d c$) has been considered in the calculations. Radioactive decay was taken into account by correcting the measured activities back to the beginning of the experiment. As an example, Fig.2 shows results for $^{22}$Na$^+$. Activities in the circulation system decrease with time. However, the large activity in solution in the circulation system at about 32 days may be an experimental artifact due to poor mixing. Tracer profiles in the rock show anomalous large activities very close to the borehole (d = 0.1 cm). These points correspond to the BDZ (stronger sorption compared with the bulk rock).

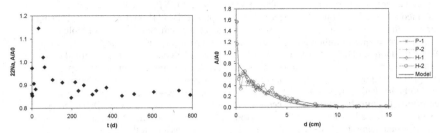

**Figure 2.** $^{22}$Na$^+$. Experimental data and model results. Left: Relative activities in solution in the circulation system (experimental data measured at PSI, Switzerland). Right: Tracer profiles in the rock (relative activities vs. distance from borehole wall). P-1 and P-2 refer to profiles measured by UJV-Rez; H-1 and H-2 refer to profiles measured by Univ. of Helsinki; model curve corresponds to activities in the porewater. Model parameters – rock: $D_e = 2\times10^{-12}$ m$^2$/s, $\alpha =$ 0.2 ($K_d = 7.3\times10^{-5}$ m$^3$/kg for a porosity $\phi = 0.0065$). Model parameters – BDZ: thickness = 1.5 mm, $D_e = 3\times10^{-12}$ m$^2$/s, $\alpha = 3$ ($K_d = 1.1\times10^{-3}$ m$^3$/kg for a porosity $\phi = 0.0065$).

**Modeling approach by UJV-Rez**

A g77 Fortran code was written which solves the 1D cylindrical problem given by

$$\frac{\partial c_j(t,r)}{\partial t} = \frac{Dp_i}{R_i(r,Kd)}\left(\frac{\partial^2 c_j(t,r)}{\partial r^2} + \frac{1}{r}\cdot\frac{\partial c_j(t,r)}{\partial r}\right) - \lambda_j.c_j(t,r) \tag{3}$$

where $c$ is concentration in pore water (mol/m$^3$), $D_p$ is the pore diffusion coefficient (m$^2$/s), $K_d$ is the distribution coefficient (m$^3$/kg), $r$ is radius (m), $R$ is the retardation factor in the rock matrix (-), $\lambda$ is the decay constant (s$^{-1}$) and $\varepsilon$ is porosity (-). Indices $i$ and $j$ refer to chemical species and radionuclides, respectively. The 1D cylindrical model contemplated symmetry around the borehole axis and explicit calculation of radioactive decay.

**Modeling approach by JAEA**

Calculations were performed with the commercial codes GoldSim (ver. 10.1; [3]) and COMSOL Multiphysics (ver. 4.2; [4]). The GoldSim code simulated simplified 1D radionuclide transport (including symmetry around borehole axis), and COMSOL simulated 3D transport. The codes solved eq. 1, and radioactive decay was taken into account by correcting the measured activities back to the beginning of the experiment. The quantitative comparison between 1D and 3D calculations justified the use of 1D simulations for this system.

## Modeling approach by University of Poitiers

This team modeled the results of out-leaching experiments performed with rock blocks overcored after the experiment. At the University of Helsinki, [3]H and stable iodide were simultaneously leached from two overcored sub samples (called HYRL1 and HYRL2) using several small boreholes (or observation boreholes) located around the central borehole (Fig. 3).

The modeling of radionuclide diffusion through rock porosity was undertaken using the TDD (Time Domain Diffusion) method [5-9]. This is a finite volume approach solved by a Lagrangian method (particle tracking) which works under the time domain. Each particle moves into a grid according to probability transitions which are pre-calculated. Particle positions are forced to be located in the center of the square grid cells. For modeling in-situ diffusion (before overcoring), the injection borehole is centered in the simulation domain, the 2D grid being perpendicular to the borehole axis (Fig. 4). Modeling of the out-leaching phase (after overcoring) was performed using the geometries shown in Fig. 3. It was assumed that the rock was homogeneous (no BDZ), having porosity of 0.65%.

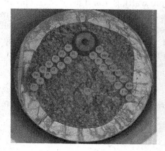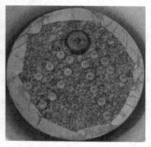

**Figure 3.** HYRL1 (left) and HYRL2 (right) sub samples were sawn from the overcored cylinder (30 cm diameter). The central injection borehole is visible in the center top of the section. Small boreholes are filled in grey and are numbered. The overcored surface is covered by impermeable foil glued to the core and the central hole was filled with resin. Diameter of the central hole: 56 mm.

## COMPARISON OF RESULTS AND CONCLUSIONS

Three different teams (NAGRA / IDAEA-CSIC, UJV-Rez, JAEA) performed modeling of the in-situ diffusion experiment based on (i) the measured drop in tracer activities in the circulation system and (ii) measured tracer distribution profiles in the rock. Additionally, the team from University of Poitiers modeled the results of out-leaching experiments ([3]H, I⁻) performed on overcored rock blocks from the in-situ diffusion experiment.

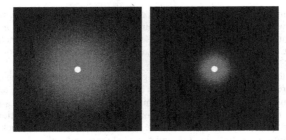

**Figure 4.** Particle clouds after 780 days of diffusion in homogeneous unlimited rock matrix (black) for $D_p = 3 \times 10^{-10}$ m²/s (left) and $D_p = 3 \times 10^{-11}$ m²/s (right), respectively. Rock porosity fixed at 0.65%. 56-mm-diameter central injection borehole is in white.

Tables I and II compile the best-fit parameters for the in-situ experiment from NAGRA / IDAEA-CSIC, UJV-Rez and JAEA. NAGRA/IDAEA-CSIC and UJV tried directly to fit the results of the in-situ experiment, while JAEA conducted a predictive modeling based on laboratory diffusion data and their scaling to in-situ conditions. This different modeling approach by JAEA could approximately reproduce the results for $^{22}$Na$^+$ and $^{134}$Cs$^+$, but it was not possible to reproduce the $^3$H results.

**Table I.** $D_e$(m²/s) and $\alpha$ values for bulk rock.

| TRACER | $D_e$ (IDAEA) | $D_e$ (UJV) | $D_e$ (JAEA) | $\alpha$ (IDAEA) | $\alpha$ (UJV) | $\alpha$ (JAEA) |
|---|---|---|---|---|---|---|
| $^3$H | $2 \times 10^{-13}$ | $1.7 \times 10^{-13}$ | $1.1 \times 10^{-12}$ | 0.0065 | 0.005 | 0.004 |
| $^{22}$Na$^+$ | $2 \times 10^{-12}$ | $6.0 \times 10^{-13}$ | $2.9 \times 10^{-12}$ | 0.2 | 0.13 | 0.062 |
| $^{134}$Cs$^+$ | $3 \times 10^{-12}$ | $4.0 \times 10^{-12}$ | $3.8 \times 10^{-12}$ | 20 | 23 | 5.2 |

**Table II.** $D_e$(m²/s) and $\alpha$ values for the BDZ (IDAEA, UJV:1.5 mm thick; JAEA: 1 mm thick).

| TRACER | $D_e$ (IDAEA) | $D_e$ (UJV) | $D_e$ (JAEA) | $\alpha$ (IDAEA) | $\alpha$ (UJV) | $\alpha$ (JAEA) |
|---|---|---|---|---|---|---|
| $^3$H | $1 \times 10^{-12}$ | $1.3 \times 10^{-12}$ | $2.2 \times 10^{-12}$ | 10 | 11.53 | 0.008 |
| $^{22}$Na$^+$ | $3 \times 10^{-12}$ | $4.8 \times 10^{-12}$ | $5.8 \times 10^{-12}$ | 3 | 5.66 | 3.6 |
| $^{134}$Cs$^+$ | $6 \times 10^{-11}$ | $3.2 \times 10^{-11}$ | $8.0 \times 10^{-12}$ | 110 | 153 | 270 |

Overall there is a relatively good agreement between the values from NAGRA/IDAEA-CSIC and UJV, due clearly to the use of the same modeling concept. The largest differences are for $^{22}$Na$^+$ in the bulk rock (a factor of about 3 for $D_e$ and about 1.5 for $\alpha$) and for $^{134}$Cs$^+$ in the BDZ (a factor of about 2 for $D_e$ and about 1.4 for $\alpha$). $\alpha$ for $^{22}$Na$^+$ also shows a difference by a factor of 2 in the BDZ.

$D_e$ values from JAEA for $^{22}$Na$^+$ and $^{134}$Cs$^+$ in the bulk rock are also in fair agreement with the values obtained by NAGRA/IDAEA-CSIC and UJV. $\alpha$ values are smaller by a factor of 2 to 4. As mentioned above, it was not possible by JAEA to obtain a good match for $^3$H data.

Results from University of Poitiers give a pore diffusion coefficient $D_p$ equal to $3\times10^{-10}$ m$^2$/s as consistent with the results from out-leaching tests for $^3$H and I$^-$ and also from previous tests with $^{36}$Cl$^-$ (Univ. of Helsinki). However, this value is not consistent with the results from NAGRA/IDAEA-CSIC and UJV, which gave $D_p$ values ($D_e$/porosity) for $^3$H about $3\times10^{-11}$ m$^2$/s in the bulk rock ($^3$H profiles measured by UJV).

Regarding the values obtained from through-diffusion experiments in the laboratory (Table III), there are some significant differences between the results from UJV and JAEA (e.g. 2 orders of magnitude for Cs$^+$ $D_e$; 1 order of magnitude for $^3$H $D_e$ and $^{22}$Na$^+$ $\alpha$), although the trend in sorption is clear, from no or very weak sorption for $^3$H to stronger sorption for Cs$^+$. Possible differences in experimental conditions (e.g. tracer concentrations or background solution compositions) and setups should be investigated in more detail.

**Table III.** $D_e$(m$^2$/s) and $\alpha$ values from laboratory-scale through-diffusion experiments.

| TRACER | $D_e$ (UJV) | $D_e$ (JAEA) | $\alpha$ (UJV) | $\alpha$ (JAEA) |
|---|---|---|---|---|
| $^3$H | $1.6 - 2.9\times10^{-11}$ | $3.3\times10^{-12}$ | <0.2 | 0.012 |
| $^{22}$Na$^+$ | $5.4 - 7.2\times10^{-12}$ | $8.7\times10^{-12}$ | 1.4 | 0.19 |
| Cs$^+$ | $1.0\times10^{-13}$ $^{134}$Cs$^+$ | $1.2\times10^{-11}$ $^{137}$Cs$^+$ | 42.4 $^{134}$Cs$^+$ | 16 $^{137}$Cs$^+$ |

Comparing laboratory and in-situ values (bulk rock), $D_e$ for $^3$H is larger in the laboratory, $D_e$'s for $^{22}$Na$^+$ are also larger in the laboratory but comparable to in-situ values, and in-situ $D_e$ for Cs$^+$ is intermediate between laboratory values from UJV and JAEA. $\alpha$ values are comparable. The largest difference is shown by $^{22}$Na$^+$ relative to UJV.

A possible effect caused by the loss of rock confining pressure after overcoring and sampling could be responsible for the observed discrepancy in $D_e$ values for $^3$H and $^{22}$Na$^+$, with larger values measured in the laboratory. Disturbed zones in the laboratory samples may also have an effect on the calculated $D_e$ and $\alpha$ values.

A still open issue is the large value of $\alpha$ for $^3$H in the BDZ (NAGRA/IDAEA-CSIC, UJV). This large value was necessary to fit the large initial drop in activity in the circulation system (uncertainty due to loss of activity in the rock profiles). Another significant observation is that all tracers sorb more strongly in the BDZ than in the bulk rock, with larger differences for the weakest sorbing tracers. This observation would be consistent with more accessible surface area in the BDZ.

The extension of the BDZ deduced from the rock measurements (maximum 2 mm), is about the same magnitude than the mean grain size of the quartz and feldspar grains (grain sizes below 5 mm; [10]). The extension of the BDZ in the PSG experiment boreholes was 3 mm [10], as indicated by a sharp decrease in the number of grain boundary pores at that distance.

**REFERENCES**

1. J.M. Soler, J. Landa, V. Havlova, Y. Tachi, T. Ebina, P. Sardini, M. Siitari-Kauppi and A. Martin, *LTD Experiment. Postmortem Modelling of Monopole I*, (Nagra Arbeitsbericht NAB 12-53, Wettingen, 2013), p.80.
2. C.I. Steefel, *CrunchFlow. Software for Modeling Multicomponent Reactive Flow and Transport. User's Manual*, (Lawrence Berkeley National Laboratory, Berkeley, 2009), p. 91.
3. GoldSim Technology Group, *GoldSim Version 10.1* (2010).
4. COMSOL Inc., *COMSOL Multiphysics Version 4.2* (2011).

5.  F. Delay, G. Porel and P. Sardini, *C. R. Geosci.* **334**, 967 (2002).
6.  P. Sardini, F. Delay, K.-H. Hellmuth, G. Porel and E. Oila, *J. Contam. Hydrol.* **61**, 339 (2003).
7.  P. Sardini, J.C. Robinet, M. Siitari-Kauppi, F. Delay and K.H. Hellmuth, *J. Contam. Hydrol.* **93**, 21 (2007).
8.  J.C. Robinet, P. Sardini, F. Delay and K.H. Hellmuth, *Transp. Porous Media* **72** (3), 393 (2008).
9.  J.C. Robinet, P. Sardini, D. Coelho, J.-C. Parneix, D. Prêt, S. Sammartino, E. Boller and S. Altmann, *Water Resour. Res.* **48**, No. 5 (2012).
10. A. Möri, *Phase VI – Pore Space Geometry (PSG) Experiment. In Situ Impregnation of Matrix Pores in Granite Rock – Study of a potential borehole disturbed zone (BDZ)*, (Nagra Arbeitsbericht NAB 05-19, Wettingen, 2005), p. 24.

Mater. Res. Soc. Symp. Proc. Vol. 1665 © 2014 Materials Research Society
DOI: 10.1557/opl.2014.633

# Quantification Of Pyrrhotiye $O_2$ Consumption By Using Pyrite Oxidation Kinetic Data

Rojo, I.[1*], Clarens, F.[1], de Pablo, J.[1], Domènech, C.[2], Duro, L.[2], Grivé, M.[2], Arcos, D.[2]

[1]Fundació CTM Centre Tecnològic, Plaça de la Ciència 2, 08243 Manresa (E)
[2]Amphos 21 Consulting, S.L. (E)

* Corresponding author: isabel.rojo@ctm.com.es

## ABSTRACT

Experiments on the dissolution kinetics of natural pyrrhotite ($FeS_{1-x}$) and pyrite ($FeS_2$) under imposed redox conditions to evaluate the oxygen uptake capacity of both minerals were carried out at 25°C and 1 bar. Experimental data indicate that in both cases, $Fe^{(II)}$ released from dissolution of these Fe-bearing sulphides is kinetically oxidized to $Fe^{(III)}$ to precipitate as $Fe^{(III)}$-oxyhydroxides. While the system is pH controlled by the extent of the sulphide oxidation, Eh is controlled by the redox pair $Fe^{2+}/Fe(III)$-oxyhydroxides. Pyrrhotite dissolution is faster than that of pyrite but generates less acidity. Consequently, the achieved redox value is more reducing. Experimental data show that the oxidation rates of both minerals (in $mol \cdot g^{-1} \cdot s^{-1}$) are equivalent under the studied conditions. This fact gives a new opportunity to quantify the reductive buffering capacity of pyrrhotite, for which no kinetic rate law has been still established.

## INTRODUCTION

The maintenance of reducing conditions in and in the vicinity of a repository for high level nuclear waste is one of the main issues of concern in any safety assessment exercise. Mobility of Uranium and other redox sensitive actinides is importantly decreased under reducing conditions. The study of processes contributing to the maintenance of the systems under reducing redox potentials has received special attention and has generated a numerous publications in the open scientific literature, as well as in specifically focused national programmes.

It is well accepted that in most deep groundwaters the redox state is governed by electron transfer between $Fe^{(II)}$ and $Fe^{(III)}$ species and/or by sulphate/sulphide reactions. The occurrence of $Fe^{(II)}$ in nature is dominated by minerals such as $Fe^{(II)}$ sulphides, pyrite and pyrrhotite being the most common natural Fe-bearing minerals (Belzile et al. 2004).

Due to the difficulty to obtain stable redox potentials from field data, Scott and Morgan (1990) proposed the use of intensive magnitudes to define the redox state of geological systems. They used the concepts of OXidising and ReDucing Capacities (OXC and RDC) defined similarly to magnitudes such as acidity or alkalinity in the case of acid/base systems. The major contribution to the ReDucing Capacity of a system will be given by those species able to react with oxidants and therefore buffer an increase in oxidant concentration.

Despite the intensive magnitudes presented above can give us an idea of the maximum redox buffering capacity of a system, it is also important to consider the rates at which the redox buffer reacts. Iron sulphides are solid phases whose oxidative dissolution is kinetically controlled. The oxidation of sulphide to sulphate implies the transfer of 8 electrons, so that kinetics is possibly going to play a role in the system. In fact, the scientific literature is full with references indicating that the oxidation of sulphide to sulphate is not reversible unless it is catalyzed by the presence of biotic activity.

Pyrrhotite dissolution rates reported in literature (Janzen et al. 2000, Wang 2008) are in the range from $10^{-10}$ to $10^{-8}$ mol m$^{-2}$ s$^{-1}$. However, although there are several studies assessing the mechanism of pyrrhotite oxidation (Belize et al. 2004; Wang 2008), the dependence of the rate on oxygen concentration or pH has not been provided. The lack of reaction rates for pyrrhotite oxidation forces the modeller to use other well-established kinetic rates for Fe-bearing sulphide minerals, being that of pyrite (modified or not) the most widely used. However, the uncertainty concerning the use of this approach is not known.

In this work we present the study of the rate at which the common iron sulphide minerals, pyrite and pyrrhotite, consume oxygen and are, therefore, able to buffer oxidant intrusions under environmental conditions common to be found in natural groundwaters. To this aim, new kinetic experimental data has been interpreted and quantitatively modelled.

## DESCRIPTION OF EXPERIMENTAL PROCEDURE

### Pyrrhotite and pyrite
The pyrrhotite and pyrite used in the experiments were natural samples from the skarn sulphide deposit of Gualba (NE Spain) and from Arnedo (N Spain), respectively. The solids were crushed and sieved to a 100-150 μm particle size.

### Experimental setup
Batch experiments to evaluate the oxygen uptake capacity of pyrrhotite or pyrite were performed at two initial dissolved oxygen concentrations, 70 and 20%, and 25±0.2 °C.

A volume of 350 cm$^3$ of $10^{-2}$ mol dm$^{-3}$ NaClO$_4$ solution was placed in the reactor of 500 cm$^3$ and was saturated with O$_2$(g) depending on the initial dissolved oxygen concentration desired (70% or 20% respectively). Between 0.5 and 3 g of pyrrhotite or pyrite were added (except in the case of the blank), and the free gas phase was eliminated. Stirring was provided by orbital shaker at 90 rpm in order to keep the solution as homogenous as possible while minimising the risk of solid grinding. Continuous monitoring of pH, Eh (Pt and Au electrodes) and dissolved oxygen (DO) was carried out with combined-glass electrodes and with an optical sensor. Calibration of pH and Au and Pt Eh electrodes (Crison, models 5221, 5262 & 5269 respectively) was performed against commercial standard buffers before and after each experiment. Additionally, Eh electrodes were mechanically cleaned before each experiment. DO concentrations were measured with an optical sensor (Ocean Optics, FoxyOR125GT & red eye) calibrated at two points: 0 % in 20 % Na$_2$SO$_3$ or N$_2$(g) and 20.9% in open air.

During the experiments aqueous samples were taken periodically and immediately filtered through 0.45 μm pore size filters for analysis. Concentrations of iron $Fe^{(II)}$ and Fe(total) were determined by the ferrozine method (Gibbs (1976)) by means of UV-Vis spectrophotometry (Shimadzu, 1600). Anion concentrations (sulphate, thiosulphate, chloride...) were determined by ionic chromatography with ion suppression (Dionex, ICS2100). Scanning electron microscopy (SEM) (Zeiss, Ultraplus) was used to characterize the solids.

## RESULTS and DISCUSSION

Figures 1 and 2 show the temporal evolution of pH and Eh (measured with both Pt and Au electrodes) measured in the pyrite and pyrrhotite kinetic experiments, as well as the aqueous concentration of $O_2$, $SO_4^{2-}$, $S_2O_3^{2-}$, $Fe^{(II)}$ and $Fe_{total}$.

In both cases, $[O_2]$ decreases and $SO_4^{2-}$, $S_2O_3^{2-}$, $Fe^{(II)}$ and $Fe_{total}$ aqueous concentrations increase with time. In the case of pyrrhotite, faster processes are observed. According to Nicholson (1994), Belzile et al. (2004) and Murphy and Strongin (2009) the presence of ordered vacancies within the Fe lattice in the non-stoichiometric pyrrhotite crystal structure ($Fe_{1-x}S$, x from 0.125 to 0), may be the reason accounting for the faster oxidation rate of pyrrhotite when compared to that of pyrite.

According to r.1 and r.2 oxidation of both pyrite and pyrrhotite generates protons and hence, causes a pH decrease. However, due to the different crystal structure of both minerals, pyrrhotite oxidation generates less acidity than pyrite oxidation (Nicholson, 1994).

$$FeS_2 (c) + 7/2\ O_2 (aq) + H_2O = Fe^{2+} + 2\ SO_4^{2-} + 2\ H^+ \hspace{2cm} r.1$$

$$Fe_{1-x}S (c) + (2 - x/2)\ O_2 (aq) + x\ H_2O = (1-x)\ Fe^{2+} + SO_4^{2-} + 2x\ H^+ \hspace{1cm} r.2$$

Figure 3 compares the redox potential measured in both experiments with that calculated by assuming a redox control of $Fe^{2+}$-$Fe(OH)_3$(am) and $Fe^{2+}$/goethite at the measured pH.

In the case of pyrite, the calculated Eh by assuming a $Fe^{2+}$-$Fe(OH)_3$(am) equilibrium satisfactorily agrees with the measured values. In the pyrrhotite case, this equilibrium also seems to control the Eh of the solution at the initial stage after the $O_2$ injection, while a transition to the redox pair $Fe^{2+}$/goethite seems to occur at longer reaction times.

From the experiments, the precipitation of Fe-oxyhydroxides onto the pyrrhotite surface can be observed. Fe-oxyhydroxides are the most common oxidation products of pyrite and pyrrhotite under near neutral and alkaline pH. These results are consistent with experimental observations and information gathered from literature (Feng and Van Deventer, 2002; Belzile et al., 2004; Wang, 2008; Murphy and Strongin, 2009).

Results shown in Figure 3 indicate that although pyrite or pyrrhotite dissolution processes are kinetically driven and control the solution pH, the redox potential is controlled by the redox pair $Fe^{2+}$/Fe(III)-oxyhydroxides. $Fe^{2+}$ generated in r.1 and r.2 is oxidized to $Fe^{3+}$ (r.3) which under the favourable conditions of neutral to alkaline pHs leads to the precipitation of amorphous Fe-oxyhydroxides which, in turn, may evolve to more crystalline phases (r.4, r.5).

$$Fe^{2+} + e = Fe^{3+} \qquad \qquad \text{r.3}$$
$$Fe^{3+} + 3\,H_2O = Fe(OH)_3(am) + 3\,H^+ \qquad \qquad \text{r.4}$$
$$Fe^{3+} + H_2O = FeOOH + 3\,H^+ \qquad \qquad \text{r.5}$$

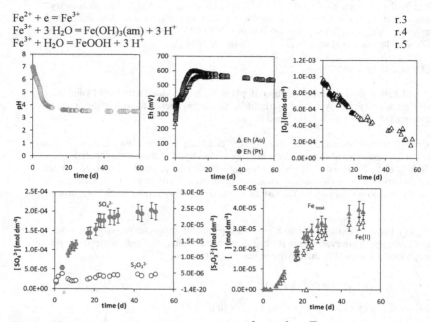

**Figure 1:** Temporal evolution of pH, Eh and $O_2$, $SO_4^{2-}$, $S_2O_3^{2-}$, $Fe^{(II)}$ and $Fe_{total}$ aqueous concentration obtained in the pyrite.

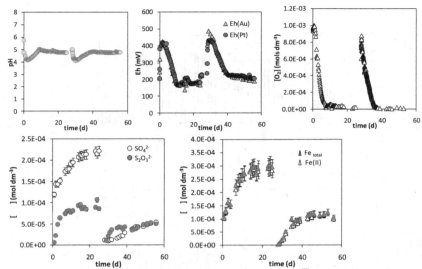

**Figure 2:** Temporal evolution of pH, Eh and $O_2$, $SO_4^{2-}$, $S_2O_3^{2-}$, $Fe^{(II)}$ and $Fe_{tota}$ aqueous concentration obtained in the pyrrhotite experiment.

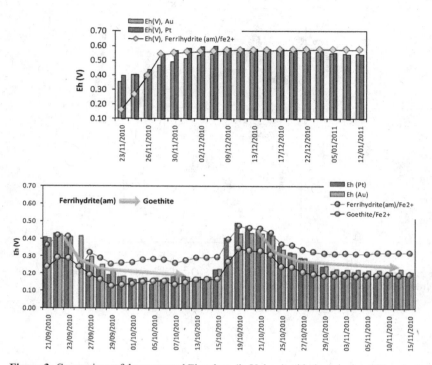

**Figure 3:** Comparison of the measured Eh values (in V, bars) with the calculated Eh values assuming a redox control by Ferrihydrite (am)/$Fe^{2+}$ and goethite/$Fe^{2+}$ equilibrium at the sample pH. Top: pyrite experiment. Bottom: pyrrhotite experiment.

Oxidation of $Fe^{(II)}$ to $Fe^{(III)}$ (r.3) is kinetically controlled. Singer and Stumm (1970) proposed a reaction rate of second order with respect to $OH^-$ for this reaction, according to which, for pH above pH 4, the abiotic oxidation rate of Fe(II) is high enough to compete with the biotic oxidation rate. For pH below 4, the abiotic oxidation rate is not significant and $Fe^{2+}$ can only be oxidized to $Fe^{3+}$ biotically.

Figure 4 shows that the redox potential determined in the pyrite experiment is in agreement with the oxidation of ferrihydrite, whereas in the case of pyrrhotite the control exerted by goethite prevails with time.

The above derived observations, by reproducing the experimental values with the code PHREEQC (Parkhurst and Appelo, 1999) and ThermoChimie data base (Duro, 2010). In the case of pyrite it was assumed that pyrite dissolves kinetically according to the rate of Williamson and Rimstidt (1994) and that the generated $Fe^{2+}$ oxidises kinetically to $Fe^{3+}$ according to the kinetic law of Singer and Stumm (1970). Ferrihydrite was allowed to precipitate in equilibrium. Initial $O_2$ aqueous concentration corresponded to that in equilibrium with 70% $O_2(g)$ in the gas phase. The reactive area of pyrite was set to 0.06 $m^2g^{-1}$. As it is illustrated in Figure 5, the numerical model satisfactorily matches the experimental data.

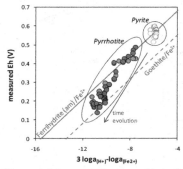

**Figure 4:** Measured Eh values (blue: Au electrode; red: Pt electrode) vs 3 $loga_{(H+)}$-$loga_{(Fe2+)}$ obtained in the pyrite and pyrrhotite experiments. Lines show the Eh values corresponding to the equilibrium $Fe(OH)_3(am)/Fe^{2+}$ and goethite/$Fe^{2+}$.

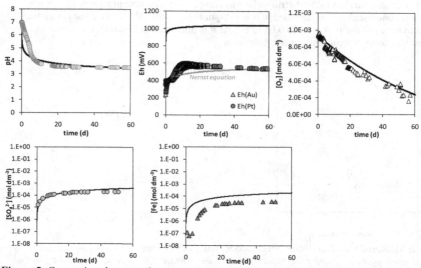

**Figure 5:** Comparison between the temporal evolution of the experimental (symbols)and the calculated (black solid lines)values of pH, Eh and $O_2$, $SO_4^{2-}$, $S_2O_3^{2-}$, Fe(II) and $Fe_{tota}$ aqueous concentration obtained in the pyrite experiment.

In this paper it is shown that the processes occurring in both experiments are similar and that the observed differences area consequence of the extent of mineral oxidation. Figure 6 compares the $O_2$ consumption rates (in $mol \cdot g^{-1} \cdot s^{-1}$) calculated from the $O_2$ concentration data measured in each experiment. In the case of pyrrhotite only data for the first cycle was used. It is observed, once the transient state finished, the consumption rate in mols $O_2$ $g^{-1} \cdot m^{-2}$ of pyrrhotite and pyrite is equivalent. This observation opens a new door to estimate the effects that pyrrhotite oxidation may exert on the pH and Eh of a host rock of a HLNWR. Presented results reveal that the processes related to oxidation of pyrite and pyrrhotite can be quantified using the same oxidation rates in mol $O_2$ $g^{-1}$ $m^{-2}$ at the mid – term.

## ACKNOWLEDGMENTS

The European Atomic Energy Community Seventh Framework Programme [FP7/2007-2013] under grant agreement n° 212287, Collaborative Project ReCosy is thanked for financial support.

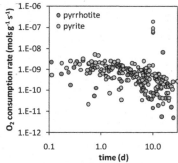

**Figure 6:** $O_2$ consumption rate (in mol $g^{-1}$ $s^{-1}$) calculated from pyrite and pyrrhotite kinetic experiments.

## REFERENCES

1.  Belzile, N., Chen, Y.W., Cai, M.F., Li, Y. (2004) A review on pyrrhotite oxidation. Journal of geochemical exploration 84, 65-76.

2.  Feng, D., van Deventer, J.S.J. (2002) Leaching behaviour of sulphidesin ammoniacalthiosulphate systems. Hydrometallurgy 63, 189– 200.

3.  Gibbs C. (1976). Characterization and application of ferrozine iron reagent as ferrous iron indicator. Anal. Chem. 48, 1197-1200.

4.  Janzen, M.P., Nicholson, R.V., Scharer, J.M. (2000) Pyrrhotite reaction kinetics: Reaction rates for oxidation by oxygen, ferric iron, and for nonoxidative dissolution. GeochimicaetCosmochimicaActa, 64 (9) 1511–1522.

5.  Murphy, R.;Strongin, D.R. (2009) Surface reactivity of pyrite and related sulfides. Surface Science Reports 64, 1-45.

6.  Nicholson, R.V. (1994) Iron-sulfide oxidation mechanisms: laboratory studies. In J.L. Jambor, D.W. Blowes (eds). Short course handbook on environmental geochemistry of sulfide mine wastes 22, Mineralogical Association of Canada, 163-183.

7.  Parkhurst, D.L.; Appelo, C.A.J. (1999) User's guide to PHREEQC ((v 2.17.5))-A computer program for speciation, batch-reaction, one-dimensional transport and inverse geochemical calculations. Washington D.C., USGS, Water resources investigations report 99-4259, 326p.

8.  Scott, M.J., Morgan, J.J. (1990) Energetic and conservative properties of redox systems. In (eds)      American Chemical Society, 368-378.

9.  Singer, P.C. and Stumm, W. (1970) Acidic mine drainage: the rate determining step. Science 167, 1121-1123.

10. Wang, H. (2008) A review on process-related characteristics of pyrrhotite. Mineral Processing and Extractive Metallurgy Reviews, 29, 1 — 41.

11. Williamson, M.A., Rimstidt, J.D. (1994) The kinetics and electrochemical rate-determining step of aqueous pyrite oxidation. Geochimica et CosmochimicaActa 58 (4), 5443-5454.

Figure 12.4 ...

References

Mater. Res. Soc. Symp. Proc. Vol. 1665 © 2014 Materials Research Society
DOI: 10.1557/opl.2014.634

# Dose Estimate in Treatment and Disposal of Contaminated Materials due to the Accident at the Fukushima Nuclear Power Plant

Seiji Takeda[1], Takuma Sawaguchi[1], and Hideo Kimura[1]
[1]Japan Atomic Energy Agency, Nuclear Safety Research Center, Environmental Safety Research Group, Tokai-mura, Naka-gun, Ibaraki pref., 319-1195, Japan

## ABSTRACT

Some kinds of material in the environment due to the accident at the Fukushima Nuclear Power Plant have been contaminated by radioactive cesium ([134]Cs and [137]Cs), which are represented by dehydrated sludge, surface soil and disaster wastes generated by the Great East Japan Earthquake. Treatment (transportation, temporary storage and incineration) and disposal of the contaminated materials should be carried out while ensuring the safety of radiation for the workers and the public. In this study, in order to provide the technical information for making the criteria, the dose estimation for scenarios on the treatment and disposal is conducted, based on the method used for driving the clearance levels in Japan. Minimum radioactive cesium concentration in contaminated material, that is, limiting activity concentration which is practicable for ordinary treatment and/or disposal, is calculated from the dose results, corresponding to the effective dose criteria indicated by the Nuclear Safety Commission of Japan. From the calculation result, it is suggested that it is necessary to forbid reusing the disposal site as construction, resident and agriculture in which the calculated doses for the public are higher than those in the other exposure pathways. Limiting concentration of radioactive cesium ([134]Cs and [137]Cs) is derived to be 8,900Bq/kg for the external exposure pathway in landfill work under the condition of limited reuse of the site. In the case of the concentration below 8,900Bq/kg, the calculated dose of the resident due to direct and sky-shine radiation scattered in the air and ground from the interim storage place is less than 1mSv/y, irrespective of the distance from the storage place.

## INTRODUCTION

Radionuclides were released to surface environment over Fukushima prefecture due to the severe accident at the Fukushima Nuclear Power Plant. A portion of the disaster wastes such as concrete waste, scrap metal, wood waste and so on, which are generated from the Great East Japan Earthquake, was contaminated by radioactive cesium. Soil particles adsorbing radioactive cesium flow into water resources which are treated at sewage plant or purification plant. Dehydrated sludge containing radioactive cesium of high concentration was also generated through the sewage and water supply processing. Treatment and disposal of thus various contaminated materials should be carried out while ensuring the safety of radiation for the workers and the public. However, in Japan, there was no criterion, namely, limiting activity concentration practicable for ordinary treatment and/or disposal with the contaminated materials. Dose estimation for scenarios on the treatment and disposal was conducted to provide the technical information for making the criteria. Based on the results of calculated dose, we derived the limiting radioactive cesium concentration practicable for ordinary treatment and/or disposal, corresponding to the effective dose criteria indicated by Nuclear Safety Commission of Japan

(1mSv/y for engaging in treatment of materials contaminated by the severe accident, and 10μSv/y for post-closure in disposal) [1].

## ANALYTICAL METHOD

### Scenario description

Most of the actual treatment for disaster wastes or sludge is similar to a series of scenarios such as transportation, incineration, disposal and reuse in the previous clearance level estimation in Japan[2]. Considering the peculiarity of the treatment for the disaster wastes or sludge, we rebuilt the scenario and exposure pathways based on the scenario description in the clearance level estimation. First, the pathways of reuse as recycled materials are excluded in this analysis, because of prohibition of reusing dehydrated sludge, ashes and so on after the accident.

For this dose estimation, it is important to consider comprehensive external exposure pathways in the initial treatment, because the calculated dose results of $^{134}$Cs and $^{137}$Cs for external exposure pathways tend to be about two orders of magnitude higher than those for the other internal exposure pathways in the previous clearance level estimation[2]. For the disaster wastes, the additional external exposure pathways are picked up as the removal and separation of piled disaster wastes, dismantling of concrete wastes and the dismantling and separation of metal wastes under the consideration of large volume size as a radiation source, For the sludge, we added the external exposure pathways for a worker relating to transportation by a special container car with opening door and for a worker relating to cleaning the inside of container.

Since the disaster wastes, sludge and their ashes are stored temporally, the pathways to estimate the effect of direct and skyshine radiation scattered in the air and ground for a worker in the interim storage place for the public living near there are considered in this analysis. We also added an external exposure pathway for residence near the road of transportation pass to the interim storage place. The list of scenarios in this analysis is shown in Table 1. The scenarios are categorized by four kinds of treatment of disaster waste, incineration plant, waste disposal and interim storage. The type of exposure pathways for the workers and the public (adult and children) is not only external exposure pathways but also internal exposure pathways by inhalation of dust and ingestion of water, foods and so on.

### Model and code

The dose estimation models used to derive the clearance level in Japan[2] are applied to three types of exposure pathways (external exposure pathway and internal exposure pathways by inhalation and ingestion), which are equivalent to the models in IAEA Safety Guide No.RS-G-1.7[3]. In the scenario on radioactive cesium migration in groundwater, the radionuclide, which is instantaneously released from a disposal facility, flows into an aquifer under the facility after the closure of disposal. The calculation of radioactive cesium migration to the well, which is located at the downstream end of the facility conservatively, was carried out by 1-D advection-dispersion model. The radioactive cesium migration and the dose for the worker and the public except for the scenario on interim storage are estimated using PASCLR code[4] from which the clearance levels for various radionuclide including Cs isotopes were derived. We apply MCNP-4C code[5] to the scenario on interim storage for estimating the effect of scattered direct and skyshine radiation.

**Table I.** List of scenarios and exposure pathways in this dose estimation

| | Scenario description | | Radioactive source | Exposed individual*1 |
|---|---|---|---|---|
| Incineration plant | Treatment of disaster wastes | Removal and separation of piled disaster wastes | Piled disaster wastes | Worker |
| | | Dismantling of broken concrete building | Concrete waste | Worker |
| | | Dismantling and separation of metal wastes of large size such as a car | Metal Waste | Worker |
| | Operation of incineration plant | Maintenance of incinerator | Incineration ashes in the incinerator | Worker |
| | | Loading and unloading | Incineration ashes | Worker |
| | | Transportation | | |
| | The public near the incinerator | Residence near the incinerator | Dust in downwind plume released from the incinerator | Public*2 |
| | | | Soil in which dust deposited | Public |
| | | Ingestion of crops | Crops cultivated in the soil in which dust deposited | Public |
| | | Ingestion of livestock | Livestock grown with the feeds cultivated with the soil | Public |
| Waste disposal | Operation of waste disposal | Loading, unloading and transportation | Disaster wastes, sludge and incineration ashes | Worker |
| | | Landfill | | Worker |
| | Reuse of disposal site after the closure | Construction of a house | Mixed soil of covered soil and the wastes | Worker |
| | | Residence | | Public |
| | | Agriculture | | Worker |
| | | Ingestion of crops | Crops cultivated in the disposal site | Public |
| | | Ingestion of livestock | Livestock grown with the feeds cultivated in the disposal site | Public |
| | | Use of park | Wastes covered with non-contaminated soil | Public |
| | Groundwater migration | Ingestion of water | Well water | Public |
| | | Agriculture | Soil irrigated with well water | Worker |
| | | Ingestion of crops | Crops cultivated with well water | Public |
| | | Ingestion of livestock | Livestock grown with the feeds cultivated with well water | Public |
| | | Ingestion of livestock | Livestock grown with well water | Public |
| | | Ingestion of fishery products | Fishery products farmed with well water | Public |
| Interim storage | | Transportation | Flexible containers loading on a truck | Worker |
| | | Residence near the road of transportation pass | | Public |
| | | Work in interim storage place | Flexible containers piled at interim storage place | Worker |
| | | Residence near interim storage place | | Public |

(*1) Three types of exposure pathways are considered as external exposure pathway and internal exposure pathways by inhalation and ingestion.

(*2) Two kinds of adult and children are set up as exposed individual basically.

## Parameter setting

In this analysis, the parameter values are basically cited from the clearance level estimation[2], and are decided from the realistic values such as the mean value or most probable value of statistical data in Japan.In the previous uses of this approach, if it was difficult to estimate the realistic value, the reasonably conservative value was selected. For additional scenarios and exposure pathways, however, we investigate the information on actual work condition and treatment record of targeted material. Based on the information, the suitable parameters in each pathway were set for the realistic estimation. Since most of contaminated disaster wastes are generated from Fukushima Prefecture, the parameters characterized by the existing incineration plant and disposal facility in Fukushima are selected. In the incineration scenario, we add the analytical case of mono-fuel combustion of disaster wastes at newly-established incinerator. Effective dose conversion factor (μSv/h per Bq/g) of each external exposure pathway for unit concentration of radioactive cesium in targeted material is estimated using QAD-CGGP2R code[6]. It is assumed that the distribution of radioactive cesium concentration inside the material is uniform. The buildup factor of air is used conservatively[7]. According to the previous clearance level estimation, effective dose conversion factor for children is assumed to be 1.3 times higher than that calculated for adult by QAD-CGGP2R code. For the scenario on interim storage, it is assumed that a large amount of flexible containers is stored temporally. A single cluster is piled with 900 flexible containers and 50 clusters are arranged in 5x10 rows at intervals of 5m. Evaluation point of external exposure for the worker is the center of 50 clusters arrangement. The distance from the source to an evaluation point for residence near the interim storage place changes in the range of 2m to 100m.

## RESULTS AND DISCUSSION

The result of dose estimation for four kinds of scenarios is shown in Figure 1, respectively. The results of maximum annual effective dose for targeted material of 1Bq/g are compared among main exposure pathways in each scenario. The ratio of $^{134}$Cs to $^{137}$Cs is determined to be $^{134}$Cs/$^{137}$Cs$=0.806$ based on the measured radioactive cesium concentration data in the contaminated soil. In the scenario on treatment of disaster waste, the dose of external exposure pathways for the workers is two orders magnitude higher than those of inhalation and ingestion pathways. The dose of external exposure pathway for dismantling of broken concrete building is the highest. In the incineration scenario of disaster wastes, the dose for the incineration scenario is calculated under both the condition of mixed combustion with domestic refuse at the existing incinerator and the condition of mono-fuel combustion of disaster wastes at newly-established incinerator. The former result is higher than the latter as indicated in Figure 1(b), because of higher incineration capacity of the existing incinerator. The dose for a worker loading and unloading of ashes is the highest in the scenario on incineration plant. The dose for the public near the incinerator is about two order magnitude lower than for operation workers in incineration plant. As shown in Figure 1(c), the external exposure pathways for landfill worker and for the public (child) dwelling at the disposal site indicate more dominant dose than the other pathways in scenario on waste disposal. The reason of higher dose in the residence at the site is that the public comes close to the wastes through excavating the site. However, the dose in the case of park reuse is three orders magnitude lower than that for the other pathways on the site reuse, because of using the covered soil in park reuse to shield the radiation from the disposed wastes. The dose for the pathways on groundwater migration tends to be lower in the scenario on waste disposal. The dose for transporter of flexible containers is the highest in the scenario on interim storage as shown in Figure 1(d). It is assured that the dose for the public dwelling near interim storage place is lower than that for transporter of flexible containers under the condition of the closest distance 2m from the place. Based on the results of calculated dose, we derived the radioactive cesium concentration, corresponding to the effective dose criteria indicated by NSC[1]. As shown in Figure 2, minimum radioactive cesium concentration, that is, limiting activity concentration practicable for ordinary treatment and/or disposal, is derived to be 8,900Bq/kg for exposure pathway in landfill work under the condition of limited reuse of the site such as construction, resident and agriculture.

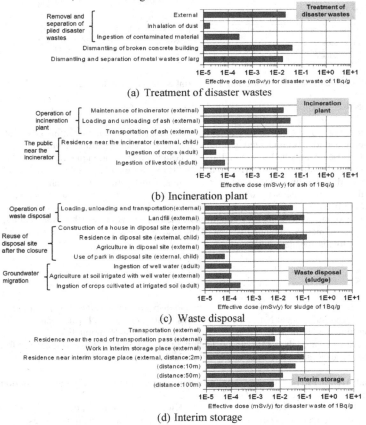

(a) Treatment of disaster wastes

(b) Incineration plant

(c) Waste disposal

(d) Interim storage

**Figure 1.** Result of dose estimation for 1Bq/g of the mixed $^{134}$Cs and $^{137}$Cs ($^{134}$Cs/$^{137}$Cs$=0.806$) **in four kinds of scenarios**

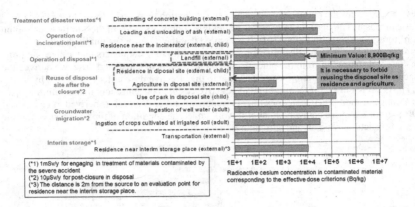

**Figure 2.** Radioactive cesium concentration in material corresponding to effective dose criteria

## CONCLUSIONS

The dose estimation for scenarios on the treatment and disposal for contaminated disaster wastes or dehydrated sludge is conducted to provide the technical information for making the criteria. Form the dose results, the limiting radioactive cesium concentration practicable for ordinary treatment and/or disposal is derived to be 8,900Bq/kg for exposure pathway in landfill work under the condition of limited reuse of the site such as construction, resident and agriculture. In the case of the activity concentration below 8,900Bq/kg, the dose by direct and sky-shine radiation from the interim storage place is less than 1mSv/y, irrespective of the distance from the storage place. Under the condition of the concentration less than 8,000Bq/kg, it is justified that general incineration and disposal of the sludge is possible under the restriction of disposal site reuse by the Japanese government.

## REFERENCES

1. Near-term policy to ensure the safety for treating and disposing contaminated waste around the site of Fukushima Dai-ichi Nuclear Power Plants, NSC, June 3, 2011. [in Japanese]
2. Special Committee on Radioactive Waste and Decommissioning, Radionuclide Concentrations for Materials not Requiring Treatment as Radioactive Wastes Generated from Dismantling etc. of Reactor Facilities and Nuclear Fuel Use Facilities, NSC, 2005. [in Japanese]
3. International Atomic Energy Agency: Application of the Concepts of Exclusion, Exemption and Clearance, Safety Guide No.RS-G-1.7, IAEA, 2004.
4. S. Takeda, M. Kanno, T. Sasaki, N. Minase and H. Kimura: Development of PASCLR Code System Version 2 to Derive Clearance Levels of Uranium and Trans Uranium Wastes, JAEA-Data/Code 2006-003, JAEA, 2006. [in Japanese]
5. J. F. Briesmeister (Ed.), MCNP - A General Monte Carlo N-Particle Transport Code, Version 4C, LA-13709-M, Los Alamos National Laboratory, 2000.

6. Y. Sakamoto and S. Tanaka, QAD-CGGP2 and G33-GP2: Revised Versions of QAD-CGGP and G33-GP, JAERI-M 90-110, JAERI, 1990.
7. M. Watanabe, S. Takeda and H. Kimura, External Effective Dose Conversion Factors for Deriving Clearance Levels of Uranium and Transuranium Wastes, JAEA-Data/Code 2008-001, JAEA, 2008. [in Japanese]

Mater. Res. Soc. Symp. Proc. Vol. 1665 © 2014 Materials Research Society
DOI: 10.1557/opl.2014.635

# A tool to draw chemical equilibrium diagrams using SIT: Applications to geochemical systems and radionuclide solubility

I. Puigdomènech[1], E. Colàs[2], M. Grivé[2], I. Campos[2], D. García[2].

[1]Swedish Nuclear Fuel & Waste Management Co. (SKB), Stockholm, Sweden.
[2]Amphos 21, Barcelona, Spain.

## ABSTRACT

A set of computer programs has been developed to draw chemical-equilibrium diagrams. This new software is the Java-language equivalent to the Medusa/Hydra software (developed some time ago in Visual basic at the Royal Institute of Technology, Stockholm, Sweden). The main program, now named "Spana" calls Java programs based on the HaltaFall algorithm. The equilibrium constants that are needed for the calculations may be retrieved from a database included in the software package ("Database" program). This new software is intended for undergraduate students as well as researchers and professionals.

The "Spana" code can be easily applied to perform radionuclide speciation and solubility calculations of minerals, including solubility calculations relevant for the performance assessment of a nuclear waste repository. In order to handle ionic strength corrections in such calculations several approaches can be applied. The "Spana" code is able to perform calculations based on three models: the Davies equation; an approximation to the model by Helgeson et al. (HKF); and the Specific Ion-Interaction Theory (SIT). Default SIT-coefficients may be used, which widens the applicability of SIT significantly.

A comparison is made here among the different ionic strength approaches used by "Spana" (Davies, HKF, SIT) when modelling the chemistry of radionuclides and minerals of interest under the conditions of a geological repository for nuclear waste. For this purpose, amorphous hydrous Thorium(IV) oxide ($ThO_2(am)$), Gypsum ($CaSO_4 \cdot 2H_2O$) and Portlandite ($Ca(OH)_2$) solubility at high ionic strengths have been modelled and compared to experimental data from the literature. Results show a good fitting between the calculated values and the experimental data especially for the SIT approach in a wide range of ionic strengths (0-4 M).

## INTRODUCTION

The understanding of both the release of radionuclides from nuclear waste and their transport/retention through engineered barriers and the geosphere requires of a detailed study of their chemical behaviour, especially of their solubility and speciation, under the different prevalent (geo)chemical conditions. Several computer codes have been developed in the past to interpret field geochemical data and laboratory results (see for example Nordstrom 2005 [11]), and some of these codes may also be used to estimate radionuclide speciation and mineral solubilities. It is worth stressing that calculation results obtained with those codes should be similar if both the equilibrium constant databases and the used activity coefficient models are equivalent.

Chemical equilibrium diagrams give a quick overview of calculation results, for example speciation as a function of some ligand concentration or of pH. Predominance area diagrams present in two dimensions the main species in solution and the conditions for which solids have solubilities lower than a given limit. Pourbaix diagrams (Eh/pH) are typical examples of predominance area diagrams. The Fortran program SolGasWater [5] was perhaps the first where predominance area diagrams could be created with the total concentration of a component as the variable in an axis. It was then adapted to be used by undergraduate students [12] using also concepts from HaltaFall [10]. With the development of operating systems the Windows programs Medusa and Hydra were developed using the Visual Basic programming language [13]. These programs called the early Fortran software to make the calculations, but the user interface was tailored for undergraduate students, with a standard thermodynamic database, and simple frames to change the chemical conditions and quickly display the results graphically. The software has been extensively used in undergraduate and graduate courses in Sweden.

With non-Windows users in mind, the software has now been thoroughly translated to the Java programming language, as the Java runtime environment is available to a wide range of operating systems. As a result of this translation, two main programs named "Spana" and "Database" have been generated. The "Database" program has been supplemented with the equilibrium constants selected by the OECD-NEA and therefore the "Spana" code can be easily applied to perform speciation and solubility calculations relevant for the performance assessment of nuclear waste repositories.

## METHODOLOGY

### Software development

The software was developed using Java 1.5 and the Netbeans IDE. A HaltaFall class was written implementing the Algol algorithm published in 1967 [10, 16], later translated to Fortran [4]. HaltaFall is able to calculate the composition of an aqueous solution in equilibrium with an unknown number of solids. Java classes where created to store data on the chemical system (equilibrium constants, stoichiometric coefficients, etc). These classes may be coupled to any software requiring the calculation of chemical equilibrium compositions.

Two program components using these classes (SED and Predom) read input text data files, perform calculations, and finally the resulting equilibrium diagram is displayed and stored in an output file. A user interface, called "Spana" allows the user to define what kind of diagram should be created, and the variables and range of variables values in the axes, etc. The user starts by selecting what chemical components should be included in the calculations through a "Database" program. The set of Java programs is similar in its structure and appearance to the Visual Basic programs Hydra and Medusa previously developed [13].

The original HaltaFall algorithm had no provision to calculate medium effects, so an additional loop was added: once the equilibrium composition has been obtained, activity coefficients are calculated using a method from a class named Factor. Several activity correction methods are described in the literature to account for the effects of background electrolytes, e.g in saline groundwaters. The Davies equation [3], a simplified version of the Helgeson, Kirkham and Flowers equations [8] and the SIT [7] have been implemented in "Spana"; as well as a class to store specific ion interaction coefficients.

The equations accounting for the different ionic strength ($I$) correction models in "Spana" are summarized in Table 1.

112

**Table 1.** Equations for different ionic strength correction approaches in "Spana".

| Name | Equation |
|------|----------|
| Davies | $\log(\gamma_i) = -Az_i^2\left(\dfrac{\sqrt{I}}{1+\sqrt{I}}\right) + 0.3I$ |
| HKF | $\log(\gamma_i) = \dfrac{-Az_i^2\sqrt{I}}{1+B\sqrt{I}} + \Gamma_\gamma + b_{\gamma,MX}I$ |
| SIT | $\log(\gamma_i) = -z_i^2\left(\dfrac{A\sqrt{I}}{1+B\sqrt{I}}\right) + \sum_k \varepsilon(i,k)\,m_k$ |

The Davies equation is an empirical extension of Debye-Hückel theory, where $z_i$ is the charge of the ion $i$, $A$ is the temperature dependent Debye-Hückel parameter. In the Davies approach, there are no specific parameters for an aqueous species, except the charge ($z_i$). In the simplified HKF model used in "Spana" the $b_{\gamma,MX}$ parameter is that of NaCl listed in [8], and $\Gamma_\gamma$ is the mole fraction to molality conversion factor ($\Gamma_\gamma = -\log(1 + 0.0180153 \cdot m)$) [8].

The SIT equation accounts for both the electrostatic long-range interactions in dilute solutions and the short-range non-electrostatic interactions occurring between ions. In the SIT equation [7], $m_k$ is the molality of an ion $k$ present in the solution such that $z_i \times z_k < 0$ and $\varepsilon(i,k)$ is a temperature and pressure dependent specific ion interaction parameter between the ions $i$ and $k$. In the SIT the $\varepsilon$ values are considered to be approximately independent of the ionic strength.

The SIT approach has been shown to provide a good estimation of salinity effects in the background concentration range 0.1 to 4 molal [7]. However, the use of this approach in geochemical calculations related to nuclear waste disposal has been usually limited due to the scarce amount of geochemical modelling codes able to support this type of ionic strength corrections. The implementation in "Spana" of the SIT, through the class Factor, includes the possibility to read a text file containing specific ion-interaction coefficients ($\varepsilon(i,k)$ in Table 1), while the default values reported in [9] are used for cases where no information is available.

To allow working at different temperatures, the HKF and SIT models use molality as a concentration scale, including the ionic strength.

**Applications**

Three chemical systems, amorphous Thorium(IV) oxide ($ThO_2(am)$), Gypsum ($CaSO_4 \cdot 2H_2O$) and Portlandite ($Ca(OH)_2$), have been selected in order to demonstrate the capabilities of the "Spana" code to calculate solubilities and draw diagrams in systems with relatively high ionic strengths using the SIT model.

The following methodology has been followed:

1. Selection of experimental datasets from literature for the three systems studied. The following literature sources were selected: a) $CaSO_4 \cdot 2H_2O$ solubility data at T = 25°C in aqueous NaCl solutions from Azimi and references therein [1], b) $Ca(OH)_2$ solubility data at ~25°C in NaCl solutions from Christov and Moller and references therein [2], and

c) $ThO_2(am)$ solubility data at T=25 °C in 3m NaCl aqueous solutions as a function of pH from Felmy et al. [6]

2. Selection of the most appropriate thermodynamic data; in this work the thermodynamic data from the SIT.dat database released within the geochemical code PhreeqC [14] was selected. No additional complexes or ion-pairs were included in the calculations.

3. Ion interaction coefficients required for the SIT calculations are those selected by the NEA [15].

4. Modelling of data selected in step 1 by using the different three different ionic strength approaches that "Spana" included.

5. Comparison between modelling results obtained with "Spana" against the experimental data from step 1.

## RESULTS AND DISCUSSION

Figure 1 shows the modelling results obtained for the systems studied in the present work. As can be seen, Gypsum ($CaSO_4 \cdot 2H_2O$) solubility (Figure 1a) slightly increases when increasing the ionic strength, due to the specific ion interactions between $Ca^{2+}$ and $Cl^-$ and between $SO_4^{2-}$ and $Na^+$ ions (from the NaCl medium).

It is also observed, that the SIT approach fairly well reproduces gypsum solubility over a wide range of ionic strengths (0-4M). In the other hand, the application of Davies and HKF approaches produces much less accurate results above $I > 0.2M$.

Portlandite ($Ca(OH)_2$) solubilities at 25°C in aqueous NaCl solutions up to ~4.5 M are shown in Figure 1b. As in the case of gypsum discussed above, portlandite solubility can be better described with the SIT approach than with the Davies or HKF approaches. In the case of portlandite solubility, the Davies model is able to reproduce the experimental data up to NaCl concentrations ~2M, while the HKF model only produces satisfactory results below $I < 0.2M$.

Figure 1c shows $ThO_2(am)$ solubility results at $T \approx 25°C$ in aqueous NaCl solutions at different pH values. From this figure it can be seen that $ThO_2(am)$ solubility decreases with increasing pH below pH~6.5 and that above this value it is pH-independent. At pH > 6.5, $ThO_2(am)$ solubility could be well described by using Davies, HKF or SIT approaches. The reason behind this is that the neutral aqueous species $Th(OH)_4(aq)$ is the predominant Th species above pH~6.5, so that the solubility of thorium is governed by reaction (1). This reaction does not include any charged species and therefore it is practically not affected by the ionic strength.

$$ThO_2(am) + 2H_2O(l) \leftrightarrow Th(OH)_4(aq) \qquad (1)$$

At pH < 6.5 the presence of charged hydrolytic species in solution (e.g. $Th(OH)^{3+}$) causes the calculated $ThO_2(am)$ solubility to increase with decreasing pH. In this pH range, and given that charged species (e.g. $Th(OH)^{3+}$) are formed, the solubility of the solid is affected by the ionic strength, and the SIT approach led to the best model results. It should be mentioned that below pH 5.5 the presence of highly charged polynuclear hydrolysis species (e.g. $Th_4(OH)_{12}^{4+}$) produced larger deviations for the solubilities calculated using the Davies and HKF models.

It must be noticed that the solubility of $ThO_2(am)$ has a relatively large uncertainty [15]. This uncertainty, which accounts for solid crystallinity variations, explains the apparent discrepancies between our modelled results and the experimental data.

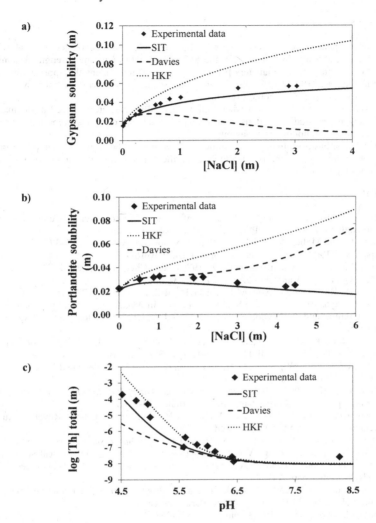

**Figure 1.** a) Gypsum solubility in NaCl solution. Experimental data from [1], b) Portlandite solubility in NaCl solution at ~25°C. Experimental data from [2], c) Amorphous Thorium oxide solubility in 3m NaCl solution at 25°C, as a function of pH. Experimental data reported in [6].

## CONCLUSIONS

A platform-independent computer program, "Spana", has been developed that allows drawing chemical equilibrium diagrams using the SIT to deal with high ionic strength systems. Default specific ion interaction parameters [9] may be used, which widens the applicability of the SIT. The Davies and HKF models, also included in "Spana", reproduce reasonably well the tested experimental systems at low ionic strengths.

Furthermore, it has been shown that when highly charged species are involved in the main chemical reactions of the studied systems, ionic strength effects are of utmost importance for a proper understanding of radionuclide behaviour.

The "Spana" code has been found to be a powerful tool that can be easily applied to perform radionuclide speciation and solubility calculations for the performance assessment of nuclear waste repositories using the SIT.

## REFERENCES

1. Azimi, G. (2010). Evaluating the Potential of Scaling due to Calcium Compounds in Hydrometallurgical Processes. PhD Thesis, University of Toronto.
2. Christov, C. and Moller, N. (2004). Geochim. Cosmochim. Acta 68, 3717–3739.
3. Davies, C.W. (1962). Ion association. London, Butterworths.
4. Ekelund R., Sillén L. G. and Wahlberg O. (1970). Acta Chem. Scand. 24, 3073.
5. Eriksson, G. (1979). Anal. Chim. Acta 112, 375-383.
6. Felmy, A.R., Rai, D. and Mason, M.J. (1991). Radiochim. Acta 55, 177-185.
7. Grenthe I., Plyasunov A. V. and Spahiu K. (1997) In *Modelling in Aquatic Chemistry* (eds. I. Grenthe and I. Puigdomenech), pp. 325-426. OECD Nuclear Energy Agency, Paris, France.
8. Helgeson, H. C., Kirkham, D. H. and Flowers, D. C. (1981). Am. J. Sci. 281, 1249–1516.
9. Hummel W. (2009) Report PSI-TM-44-09-01, Paul Scherrer Institut.
10. Ingri N., Kakolowicz W., Sillén L. G. and Warnqvist B. (1967) Talanta 14, 1261-1286. Errata: 15(3) (1968) xi-xii.
11. Nordstrom, D. K., (2005). In *Surface and Ground Water, Weathering, and Solids* (ed. J.I. Drever), pp. 37–72. Amsterdam, The Netherlands: Elsevier.
12. Puigdomenech I. (1983) Report TRITA-OOK-3010, Dept. Inorg. Chem., Royal Institute of Technology, 100 44 Stockholm, Sweden.
13. Puigdomenech I. (2000) In "*219th ACS National Meeting. San Francisco, CA, March 26-30, 2000. Abstracts of Papers*", Vol. 1, abstract I&EC-248. American Chemical Society.
14. Parkhurst, D.L. and Appelo, C.A.J. (2013). U.S. Geological Survey Techniques and Methods, Book 6, Chap. A43.
15. Rand., M., Fuger, J., Grenthe, I., Neck, V., Rai, D. (2009) Chemical Thermodynamics of Thorium. OECD Nuclear Energy Agency, Paris, France.
16. Warnqvist B. and Ingri N. (1971) Talanta 18, 457-458.

Mater. Res. Soc. Symp. Proc. Vol. 1665 © 2014 Materials Research Society
DOI: 10.1557/opl.2014.636

# Experimental adsorption studies on different materials selected for developing a permeable reactive barrier for radiocesium retention

Miguel García-Gutiérrez[1], Tiziana Missana[1], Ana Benedicto[1], Carlos Ayora[2] and Katrien DePourcq[2]
[1]CIEMAT, Departamento de Medioambiente, Av. Complutense 40, 28040 Madrid (SPAIN).
[2]IDAEA-CSIC, Jordi Girona 18-26, 08034 Barcelona (SPAIN)

## ABSTRACT

Cs-137 was accidentally spilled in an industrial waste repository located in a salt marsh in southern Spain, and a permeable reactive barrier was proposed to retain it. Cs adsorption properties of different natural clayey materials were analyzed. The salt marsh waters show high salinity and high chemical variability, therefore Cs adsorption was also analyzed in the presence of competitive ions, especially $K^+$ and $NH_4^+$.

Cs adsorption was non-linear in all the analyzed materials, indicating more than one adsorption sites with different selectivity. It was shown that in mixed clay systems with illite, montmorillonite and kaolinite, the presence of illite favors Cs retention at low and medium Cs loadings and montmorillonite at high Cs loadings. In the presence of illite and montmorillonite, kaolinite plays almost no role in Cs retention. The presence of $K^+$ and $NH_4^+$ significantly hinders cesium adsorption.

## INTRODUCTION

Cs-137 is a major radionuclide in spent nuclear fuel and in the global radioactive waste inventory. It has been accidentally introduced in the environment by nuclear accidents and from nuclear weapons testing. Radiocesium is particularly relevant from an environmental point of view because it exists predominantly as the monovalent highly soluble cation $Cs^+$. Geochemical barriers can be designed to retard or stop its migration in groundwater from contaminated zones: many solids are still under study for cesium retention [1,2] but the barriers comprised of argillaceous materials, are reported to be the most effective for its retardation [3].

Thus in this work, for the development of geochemical reactive barrier for cesium, different clayey materials were analyzed to measure their sorption capacity and to understand which minerals contribute most to cesium retention. In order to make the barrier permeable (clays usually have very low permeability) the sorbing fraction will be mixed with other materials, for example, wood shavings with a null sorption capability.

Cesium retention occurs in clays mainly by ionic exchange; therefore the salinity of the waters and ion competition are expected to hinder its retention. The Spanish region, where Cs-137 was accidentally spilled, was located in a salt marsh. The salt marsh waters, affected by the tide, show very high salinity and high chemical variability, therefore Cs adsorption was analyzed under saline condition and in the presence of ions like potassium and ammonium potentially competitive for Cs adsorption [4,5]

## EXPERIMENTAL

Sorption on three different clayey materials (Rojo Carboneros (RC), San Juan (SJ) and Canal de Drenaje (CD)) will be analyzed in this study using 2 different natural saline waters and 0.5 M NaCl. Table 1 shows the main mineralogical composition of the three clayey materials. The main clay minerals are illite, kaolinite and montmorillonite, which is present only in SJ. Table 2 shows the main chemical composition of the two natural waters (W-1 and W-2) used in sorption tests.

**Table I.** Main minerals present in the three materials under study.

| | Clay 1 (RC) | Clay2 (SJ) | Clay 3 (CD) |
|---|---|---|---|
| Illite | 57 | 40 | 38 |
| Kaolinite | 1 | 12 | 12 |
| Montmorillonite | --- | 9 | --- |
| Quartz | 27 | 18 | 26 |
| Calcite/Dolomite | 8 | 4 | --- |
| Chlinoclore | --- | 14 | 14 |
| Other Minor | 7 | 3 | 10 |

**Table II.** Main chemical composition of the two natural waters used for sorption tests

| Element | W-1 | W-2 |
|---|---|---|
| $F^-$ | 5 | 33 |
| $Cl^-$ | 17250 | 60004 |
| $NO_3^-$ | 58 | 1295 |
| $PO_4^{3-}$ | 2280 | 6959 |
| $NH_4^+$ | 9 | n.a |
| $Ca^{2+}$ | 840 | 1056 |
| $Mg^{2+}$ | 950 | 3481 |
| $Na^+$ | 8500 | 26555 |
| $K^+$ | 286 | 1179 |
| pH | 5 | 5 |
| Cond. (mS/cm) | 42 | 106 |

The natural waters present high salinity and the main cations are Na, Mg, Ca and K. The ionic strength of W-2 is approximately 2 times higher than that of W-1.

All the experiments were performed at a room temperature (22-25 °C) and under oxic conditions. The radionuclide used in this study was $^{137}Cs$ and its activity in solution was measured by $\gamma$-counting with a NaI detector (Packard, Cobra II).

Batch sorption experiments were carried out with a solid to liquid ratio of 10 g/L. Kinetic experiments (1 to 35 days) were carried out to determine the equilibrium time. $K_d$ measurements were carried out with [Cs] = $3 \cdot 10^{-9}$ M and sorption isotherms were carried out by varying the radionuclide concentration from [Cs] = $1 \cdot 10^{-10}$ M to [Cs] = $1 \cdot 10^{-3}$ M and a contact time of 14 days. For the experiment with high Cs concentrations (higher than $1 \cdot 10^{-6}$ M), a non-radioactive chemical of high purity (CsCl, Merck) was used in addition to the radiotracer. The solid and liquid phases were separated by centrifuging (22500g, 20 min).

The distribution coefficient, $K_d$ (mL·g-1), is calculated by:

$$K_d = \frac{C_{in} - C_{fin}}{C_{fin}} \cdot \frac{V}{m} \tag{1}$$

$C_{in}$ and $C_{fin}$ are the initial and final concentration of cesium in the liquid phase (Bq·mL$^{-1}$), m the mass of the clay (g) and V the volume of the liquid (mL). Sorption onto vessels was always lower than <2 %, therefore it was not accounted for $K_d$ calculations.

## RESULTS AND DISCUSSION

Figure 1 shows the results of the kinetic sorption experiments carried out with the three clays (RC, SJ and CD) in the two natural waters. The sorption equilibrium is reached around two weeks of contact time in all the cases. At the equilibrium the distribution coefficients range between 80 and 180 mL/g in W-2 and between 480 and 650 mL/g in W-1. The material presenting the lower adsorption capacity was RC, the other two showed similar $K_d$ values.

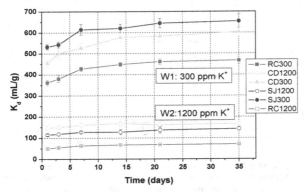

**Figure 1.** Cesium adsorption kinetic in three different clays (RC, SJ and CD) and two different natural waters (W-1 and W-2).

The chemical composition of the water has then an effect of almost one order of magnitude on the retention capability of the material. Cesium in fact adsorbs mainly by ionic exchange, therefore the higher the salinity the lower the adsorption, as shown in our results. However it is quite important to understand which ions can compete more effectively with cesium for sorption. One of the main difference between the waters that can be mainly responsible for the different sorption behavior is the potassium content: in fact, W-2 has a significantly higher potassium content (about 1200 ppm ) than W-1 (about 300 ppm).

Potassium and ammonium are known to be ions especially important in cesium adsorption, because of the existence of sorption sites, related to the presence of micaceous minerals, such as illite, which dominate cesium sorption. For this reason the effect of their presence was especially evaluated and quantified in this study. Thus, the possible effects of the mineralogy of the sorbents, cesium concentration and potassium and ammonium as competitive ions were analyzed more in depth using the SJ and RC clays, simulating saline conditions with 0.5 M NaCl and performing sorption isotherms.

Figure 2 (left) shows the comparison of the sorption isotherms for the SJ and RC clays in NaCl. In both materials, $K_d$ varies with cesium concentration, i.e. cesium adsorption is not linear. This behavior is very typical for cesium above all in the presence of micaceous minerals. Illite, in addition to planar sites, possess frayed edge sites, FES, arising from the weathering of the clay particle's edges, and especially important for alkaline cations adsorption. These sites have very small density (<1% of the CEC of the clay) but adsorb these cations selectively. Cations including $Cs^+$, $Rb^+$, $Li^+$ , $NH_4^+$, characterized by low hydration energy and small dehydrated radius, can replace $K^+$ at these weathered edges. Other monovalent ions, such as $Na^+$, can access these sites only when their concentration is high enough. Yet at these sites, divalent cations with larger sizes and high hydration energy are improbably found.

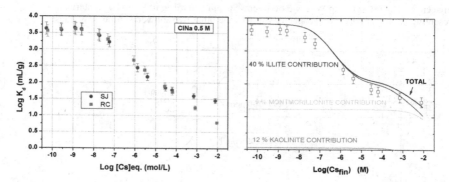

**Figure 2.** Left: sorption isotherms of cesium onto SJ and RC clays in 0.5 M NaCl. Right: simulation of cesium sorption isotherm in a mixed clay system with 40% illite, 9% smectite and 12 % kaolinite in 0.5 M NaCl. The simulation obtained with the parameters of Missana et al. (2013) is superimposed to the experimental points obtained for the SJ clay, for comparison.

120

Thus, cesium concentration is a very important parameter in sorption experiments as the variation in $K_d$ values can be more than two orders of magnitude, depending on the selected concentration. $K_d$ values are quite similar in both clays, except for very high Cs concentrations. At these high Cs concentrations, $K_d$ values for SJ are slightly higher than those for RC. This can be due to the presence of montmorillonite in the SJ clay.

To evaluate the presence of potassium and ammonium in the system, sorption isotherms were carried out in both the clays, with 0.5 M NaCl and 782 ppm of $K^+$ and 360 ppm of $NH_4^+$ (the molar concentration of both is 0.02 M) as shown in Figure 3. The concentration represents the average value in the contaminated area. Figure 3 (left) shows the results obtained with clay RC and Figure 3 (right) the results obtained with clay SJ.

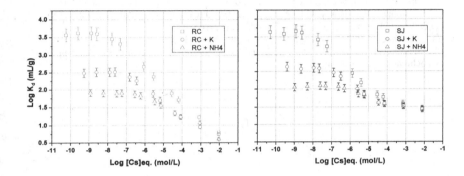

**Figure 3.** Sorption isotherms of cesium in 0.5 M NaCl and with the additions of $K^+$ (782 ppm) and $NH_4^+$ (360 ppm). Left: RC clay and Right: SJ clay.

The effects of potassium and ammonium are evident and similar in both clays, above at low-medium cesium loading, where adsorption occurs mainly in the FES sites of illite. Here, the decrease in $K_d$ values is one order of magnitude or more. Ammonium is more selectively adsorbed than potassium in FES, in fact its presence hinders more effectively cesium adsorption.

## CONCLUSIONS

Cesium adsorption has been analyzed onto different clayey rocks under very saline conditions typical of a marsh zone.

The effects of the mineralogy of the sorbent, chemistry of the water and cesium concentration have been evaluated. In particular, the competitive effects of potassium and ammonium (even in such saline waters) have been assessed.

Cesium sorption in all the materials was non-linear, due to the presence of illite, providing highly selective sites for cesium adsorption. Cesium sorption is always dominated by illite at low and medium Cs loadings whereas montmorillonite starts to be relevant at high Cs concentration. The contribution of kaolinite in the presence of illite and montmorillonite is almost irrelevant

## ACKNOWLEDGMENTS

This work has been partially supported by the CeluCem project (CTQ2011-28338) of the Spanish Ministry of Economy and Competitiveness and the CIEMAT-ENRESA association.

## REFERENCES

1. E.H Borai, R. Hariula, L. Malinen, A. Paajanen (2009), Journal of Hazardous Materials, 172(1), 416-422
2. Y. Park, W.S. Shin, S. Choi (2012), Journal of Radioanal. Nucl. Chem, 292, 837-852.
3. J.L Krumhans, P.V. Brady, H.L. Anderson (2001). J. Contaminant Hydrology, 233-240.
4. D.D. Eberl (1980), Clay and Clay Minerals, 28(3), 161-172.
5. B.L. Sawhney (1972), Clays and Clay Minerals, 20, 93-100.
6. T. Missana, M. García-Gutiérrez, A. Benedicto, C. Ayora, K. De-Pourcq (2013), Applied Geochemistry, to be submitted.

Mater. Res. Soc. Symp. Proc. Vol. 1665 © 2014 Materials Research Society
DOI: 10.1557/opl.2014.637

# Diffusion Modeling in Compacted Bentonite Based on Modified Gouy-Chapman Model

Kenji Yotsuji[1], Yukio Tachi[1] and Yuichirou Nishimaki[2]
[1] Japan Atomic Energy Agency, 4-33, Muramatsu, Tokai, Ibaraki, 319-1194, Japan
[2] Visible Information Center, Inc., 440, Muramatsu, Tokai, Ibaraki, 319-1112, Japan

## ABSTRACT

The integrated sorption and diffusion (ISD) model has been developed to quantify radionuclide transport in compacted bentonite. The current ISD model, based on averaged pore aperture and the Gouy-Chapman electric double layer (EDL) theory can quantitatively account for diffusion of monovalent cations and anions under a wide range of conditions (e.g., salinity, bentonite density). To improve the applicability of the current ISD model for multivalent ions and complex species, the excluded volume effect and the dielectric saturation effect were incorporated into the current model, and the modified Poisson-Boltzmann equations were numerically solved. These modified models had little effect on the calculation of effective diffusivity of $Sr^{2+}/Cs^+/I^-$. On the other hand, the model, modified considering the effective electric charge of hydrated ions, calculated using the Gibbs free energy of hydration, agreed well with the diffusion data including those of $Sr^{2+}$.

## INTRODUCTION

Diffusion and sorption of radionuclides in compacted bentonite are key processes in the safety of geological disposal of radioactive waste. The ISD model [1−4] gives consistent consideration to porewater chemistry, sorption and diffusion processes in compacted bentonite. The diffusion component based on the Gouy-Chapman EDL theory in the ISD model accounts consistently for cation $D_e$ overestimation and anion exclusion in narrow pores. The key parameter of the diffusion model is an electrostatic constrictivity $\delta_{el,i}$ [-], related to $D_{e,i}$ [m$^2$ s$^{-1}$] of species $i$ by:

$$D_{e,i} = \phi \frac{\delta_g \delta_{el,i}}{\tau^2} D_{w,i} \qquad (1)$$

where $\phi$ is porosity [-], $\tau$ is tortuosity [-], $\delta_g$ is geometrical constrictivity [-] and $D_{w,i}$ is tracer diffusivity of species $i$ in bulk liquid water [m$^2$ s$^{-1}$]. The electrostatic constrictivity is the averaged ratio between the ionic concentration in the diffuse layer and in the bulk water, by taking into account the enhanced water viscosity by viscoelectric effects in the interlayers:

$$\delta_{el,i} := \frac{1}{d} \int_0^d \frac{\eta_0}{\eta(x)} \cdot \frac{n_i(x)}{n_{b,i}} \, dx = \frac{1}{d} \int_0^d \frac{1}{1 + f_{ve}(d\psi/dx)^2} \exp\left(-\frac{ez_i\psi(x)}{kT}\right) dx \qquad (2)$$

where $n_i(x)$, $\eta(x)$ and $\psi(x)$ are the local number density of ionic species $i$ [m$^{-3}$], the viscosity of the water [N s m$^{-2}$] and the electric potential [V] in the interlayer at distance $x$ [m] from the clay basal surface, respectively; $n_{b,i}$ and $\eta_0$ are the corresponding number density and the viscosity of the water in the bulk, respectively; $d$ is the interlayer width [m], $z_i$ is the valence of ionic species $i$ [-], $e$ is the absolute value of the elementary electric charge [C], $k$ is the Boltzmann constant [J

$K^{-1}$], $T$ is the temperature [K], and $f_{ve}$ is a viscoelectric constant (= $1.02 \times 10^{-15}$ [$m^2$ $V^{-2}$], [5]).

The Poisson-Boltzmann (P-B) equation, which describes the distribution of the electric potential $\psi(x)$ in the EDL, is the following (e.g., [6]):

$$\frac{d^2\psi}{dx^2} = -\frac{1}{\varepsilon_b \varepsilon_0} \sum_i e z_i n_{b,i} \exp\left(-\frac{e z_i \psi(x)}{kT}\right), \quad \left.\frac{d\psi}{dx}\right|_{x=0} = -\frac{\sigma_0}{\varepsilon_b \varepsilon_0}, \quad \left.\frac{d\psi}{dx}\right|_{x=d/2} = 0 \quad (3)$$

where $\varepsilon_b$ is the relative permittivity of water (= 78.36 [-] at 298.15 [K]), $\varepsilon_0$ is the vacuum permittivity [$C$ $V^{-1}$ $m^{-1}$]. Surface charge density $\sigma_0$ is −0.129 [$C$ $m^{-2}$], calculated from the *CEC* (= 108 [meq/100g]) and specific surface area (= 810 [$m^2$ $g^{-1}$]) for montmorillonite [7]. The entire pore volume is assumed to be distributed in slit-like pores with an average aperture, $d = 2.23 \times 10^{-9}$ [m], calculated for the compacted montmorillonite at dry density of 800 [kg m$^{-3}$]. $D_{e,i}$ values were evaluated based on Eq. (1) using the same porosity $\phi$ (= 0.723) and the same geometric factor $\delta_g/\tau^2$ (= 0.0989) both for cations and anions.

The current ISD model [4] can quantitatively account for diffusion of monovalent cations and anions, however, the model predictions disagree with diffusion data for multivalent cation and complex species [2, 3]. The main basis of the current ISD diffusion model is the classical Gouy-Chapman (G-C) model of the EDL theory and the homogeneous pore model. To improve the applicability of the model, it is necessary to consider the atomic level interactions between solute, solvent or clay mineral. For instance, the short-range interaction caused by the quantum mechanical effect between the solute particles appears as the ionic volume exclusion effect. For another example, the atomic level interaction between the polar solvent and the charged interface reduces the relative permittivity of the solvent due to the dielectric saturation effect, although the relative permittivity of the solvent is constant in the classical G-C model (e.g., [8, 9]).

Based on the previous studies (e.g., [8–11]), this paper focuses on the excluded volume effect and the dielectric saturation effect. These effects are incorporated into the current ISD model and the modified P-B equations numerically solved. The modified model considering the effective electric charge of hydrated ions, calculated using the Gibbs free energy of the hydration, is also investigated.

## MODIFIED ISD DIFFUSION MODEL

### Excluded volume effects

The excluded volume effect caused by quantum mechanical short range repulsive force of inter-particle is firstly incorporated to the ISD model. The modified P-B equation considering the excluded volume effect can be expressed by the following equation, according to [10]:

$$\frac{d^2\psi}{dx^2} = -\frac{1}{\varepsilon_b \varepsilon_0} \sum_i \frac{e z_i n_{b,i}^* \exp[-e z_i \psi(x)/kT]}{1 + v \sum_j n_{b,j}^* \exp[-e z_j \psi(x)/kT]}, \quad n_{b,i}^* := \frac{n_{b,i}}{1 - v \sum_k n_{b,k}} \quad (x \geq a/2) \quad (4)$$

where $a/2$ is the closest distance between the center of a counter-ion and the interface, and $v$ is the average excluded volume which is inaccessible to ion. The differential equation (4) was numerically solved using the Runge-Kutta method and the Shooting method together [12].

As shown in Figure 1(a), evaluated concentration distributions of Na$^+$ in the modified model are shifted to outer side in the diffused layer. However, the total quantity of electric

124

charge in the interlayer is controlled by the total surface charge because of the electric neutral condition. Although the concentration distributions of counter-ion are influenced by the excluded volume effect, the total quantity of counter-ion almost does not change.

Effective diffusivity of $Sr^{2+}/Cs^+/I^-$ calculated by the modified model and the current ISD model are shown in Figure 1(b) together with measured data [3,4]. The difference in the effective diffusivity between the two models is caused by the difference in the electrostatic constrictivity ($\delta_{el}$). The difference in $\delta_{el}$ can be influenced by the excluded volume factor and electrostatic potential. The calculated excluded volume factor is 0.7 under salinity condition = 0.01 [mol $L^{-3}$]. On the other hand, the Boltzmann factor: $\exp\{-z_ie\psi(x)/kT\}$ for monovalent cation in the modified model is approximately 2.3 times higher than that for current model at salinity = 0.01 [mol $L^{-3}$]. As coupled effect of these factors, $\delta_{el}$ in the modified model can be evaluated to be approximately 1.6 times of that for current model in the case of a monovalent cation under salinity = 0.01 [mol $L^{-3}$]. As shown in Figure 1(b) by similar calculations for all tracer ions and salinity conditions, it can be seen that the excluded volume effect hardly influence the effective diffusivity.

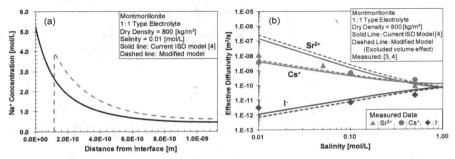

**Figure 1.** (a) $Na^+$ concentration as a function of distance from interface (salinity; 0.01 mol/L), (b) Effective diffusivity of $Sr^{2+}/Cs^+/I^-$ as a function of salinity, calculated by the modified model considering the excluded volume effect and the current ISD model. Stokes radius (= $1.84 \times 10^{-10}$ [m], [13]) as ionic radius in a solution and crystal radius (= $1.16 \times 10^{-10}$ [m], [14]) at the interface were used.

## Dielectric saturation effects

The dielectric saturation effect caused by strong external electric field in the vicinity of a negatively charged clay surface is incorporated to the ISD model. The modified P-B equation considering the dielectric saturation effect can be expressed by [8, 9]:

$$\frac{d}{dx}\left(\varepsilon_r(x)\frac{d\psi}{dx}\right) = -\frac{1}{\varepsilon_0}\sum_i ez_in_{b,i}\exp\left(-\frac{ez_i\psi(x)}{kT} + \frac{\Delta G_i \cdot \varepsilon_b}{N_AkT(\varepsilon_b-1)}\left\{\frac{1}{\varepsilon_r(x)}-\frac{1}{\varepsilon_b}\right\}\right) \quad (5)$$

where $\Delta G_i$ is the Gibbs free energy of the hydration of ionic species $i$ [J $mol^{-1}$] and $N_A$ is the Avogadro constant [$mol^{-1}$]. $\Delta G_i$ data are taken from [15]. As the electric field dependence of the relative permittivity, the Booth equation was adopted according to the previous studies [8, 9, 16]:

$$\varepsilon_r(E) = n^2 + (\varepsilon_b - n^2)\frac{3}{\beta E}\left[\coth(\beta E) - \frac{1}{\beta E}\right] \; , \quad \beta = \frac{5\mu(n^2 + 2)}{2kT} \tag{6}$$

where $\varepsilon_r(E)$ is a relative permittivity of the solution [-] under external electric field $E$ [V m$^{-1}$], $n$ is a refractive index of the water (= 1.33 [-] at 298.15 [K]) and $\mu$ is the absolute value of the permanent dipole moment vector of the water molecule in the vapor phase (= $6.19 \times 10^{-30}$ [C m]). The differential equations (5) and (6) were numerically solved using the Runge-Kutta method and the Shooting method together.

As shown in Figure 2(a), counter-ion Na$^+$ seems to be strongly eliminated from the interface, because the existence of ion becomes unstable as the relative permittivity of the solvent decreases. On the other hand, the counter-ions receive attractive force from an interfacial negative charge. The peak in Na$^+$ concentration in the case of the modified model can be explained by the competition between repulsive and attractive forces.

Effective diffusivity of Sr$^{2+}$/Cs$^+$/I$^-$ calculated by the modified model and the current ISD model is shown in Figure 2(b) together with measured data [3, 4]. For Cs$^+$ as an example, electrostatic potential term in Boltzmann factor of the modified model is larger than that of the current model, however hydration free energy term in Boltzmann factor cancels out this increment. Consequently, $D_e$ obtained by the modified model did not show significant changes in respect to the current model. As other ionic species can be considered as well as Cs, it can be seen that the dielectric saturation effect hardly influence the effective diffusivity.

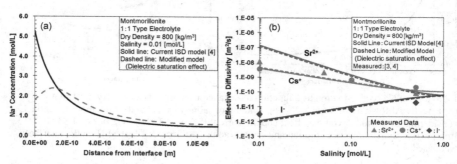

**Figure 2.** (a) Na$^+$ concentration as a function of distance from interface (salinity; 0.01 mol/L), (b) Effective diffusivity of Sr$^{2+}$/Cs$^+$/I$^-$ as a function of salinity, calculated by the modified model considering the dielectric saturation effect using the Booth equation and the current ISD model.

## Consideration of the effective electric charge

As additional approach based on the effective electric charge of aqueous ions, is investigated. It can be seen that this model gives consideration of the interaction between the solute - solvent. To keep the consistency with the P-B equation based on the classical electromagnetics, we model the aqueous ions hydration Gibbs free energies classically; the modified Born equation for the Gibbs free energies of hydration based on [17] was expressed by the following equation:

$$\Delta G_i = -\frac{N_A e^2}{8\pi\varepsilon_0}\left(\frac{z_i^2}{r_{v,i}} - \frac{z_{\text{eff},i}^2}{\varepsilon_{\text{eff},i} r_{c,i}}\right), \qquad \frac{1}{\varepsilon_{\text{eff},i}} := \frac{1}{2}\left(\frac{1}{\varepsilon_{\text{surf},i}} + \frac{1}{\varepsilon_b}\right) \tag{7}$$

where $z_i$ is the valence of ionic species $i$ [-] based on a formal electric charge, $r_{v,i}$ is the van der Waals radius of ionic species $i$ [m] and $r_{c,i}$ is the crystal radius of ionic species $i$ [m]. The van der Waals radii of isolated ions of the noble gas type electronic structure in vacuum can be calculated from those of the isoelectronic noble gases using the quantum mechanical scaling principle. According to this principle, any characteristics of ions or atoms along isoelectronic sequences are inversely proportional to the effective atomic number [17]. $\varepsilon_{\text{eff},i}$ is a effective relative permittivity of the water surrounding ionic species $i$ [-]. $z_{\text{eff},i}$, is the effective electric charge, which is calculated by Eq. (7) using experimental data of the Gibbs free energy of the hydration, $\Delta G_i$. We used Marcus's data [15] for experimental data of the Gibbs free energy of the hydration and the crystal radius of ionic species $i$. In addition, we used the Booth equation (6) in calculation of the surface relative permittivity, $\varepsilon_{\text{surf},i}$.

The effective electric charge calculated by this method is listed for some ions in Figure 3(a). The effective electric charge for monovalent cations is approximately equal to its formal electric charge, but for multivalent cations the effective electric charge became smaller because of the strong interaction with water molecules. In addition, absolute value of the effective electric charge for monovalent anions is generally smaller than that for monovalent cations. When water molecules hydrate cations, the electrostatic interaction between cations and the electric dipole of the water molecules, and the interaction by electron donation from water molecules to cations must be considered. On the other hand, when water molecules hydrate anions, the interaction by electron donation from anions to water molecules must be considered in addition to the electrostatic interaction. Assuming the electrostatic interaction between water molecules and cations/anions which have the same valence and radius, it is thought that water molecules strongly hydrate to cations rather than anions because water molecules have stronger electron acceptability than electron donating capability [18].

Effective diffusivity of $Sr^{2+}/Cs^+/I^-$ calculated by the modified model and the current ISD model is shown in Figure 3(b) together with measured data [3, 4]. The effective electric charge for these tracer ions is used, however the formal electric charge is applied for the electrolyte ions which decided an electrostatic field because of the electric neutral condition. The modified model considering the effective electric charge satisfactorily reproduces measured data, especially for multivalent cation, $Sr^{2+}$.

**(a)**

| Ion (Formal Charge) | Effective Charge ($\times 1/e$) [−] |
|---|---|
| Na(1+) | 0.912 |
| K(1+) | 0.978 |
| Cs(1+) | 0.975 |
| Mg(2+) | 1.538 |
| Ca(2+) | 1.494 |
| Sr(2+) | 1.514 |
| Al(3+) | 2.223 |
| Cl(−) | −0.754 |
| Br(−) | −0.778 |
| I(−) | −0.877 |

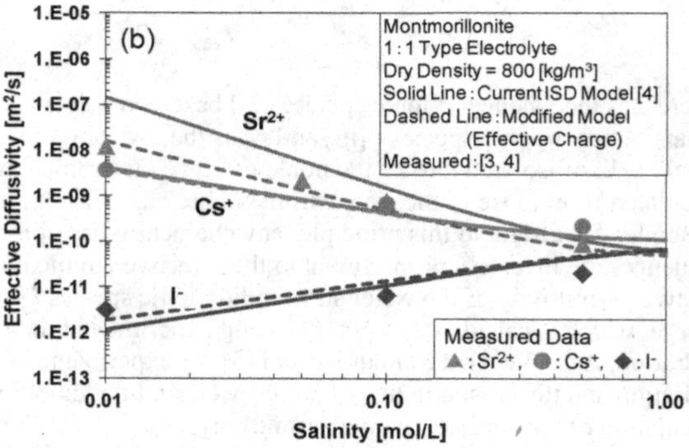

**(b)** — Effective Diffusivity [m²/s] vs Salinity [mol/L]

Montmorillonite
1 : 1 Type Electrolyte
Dry Density = 800 [kg/m³]
Solid Line : Current ISD Model [4]
Dashed Line : Modified Model (Effective Charge)
Measured : [3, 4]

Measured Data
▲ : $Sr^{2+}$, ● : $Cs^+$, ◆ : $I^-$

**Figure 3.** (a) The effective electric charge calculated by Eq. (7) for some cations and anions. (b) Effective diffusivity of $Sr^{2+}/Cs^+/I^-$ as a function of salinity, calculated by the modified model considering the effective charge and the current ISD model.

## DISCUSSION AND CONCLUSIONS

In this study, potential key factors influencing diffusion model in narrow charged pores in compacted bentonite were investigated. The modified diffusion models considering the excluded volume effect and the dielectric saturation effect were solved numerically and gave small influence on the calculated effective diffusivity of $Sr^{2+}/Cs^+/I^-$. Therefore it was concluded that the disagreements with experimental data observed in current ISD model cannot be improved by these two factors. The reasons are considered: 1) the ionic concentration distributions change depending on the models, however the total quantity of counter-ion does not almost change in the interlayer because of the electric neutral condition, 2) the mean electrostatic potential distributions change depending on the models, however the correction factor seems to cancel the influence of the change of the mean electrostatic potential. On the other hand, the modified model based on the effective electric charge gave good representation of measured data. The hydrated ions in solution would behave like the particle with the effective electric charge. This may be explained by some mechanisms, e.g., the decrease in the electric charge by forming "the solvent-shared ion-pair" and/or "the contact ion-pair" in the domain where concentration of ions become very higher like the electric double layer. To verify the idea of the effective electric charge, it is thought that the electrophoresis experiment with higher electrolyte concentration or analysis using the molecular dynamics simulation is effective.

## ACKNOWLEDGMENTS

This study was partly funded by the Ministry of Economy, Trade and Industry of Japan. We would like to thank M. Ochs, T. Ohe and H. Kato for useful discussions at the early stage of this work.

**REFERENCES**

1. M. Ochs, B. Lothenbach, H. Wanner, H. Sato, M. Yui, *J. Contam. Hydrol.* **47**, 283–296 (2001).
2. Y. Tachi, T. Nakazawa, M. Ochs, K. Yotsuji, T. Suyama, Y. Seida, N. Yamada, M. Yui, *Radiochim. Acta* **98**, 711–718 (2010).
3. Y. Tachi and K. Yotsuji, in *abstract book of the 5th meeting on international meeting "Clays in Natural and Engineered Barriers for Radioactive Waste Confinement"* (2012).
4. Y. Tachi and K. Yotsuji, *Geochim. Cosmochim. Acta*, accepted (2013).
5. J. Lyklema and J. Th. G. Overbeek, *J. Colloid Sci.* **16**, 501 (1961).
6. E.J.W. Verwey and J.Th.G. Overbeek, *Theory of the Stability of Lyophobic Colloids*, Elsevier, Amsterdam (1948).
7. D.L. Carter, M.D. Heilman and C.L. Gonzalez, *Soil Sci.* **100**, 356 (1965).
8. Y. Gur, I. Ravina and A. J. Babchin, *J. Colloid Interface Sci.* **64**, 326, 333 (1978).
9. S. Basu and M.M. Sharma, *J. Colloid Interface Sci.* **165**, 355 (1994).
10. V.N. Paunov, R.I. Dimova, P.A. Kralchevsky, G. Broze and A. Mehreteab, *J. Colloid Interface Sci.* **182**, 239 (1996).
11. J. Lehikoinen, A. Muurinen and M. Olin, *Mater. Res. Soc. Proc.* **506**, 383 (1998).
12. C.F. Gerald and P.O. Wheatley, *Applied Numerical Analysis (5th ed.)*, Addison-Wesley Publishing Company, Inc. (1994).
13. E.R. Nightingale Jr., *J. Phys. Chem.* **63**, 1381 (1959).
14. R.D. Shannon, *Acta Cryst.* **A32**, 751 (1976).
15. Y. Marcus, *Ion Properties*, Marcel Dekker, Inc., New York (1997).
16. F. Booth, *J. Chem. Phys.* **19**, 391, 1327, 1615 (1951).
17. R.H. Stokes, *J. Am. Chem. Soc.* **86**, 979 (1964).
18. H. Ohtaki, *Hydration of the Ion*, Kyoritsu Shuppan, Tokyo (1990) [in Japanese].

Mater. Res. Soc. Symp. Proc. Vol. 1665 © 2014 Materials Research Society
DOI: 10.1557/opl.2014.638

# Addition of Al₂O₃ nanoparticles to bentonite: effects on surface charge and Cd sorption properties

Natalia Mayordomo, Ursula Alonso, Tiziana Missana, Ana Benedicto, Miguel García-Gutiérrez

CIEMAT, Department of Environment, Avenida Complutense 40 28040 Madrid (SPAIN)

## ABSTRACT

Compacted bentonite barrier in radioactive waste repositories is expected to prevent radionuclide migration, due to its high sorption capability for many radionuclides. This study analyses whether the addition of Al₂O₃ nanoparticles (NPs) enhances the sorption properties of bentonite. The study was carried out with ¹⁰⁹Cd, highly pollutant heavy metal and divalent fission product. Sorption experiments were conducted in NaClO₄ at different ionic strengths ($5 \cdot 10^{-4}$ to $10^{-1}$ M) and pH (2 to 10), using mixtures of sodium homoionised bentonite and Al₂O₃ in different proportions.

It has been probed that addition of Al₂O₃ NPs to bentonite enhances Cd sorption at pH higher than 6. The effect of Al₂O₃ NPs addition on the surface properties of bentonite colloids was also analyzed by measuring particle size and surface charge in all studied systems.

## INTRODUCTION

Bentonite clay, mainly composed by smectite, is a buffer and backfill material considered in nuclear waste repositories, owing to its swelling properties and high sorption capability for many radionuclides. When hydrated, smectite particles take on a permanent negative surface charge that promotes retention of positively charged ions, while radionuclides whose dominant aqueous species were anionic can be mobile. The charge on the edge of bentonite particles is positive or negative, depending on pH conditions.

The aim of this study is to evaluate whether the addition of Al₂O₃ nanoparticles (NPs) enhances the sorption properties of bentonite. The addition of nanoparticles (diameters < 50 nm) positively charged may favour the sorption of negatively charged species or modify sorption of cations. The enhancement of bentonite sorption properties has been tested with magnetic Fe nanoparticles, promoting contaminant reduction[1, 2], but no previous study with Al₂O₃ was reported.

FEBEX clay, a Spanish Ca-Mg bentonite, was selected for the study [3]. The FEBEX clay was homoionised in Na (Na-bentonite) and mixed at different weight proportions with Al₂O₃ NPs (nominal size < 50 nm), in NaClO₄ at different ionic strengths. The surface properties (size and surface charge) of the independent Na-bentonite clay and Al₂O₃ NPs suspensions and of the bentonite /Al₂O₃ mixtures were analysed. In the binary systems interaction of particles with very different surface charge properties may promote changes in charge or particle coagulation that affects contaminant retention, as previously observed for hematite, magnetite and TiO₂ /clay mixtures [4, 5].

The sorption properties of the Na-bentonite /Al₂O₃ mixtures were evaluated with ¹⁰⁹Cd(II). Cadmium is a heavy metal of great environmental concern that has a low fission product yield, whose dominant aqueous species are positive. Previous studies on Cd sorption onto bentonite [6, 7, 8, 9] or alumina are reported [10, 11], but no previous study with bentonite / $Al_2O_3$ mixtures is available.

## EXPERIMENTAL DETAILS

### Materials

The bentonite selected is the FEBEX clay, a Ca-Mg bentonite from Spain [1]. The cation exchange capacity (CEC) of FEBEX clay is $102 \pm 4$ meq/100g and the BET surface area is 33 $m^2/g$ [12]. FEBEX bentonite was purified and homoionized with Na (Na-bentonite) by washing three times with 1 M $NaClO_4$. The colloidal fraction (size smaller than 500 nm) was obtained by centrifuging several times the suspension at 600 x g during 10 min and collecting the supernatant. Samples were equilibrated by dialysis with $NaClO_4$ electrolyte at the desired ionic strength.

The $Al_2O_3$ nanoparticles (NPs), whose phase is $\gamma$- $Al_2O_3$, were prepared in NaClO4 electrolytes without previous cleaning procedure from $Al_2O_3$ nanopowders with a nominal size < 50 nm (Aldrich) and a with a BET surface area of 136 $m^2/g$.

Four different suspensions were prepared varying the percentage in weight: Na-bentonite 100 %, Na-bentonite /$Al_2O_3$ 50:50 %, Na-bentonite / $Al_2O_3$ 10:90 %, and $Al_2O_3$ 100%, prepared in $NaClO_4$ electrolytes. For characterisation and stability studies, a solid to liquid concentration of 10 mg/L was considered, while sorption experiments were carried out with 0.5 g/L.

Cadmium solution was prepared in HCl 0.1 M from a commercial $^{109}CdCl_2$ (Ecker &Ziegler) solution bearing a carrier ([Cd]$_{TOT}$= $4.45 \cdot 10^{-3}$ M, [$^{109}$Cd] = $3.54 \cdot 10^{-6}$ M). For sorption experiments the final Cd concentration used was $4.6 \cdot 10^{-8}$ M.

Cd speciation in $NaClO_4$ was analysed with the CHESS code [13]. The predominant aqueous specie up to pH 8 is $Cd^{2+}$ and at alkaline pH the hydrolysed species $CdOH^+$ and $Cd(OH)_2$(aq) dominates. For pH >11 the anionic species $Cd(OH)_4^{2-}$ and $Cd(OH)_3^-$ may have some influence. No Cd precipitation under the experimental conditions is expected.

### Suspensions characterization

The surface characteristics of the suspensions were studied as a function of pH and ionic strength. Photon Correlation Spectrometry (PCS) technique was used to measure the mean particle size, using a Zetasizer Nano S Malvern Instrument of wavelength $\lambda = 633$ nm equipped with a photomultiplier at 173°.

To evaluate the surface charge of the particles, zeta potential ($\zeta$) was measured by Laser Doppler electrophoresis with a Zetamaster Malvern system equipped with a Spectra-Physics 2mW He-Ne laser ($\lambda = 633$ nm). The electric conductivity of the samples was always checked upon HCl or NaOH addition to detect changes in ionic strength. For $4 \geq pH \geq 10$ increase in ionic strength is expected.

### Sorption experiments

Sorption edges were carried out at room temperature under oxic conditions, in the independent suspensions and Na-bentonite /$Al_2O_3$ mixtures, by changing the pH from pH 2 to 12, adding NaOH or HCl 0.1 M. Sorbent total concentration in all experiments was 0.5 g/L, and total cadmium concentration was fixed to $4.6 \cdot 10^{-8}$ M.

Samples were maintained in polyethylene tubes (10 mL) in agitation during one week. After that samples are centrifuged (21275 g, during 1 hour) to separate the liquid from the solid. Three aliquots (2 mL) of the supernatant are sampled and [109]Cd activity is measured with a Packard Autogamma COBRA 2 counter.

Distribution coefficients (Kd) were calculated from the three aliquots of the supernatant with this equation:

$$Kd = \frac{C_i - C_f}{C_f} \cdot \frac{V}{m} \qquad \text{Eq (1)}$$

where $C_i$ is the initial activity (counts/ml), $C_f$ the final activity in the supernatant (counts/ml), m is the mass of the solid (g) and V the liquid volume (mL).

On $Al_2O_3$ system, Cd sorption on tube walls was estimated considering that sorption in oxides at lower pH (2-5.5) values should be null and 7% of the initial Cd concentration was subtracted.

## RESULTS

### Suspensions characterization

Figure 1 shows the zetapotential ($\zeta$) measured on independent Na-bentonite (100) and $Al_2O_3$ (100) suspensions and on Na-bentonite / $Al_2O_3$ mixtures (50:50 and 10:90). Only figures at $5 \cdot 10^{-4}$ M (left) and $10^{-1}$ M (right) ionic strength are shown, but similar behaviour was found for $10^{-2}$ M and $10^{-3}$ M.

In Figure 1 it can be appreciated that the zeta potential of Na-bentonite is always negative ($\zeta \approx$ -35 mV) almost independent of pH or ionic strength [14]. The zeta potential of $Al_2O_3$ depends on pH, being positive for acidic pH and negative for pH higher than the isoelectric point (IEP). The $Al_2O_3$/bentonite mixtures, showed negative $\zeta$ values, within the whole pH range that are lower than those of bentonite for pH values lower than the isolelectric point of $Al_2O_3$: This behaviour was equivalent at all studied ionic strengths. The different charge of the mixed systems may affect the retention of contaminants.

Figure 2 shows the average size of particles measured by PCS on the independent $Al_2O_3$ and Na-bentonite suspensions and on Na-bentonite / $Al_2O_3$ mixtures, as function of pH, at $5 \cdot 10^{-4}$ M (Figure 2 left) and 10-1 M (Figure 2 Right). Measurements were carried out after ultrasonication treatment for five minutes once pH was fixed. Equivalent behaviour was found for 10-2 M and 10-3 M ionic strengths (I).

The initial average size of bentonite colloids in $NaClO_4$ $5 \cdot 10^{-4}$ M is around 300 nm (Figure 3 Left) and independent of pH. This average size remains practically unvaried up to I=$1 \cdot 10^{-3}$ M, but at I=$10^{-1}$ M bentonite colloids are completely destabilized (size > 1500 nm in Figure 2 right).

As expected, the average size of the $Al_2O_3$ suspension showed dependence on pH, with the maximum size corresponding to the isoelectric point (IEP), pH $\approx$ 8-9, where aggregation is promoted [15, 16]. Far from the IEP and at low ionic strength, the average size was 425±40 nm, indicating that $Al_2O_3$ NPs are stable as nanoaggregates (nominal size of 50 nm confirmed by AFM measurements). The average particle size increases up to 540 nm at $10^{-3}$ M, 610 nm at $10^{-2}$ M and 1000 nm in $10^{-1}$ M (Figure 3 Right).

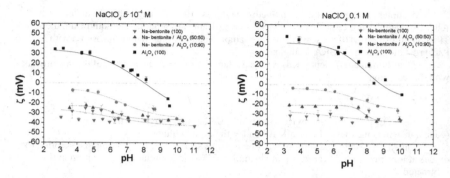

**Figure 1.** Average zeta potential as a function of the pH at **(Left)** $5 \cdot 10^{-4}$ M and **(Right)** $10^{-1}$ M, for ▼ Na-bentonite (100), ▲ Na-bentonite / $Al_2O_3$ (50:50), ● Na-bentonite / $Al_2O_3$ (10:90), ■ $Al_2O_3$ (100). Lines depict the tendency of experimental data.

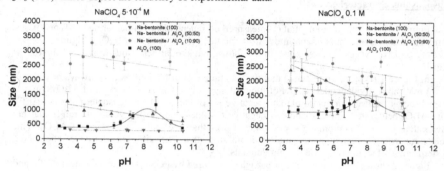

**Figure 2.** Average particle size measured by PCS as a function of the pH in $NaClO_4$ **(Left)** $5 \cdot 10^{-4}$ M and **(Right)** $10^{-1}$ M, for ▼ Na-bentonite (100), ▲ Na-bentonite / $Al_2O_3$ (50:50), ● Na-bentonite / $Al_2O_3$ (10:90), ■ $Al_2O_3$ (100). Lines show the tendency of experimental data.

In Na-bentonite /$Al_2O_3$ mixtures particles are fastly coagulated (average size > 1000 nm), even at low ionic strength (Figure 3 Left). This size increment seems to be less effective at alkaline pH, because both Na-bentonite and $Al_2O_3$ are negatively charged and repulsive electrostatic forces may reduce particle interaction. But, the particle destabilisation at low ionic strength and acidic pH is not in agreement to zeta potential measurements (Figure 2).

The charge behaviour of oxides /clay mixtures is complex since oxides and clays exhibit surface charges of different origin. While in bentonite, the structural and pH-independent permanent negative charge is due to isomorphic substitution of $Al^{3+}$ for $Si^{4+}$ in the tetrahedral sheets and $Mg^{2+}$ for $Al^{3+}$ in the octahedral sheets of the 2:1 layer. The pH-dependent charge on the edges of the lamellar bentonite particles (Al-OH and Si-OH sites), can be de/protonated, being positive or negative depending on pH conditions, but this charge is too small to overcome the structural negative charge on the lamellar faces. In oxides, surface sites (S-OH) exhibit different charge depending on pH conditions ($SOH_2^+$ for pH< $pH_{IEP}$ and $SO^-$ for pH > $pH_{IEP}$). In the oxide /clay mixtures, complex charge interactions are expected [4, 17, 18].

## Sorption experiments

Figure 3 shows the Cd sorption edges obtained as a function of the pH in NaClO$_4$ $5 \cdot 10^{-4}$ M and $10^{-1}$ M, onto the independent systems and mixtures. In the Figures, two different regions are clearly identified for pH lower or higher than pH ≈ 6.5.

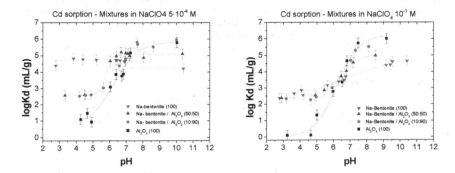

**Figure 3.** Sorption edges obtained as a function of the pH in NaClO$_4$ **(Left)** $5 \cdot 10^{-4}$ M and **(Right)** $10^{-1}$ M, for Cd sorption onto ▼ Na-bentonite, ▲ Na-bentonite / Al$_2$O$_3$ (50:50), ● Na-bentonite / Al$_2$O$_3$ (10:90) and ■ Al$_2$O$_3$ (100) when Cd concentration is kept constant ([Cd] = $4.6 \cdot 10^{-8}$ M).

Cd sorption onto Na-bentonite (100) depends on ionic strength at acidic pH, suggesting a cation exchange process with the aqueous $Cd^{2+}$, as previously described for other bivalent cations [12, 19, 20]. However, the slight dependence on pH observed at the higher ionic strength (Figure 4, right) may indicate the contribution of surface complexation processes.

Cd sorption onto Al$_2$O$_3$ (100) is independent on ionic strength and highly pH-dependent as expected for an oxide where sorption mainly takes place by surface complexation.

Cd sorption results obtained in the bentonite /Al$_2$O$_3$ mixtures suggest that at acidic pHs sorption is dominated by Na-bentonite, even though no differences amongst the two studied weight proportions (50:50 or 10:90) are observed. At higher pHs, the addition of Al$_2$O$_3$ NPs clearly enhances Cd sorption as Kd values increase with increasing Al$_2$O$_3$ NPs amount.

At present, modelling of Cd sorption results is in progress in order to verify the hypothesis. Potentiometric titrations were obtained on the independent systems, to obtain the protonation/ deprotonating constants of the surface sites. The hypothesis of Cd cation exchange process onto Na-bentonite is also being verified. Further aim of this experimental study is to simulate the enhanced Cd sorption behaviour on the Na-bentonite /Al$_2$O$_3$ mixtures, with the thermodynamic constants obtained on the independent systems.

## CONCLUSIONS

The effect of Al$_2$O$_3$ NPs addition on the surface characteristics and sorption properties of Na-bentonite was analysed. Size and zeta potential measurements carried out in the Na-bentonite /Al$_2$O$_3$ mixtures indicated that some particle destabilisation occurred, even at low ionic strength, and this has to be further investigated.

Sorption experiments indicated that the addition of Al$_2$O$_3$ NPs to bentonite, in Al$_2$O$_3$ to bentonite 50/50 and 90/10 ratios, reduces sorption when pH is lower than 6.5 and enhances Cd sorption for pH higher than 6.5, which is shown in the increase of log Kd value from log Kd = 4 to log Kd = 5.5. Modelling of Cd sorption results is in progress, to verify the sorption hypothesis and to demonstrate if enhanced Cd sorption behaviour on the Na-bentonite /Al$_2$O$_3$ mixtures can be described with the thermodynamic constants obtained on the independent systems.

The applicability for other relevant radionuclides, for example with anionic species, should be analysed more in detail.

## ACKNOWLEDGMENTS

The research leading to these results has received funding form the Spanish Government under the project NANOBAG (CTM2011-27975). N. Mayordomo acknowledges the FPI BES-2012-056603 grant from MINECO (Spain).

## REFERENCES

1. A. Mockovciaková, Z. Orolínová, J. Skvarla, *Journal of Hazardous Materials* **180** 274–281 (2010).
2. T. Shahwan, Ç. Üzüm, A.E. Eroğlu, I. Lieberwirth, *Applied Clay Science* **47** 257–262 (2010).
3. F. Huertas et al. *Report* EUR 19147 EN. European Commission, Brussels (2000).
4. E. Tombácz, Z. Libor, E. Illés, A. Majzik, E. Klumpp, *Organic Geochemistry* **35** 3, 257-267 (2004).
5 D. Zhou, A. Abdel-Fattah, A. A. Keller, *Environmental Science & Technology* **46**, 7520-7526 (2012).
6. G. Purna Chandra Rao, S. Satyaveni, A. Ramesh, K. Seshaiah, K.S.N. Murthy, N.V. Choudary, *Journal of Environmental management* **81**, 265-272 (2006).
7. D. Zhao, S. Chena, S. Yangb, X. Yangb, S. Yangb, *Chemical Engineering Journal* **166**, 1010–1016 (2011).
8. M. Hamidpour, M. Kalbasi, M. Afyuni, H. Shariatmadari, P. E. Holm, H. Ch. Brunn Hansen, *Journal of Hazardous Materials* **181**, 686–691(2010)
9. S. Kozar, H. Bilinski, M. Branica, M.J. Schwuger, *The Science of the Total Environment* **121**, 203 (1992).
10. T. K. Naiya, A. K. Bhattacharya, S. K. Das, *J. Colloid & Interface Science* **333**,14–26 (2009).
11. Ch. Kosma, G. Balomenou, G. Salahas, Y. Deligiannakis, *J. Colloid & Interface Science* **331**, 263 (2009).
12. T. Missana, M. Garcia-Gutierrez, U. Alonso, *Physics and Chemistry of the Earth*, (33), S156 (2008).
13. J. van der Lee and L. DeWindt, CHESS tutorial and cookbook. Technical Report École des Mines, LHM/RD/99/05 Paris-France (1999).
14. T. Missana, A. Adell, *Journal of Colloid and Interface* Science **230**, 150-156 (2000)..
15. S. Ghosh, H. Mashayekhi, B. Pan, P. Bhowmik and B. Xing, *Langmuir* **24**, 12385 (2008).
16. T.K. Darlington, A.M. Neigh, M.T. Spencer, O.T. Nguyen and S.J. Oldenburg, *Environ. Toxicol. Chem.***28**, 1191 (2009).
17. A. Molinard, A. Clearfield, H.Y. Zhu and E.F. Vansant, *Microporous Mater* **3**, 109 (1994).
18. D. Zhou, A. Abdel-Fattah, A. A. Keller, *Environmental Science & Technology* **46**, 7520-7526 (2012).

19. T. Missana and M. Garcia-Gutierrez, *Physics and Chemistry of the Earth* **32,** 559-567(2007).
20. B. Baeyens and M.H. Bradbury, *Journal of Contaminant Hydrology* **27,** 199-222 (1997).

Mater. Res. Soc. Symp. Proc. Vol. 1665 © 2014 Materials Research Society
DOI: 10.1557/opl.2014.639

# Improvement of Inventory and Leaching Rate Measurements of C-14 in Hull Waste, and Separation of Organic Compounds for Chemical Species Identification

Ryota Takahashi[1], Michitaka Sasoh[1], Yu Yamashita[1], Hiromi Tanabe[2], and Tomofumi Sakuragi[2]

[1]Toshiba Corporation, 8, Shisugita-Cho, Isogo-Ku, Yokohama 235-8523, Japan
[2]Radioactive Waste Management Funding and Research Center, Pacific Marks Tsukishima, 1-15-7 Tsukishima, Chuo-ku, Tokyo, 104-0052, Japan

## ABSTRACT

In order to analyze the C-14 inventory and leaching rate for safety evaluation of transuranic waste disposal, it is necessary to establish an analytical method that can measure C-14 with sufficient precision [1]. Oxidative decomposition of organic compounds containing C-14 is carried out to absorb carbon dioxide ($CO_2$) in an alkaline solution, which is mixed with a liquid scintillation cocktail, and the amount C-14 is quantified by measuring a beta ray spectrum with a liquid scintillation counter. It has been difficult to completely decompose carbon compounds in a sample, even to $CO_2$, by using conventional oxidizing agents. In the work described here, we improved the method of oxidative decomposition used to completely decompose carbon compounds using peroxydisulfuric acid ($K_2S_2O_8$). When C-14 in the form of $CO_2$ was absorbed in a sodium hydroxide (NaOH) aqueous solution, only 80% of the actually used quantity was detected. Total organic carbon measurements showed that the entire quantity of $CO_2$ was absorbed by NaOH. When NaOH aqueous solution was used, it was found that only the analytical value was 80%. The entire quantity of the actually used carbon could be measured by absorbing the $CO_2$ in Carbo-Sorb®. An anion form and a neutral molecule exist in the organic compound released from activated metals. In order to identify organic compounds efficiently, fractionation into an anion and a neutral molecule and separation by high performance liquid chromatography (HPLC) are necessary. Here, we propose the combined use of an ion exchange resin and HPLC as an improved technique for identification of the chemical species.

## INTRODUCTION

In order to evaluate the actual sources of C-14 in activated cladding metals, it is essential to measure the amount of C-14 released from radioactive cladding and to identify its chemical form. Transuranic (TRU) waste includes a large amount of C-14, which is one of the dominant radioactive nuclides. C-14 in metals such as Zircaloy, stainless steel, and Inconel is assumed to be released at a rate related to the individual corrosion rates of each metal.

We consider four main issues associated with C-14 analysis:
(1) Evaluating the oxidizing agents used in the wet oxidation method.
(2) Improving the catalysts used in the oxidation furnace.
(3) Examining the influence of the composition of the uptake liquid on liquid scintillation counter (LSC) measurements.
(4) Improving chemical species identification.

The C-14 inventory, leaching rate, and chemical form have been investigated in several studies to clarify the behavior of C-14 released from activated metals. Tables I to III summarize the measurement techniques and results of some relevant research. However, the available knowledge is not sufficient for evaluating the migration behavior of C-14, including its chemical form and sorption behavior in typical disposal conditions or geological environments, the chemical species of the C-14 form, its relationship with corrosion rate, etc.

**Table I** Pretreatment methods for measuring the C-14 inventory of irradiated metals and graphite.

| Material | Source | Measurement of C-14 inventory | Ref. |
|---|---|---|---|
| Zry | PWR Zry-4 Base metal, Oxide film | Base metal specimens were dissolved in a $HNO_3+HF+H_2O_2$ solution, and C-14 was collected in alkaline solution as vaporized carbon dioxide. C-14 activity was determined with a liquid scintillation counter (LSC). Oxide film specimens were dissolved in a $HNO_3+HF+H_2O_2$ solution. | 2 |
| | BWR Zry-2 STEP I Base metal, Oxide film | Specimens were dissolved in a $HNO_3+HF$ solution, and C-14 was collected in alkaline solution as vaporized carbon dioxide. C-14 activity was determined with an LSC. | 3 |
| | BWR Zry-2 STEP III Base metal | Specimens were dissolved in a $HNO_3+HF$ solution, and C-14 was collected in alkaline solution as vaporized carbon dioxide. C-14 activity was determined with an LSC. | 4 |
| Stainless steel (SUS) | BWR Upper grid | Specimens were dissolved in a $HNO_3+HF$ solution, and C-14 was collected in alkaline solution as vaporized carbon dioxide. C-14 activity was determined with an LSC. | 5 |
| | BWR Shroud | Specimens were dissolved in a $HNO_3+HF$ solution, and C-14 was collected in alkaline solution as vaporized carbon dioxide. C-14 activity was determined with an LSC. | 6 |
| Graphite | Moderator Reflector | The concentration of C-14 was determined by complete combustion of a powdered sample of irradiated graphite, and the inventory was calculated by multiplying the C-14 concentration and the total mass of the sample, $1.6 \times 10^6$ kg. | 7 |

**Table II** Measurement of the amount and chemical form of C-14 in samples.

| Material | Source | Abstract | Ref. |
|---|---|---|---|
| Zry | PWR Zry-4 Base metal, Oxide film | The amount of C-14 was analyzed by wet oxidation, in which organic C-14 was oxidized by $KMnO_4$. Most of the organic C-14 was neutral species (organic C-14 passed through anion exchange paper). | 2 |
| | BWR Zry-2 STEP I Base metal, Oxide film | The amount of C-14 was analyzed by wet oxidation, in which organic C-14 was oxidized by $KMnO_4$ or $K_2S_2O_8+AgNO_3$. C-14 was detected in gas and liquid phases. The amount of C-14 leaching out from an oxide film was more than that from bare Zry. The same amount of inorganic and organic C-14 was detected in a 30-month leaching test. In the leaching test conducted by the stepwise sampling method, the amount of C-14 in the sample was observed to decrease. Formate, formaldehyde, and acetate were isolated by HPLC. | 3 |
| | BWR Zry-2 STEP III Base metal | The amount of C-14 was analyzed by wet oxidation, in which organic C-14 was oxidized by $KMnO_4$ or $K_2S_2O_8+AgNO_3$. C-14 was detected in liquid and gas phases; however, it was not detected in the gas phase in a 12-month leaching test. Leaching tests were continued. | 4 |
| Stainless steel (SUS) | BWR Upper grid | Organic C-14 was not measured in the gas phase but was detected in the liquid phase. Organic C-14 was oxidized by $KMnO_4$. A C-14 chromatograph was measured by HPLC + LSC, some peaks were obtained and attributed to formate, formaldehyde, acetate, and methanol. | 5 |
| | BWR Shroud | C-14 was detected in gas and liquid phases. Organic carbon in the liquid phase was separated by an anion exchange resin to obtain neutral and anion species. | 6 |
| Graphite | Moderator Reflector | C-14 was not measured in the gas phase. The amount of C-14 was analyzed by wet oxidation, in which organic C-14 was oxidized by $KMnO_4$. A C-14 chromatograph was measured by HPLC + LSC, and peaks attributed to formate, acetate, and methanol were observed. The $K_d$ value of OPC(ordinary Portland cement) for total C-14 was 69 cm$^3$ g$^{-1}$. The $K_d$ values of organic carbon assigned to the peaks in the C-14 chromatograph were more than 30 cm$^3$ g$^{-1}$, and were not consistent with that of C-14 labeled compounds for OPC. It was suggested that organic carbon leaching out from the graphite may not have been formate, acetate, or methanol. | 7 |
| | Fe$_3$C, ZrC, Carbon Steel (Non-activated) | C-14 was not measured in the gas phase. A C-14 chromatograph was measured by HPLC + LSC, and peaks attributed to formate, acetate, and methanol were observed. The $K_d$ value of OPC for total C-14 was 69 cm$^3$ g$^{-1}$. The $K_d$ values of organic carbon assigned to the peaks in the C-14 chromatograph were more than 30 cm$^3$ g$^{-1}$, and were not consistent with that of C-14 labeled compounds for OPC. It was suggested that organic carbon leaching out from the graphite may not have been formate, acetate or methanol. | 8 |

**Table III** Chemical forms of C-14 in gas phase and liquid phases.

| Material | Source | Analysis | Gas phase | | Liquid phase | | Ref. |
|---|---|---|---|---|---|---|---|
| | | | Inorganic C-14 | Organic C-14 | Inorganic C-14 | Organic C-14 | |
| Zry Base Metal | PWR Zry-4 Base metal | Wet oxidation + LSC | N.D. ($CO_2$, CO) | N.D. ($CH_4$) | 11 months: N.D. 5.5 months: 21% | 11 months: 100%. 5.5 months: 79% Neutral | 2 |
| | BWR Zry-2 STEP I Base metal | Wet oxidation + LSC | Inorganic + Organic: N.D. (0%) to 76% 30 months: 71% | | Inorganic + Organic: 24% to 100% 30 months: Inorganic 17%, Organic 12% | | 3 |
| | BWR Zry-2 STEP III Base Metal | Wet oxidation + LSC | Inorganic + Organic 6–9 months: 50% 12 months: N.D. (0%) | | 18–20% 12 months: 72% | 30–31% 12 months: 28% | 4 |
| Zry with oxide film | PWR Zry-4 Oxide film | Wet oxidation + LSC | N.D. ($CO_2$, CO) | N.D. ($CH_4$) | 11 months: N.D. | 11 months: 100%. | 2 |
| | BWR Zry-2 STEP I Oxide film | Wet oxidation + LSC | Inorganic + Organic: N.D. (0%) to 76% 30 months: 8% | | Inorganic + Organic: 5% to 70% 30 months: Inorganic 30%, Organic 62% | | 3 |
| Stainless steel | BWR Upper grid | Wet oxidation HPLC + LSC | No data | | 23–34% | 66–75% formate 10%, acetate 43%, formaldehyde 9%, methanol 10%, ethanol 6%, Unknown 22%. | 5 |
| | BWR Shroud | Ion exchange + LSC | Inorganic + Organic : 42 months: 25% | | 42 months: 37% | 42 months: 38% Neutral: 18% Anion: 82% | 6 |
| Graphite | Moderator Reflector | HPLC + LSC | No data | | ca.20% | 80% Detected peaks of formate, acetate, and methanol at retention time | 7 |
| Fe₃C | | HPLC | No data | | 35–45% | 55–65% | 8 |
| ZrC | | HPLC | No data | | 10–12% | 88–90% | |
| Carbon Steel | | HPLC | No data | ca. 0.01% | 10–12% | 70–85% | |
| Zr | | HPLC | No data | | 15–45% | 55–85% | |

N.D.: Not Detectable

Both inorganic and organic forms have been detected in the liquid phase in which a cladding sample was immersed [3] [4]. The organic C-14 was decomposed to $CO_2$ by an oxidizing reagent and was collected in an alkali solution. In those studies, the amount of C-14 was measured with an LSC. A potassium permanganate ($KMnO_4$) aqueous solution was used as an oxidizing agent. However, the organic compound was not completely decomposed to $CO_2$.

On the other hand, using an oxidation furnace, depending on the kind of gas, the carbon compound is sometimes not decomposed to $CO_2$.

In this research, we provide a procedure that can completely decompose an organic compound, even into $CO_2$, by improving the catalyst used in the oxidation furnace and the oxidizing agents used in the wet oxidation method.

For C-14 gas collected in NaOH aqueous solution by the wet oxidation process of organic C-14, the amount of C-14 analyzed by LSC is always 80% of the amount of C-14 supplied to the experimental unit before the oxidation treatment; that is, 20% of the organic C-14 vanishes in the wet oxidation process. In order to investigate the cause of this and to improve the wet oxidation process of C-14, we investigated the presence of C-14 in the experimental unit for all wet oxidation processes.

We investigated the presence of an impurity in a labeled compound that could be oxidized easily, the change in the LSC output value due to sample acidity or alkalinity, and the mass balance of carbon in the wet oxidation process system using a total organic carbon (TOC) analyzer, and we report the analytical ability of the fixed-quantity method of performing C-14 inventory of activated metals and immersed samples.

Further, it has been reported that C-14 released from activated metal has been detected as an inorganic compound and an organic compound. The organic compound is considered to be an alcohol, aldehyde, or carboxylic acid whose carbon number is one or two [5].

The organic compound released from activated metal is identified using HPLC. When the retention value of a peak of the organic compound in a sample is in agreement with that for the chemical species of a standard reagent, the organic compound is identified as the chemical species of the standard reagent. For example, when a peak appears at a retention time of 45 min at which acetate appears, the peak is considered to be assigned to acetate. However, an organic compound cannot be identified by comparison with the retention value in HPLC.

The chemical species of C-14 released from irradiated graphite has also been measured by HPLC as well as activated metals. Isobe et al., investigated the distribution coefficient ($K_d$) of C-14 in irradiated graphite in Ordinary Portland Cement (OPC), a mixture of Low-Heat Portland Cement (LPC) and Portland Fly-Ash Cement (LPC/FA), and depleted-LPC/FA [7]. Values of $K_d$ were measured by the following two methods.

**Table IV** Issues associated with C-14 measurement.

| Parameter | Analytical method | Issues with analytical method |
|---|---|---|
| Decomposition rate of organic compound in sample solution | Wet oxidation | It is difficult to completely decompose the carbon compound in a sample, even to $CO_2$, using conventional oxidizing agents. |
| Decomposition rate of C-14 gas | Decomposition of C-14 gas to $CO_2$ by oxidation furnace using a CuO catalyst | In order to measure C-14 gas from an immersed sample, the C-14 gas is aerated in an oxidation furnace and is collected in the form of $CO_2$ in an alkaline solution. A CuO catalyst provides a $CO_2$ conversion ratio of about 80% for $CH_4$. |
| Consistency of amounts of C-14 before and after wet oxidation | Measurement of C-14 in gas and liquid phases by LSC | When C-14 in the form of $CO_2$ is absorbed in a sodium hydroxide (NaOH) aqueous solution, only 80% of the actually used quantity is detected. |
| Identification of organic C-14 | HPLC | Organic compounds cannot be identified by ion exchange or HPLC alone. |

(a) C-14 retained in the liquid phase of leaching sample was separated into inorganic and organic C-14 by oxidation of the liquid phase, and the $K_d$ values of the inorganic and organic C-14 were measured. The $K_d$ values for organic C-14 were 47 cm$^3$ g$^{-1}$ for OPC, 31 cm$^3$ g$^{-1}$ for LPC/FA, and 17 cm$^3$ g$^{-1}$ for depleted-LPC/FA[9]. We observed that $K_d$ for organic C-14 from irradiated graphite was larger than that from activated metal. The results suggest that the chemical form of organic C-14 from irradiated graphite was different from that from activated metal.

(b) The $K_d$ value for each chemical form separated by HPLC was measured by comparing the peaks in a C-14 chromatograph for the initial sample solution and that for a sample solution containing with cement powder. The initial sample solution was prepared by immersing the irradiated graphite in simulated groundwater for 2 months. The C-14 content of each sample solution separated by HPLC was measured using an LSC to obtain a C-14 chromatograph containing five independent peaks at retention times from 20 min to 70 min. Subsequently, OPC, LPC/FA, and depleted LPC/FA were each immersed in a liquid phase sample, and the liquid phase sample was separated by HPLC to obtain a C-14 chromatograph. The $K_d$ value for each peaks that represented a chemical form was calculated from the areas of the peaks before and after immersion. The C-14 chromatograph is shown in Figure 1 [7]. The $K_d$ values calculated from the C-14 chromatograph were 31 cm$^3$ g$^{-1}$ for LPC/FA and 9 cm$^3$ g$^{-1}$ for depleted LPC/FA, and the results were consistent with the $K_d$ values obtained by the standard method. The $K_d$ values for the 3rd peak were 180 cm$^3$ g$^{-1}$ for LPC/FA and 120 cm$^3$ g$^{-1}$ for depleted-LPC/FA, showing that there exists a chemical form of C-14 with a large $K_d$. The chemical form corresponding to each peak has not yet been identified.

The retention values of the peaks shown in Figure 1 are consistent with those of representative organic compounds, i.e., acetate, methanol, formate, formaldehyde, and ethanol, emitted from activated metal. As shown in Table V, the $K_d$ values of the representative organic compounds for OPC obtained with a standard method, ranging from 0.4 to 2.7 cm$^3$ g$^{-1}$, are not consistent with those calculated from the peak areas in the C-14 chromatograph. By comparing the retention value of each peak, the chemical species could not be identified. Thus, the HPLC+LSC method provides insufficient information to identify the organic compounds. Here, we propose a means of identifying an organic compound directly and a method of identifying an organic compound released from activated metal with a combination of HPLC and an ion exchange method.

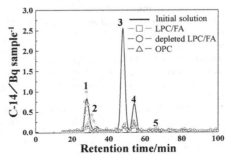

**Figure 1.** C-14 chromatograph for liquid phase of the sample solution in which cement was immersed, showing that the peak area is smaller than that of the initial solution. The $K_d$ values were estimated from the peak areas in the chromatograph [7].

**Table V** Result of C-14 sorption test on the leached liquid of irradiated graphite, Comparison of $K_d$ values of OPC related organic C-14 species / cm$^3$ g$^{-1}$.

| Method of sorption test | Test solution | Formate HCOOH Peak No.3 | Acetate CH$_3$COOH Peak No.4 | Formaldehyde HCHO – | Methanol CH$_3$OH – | Ethanol C$_2$H$_5$OH Peak No.5 |
|---|---|---|---|---|---|---|
| HPLC method | Irradiated graphite leaching liquid [7] | 200 | 34 | – | – | N.D. |
| Ordinary method | C-14 labeled solution [7] | 2.7 | 1.2 | 2.6 | 0.7 | 0.4 |
| | Irradiated Zry leaching liquid[14] | 1.9~5.8 | | | | |

## EXPERIMENTAL

(1) Evaluating the oxidizing agents used in the wet oxidation method

To select the oxidizing agent for use in the wet oxidation method, we evaluated some typical agents used in general industrial fields. We determined the oxidizing conditions, such as temperature and decomposition time, while carrying out a cold test and checking the ability of the selected oxidizing agent to decompose the organic compound.

For the selected oxidizing agent, we examined the wet oxidation reaction by using a C-14 labeled compound (tracer) while varying the oxidizing conditions, i.e., the temperature, decomposition time, and composition of the organic compound, and we measured the decomposition rate of the organic compound. From these results, we determined a suitable procedure for conducting wet oxidation of organic C-14.

As oxidizing agents, we selected $K_2Cr_2O_7$, $K_2S_2O_8$, $KIO_3$, $KMnO_4$, $H_2O_2$, $KBrO_3$, and $Ce(SO_4)_2$ [5]. The procedure of the wet oxidation technique is described below.

(a) The initial solution, the carrier such as $Na_2CO_3$, and the C-14 tracer were fed into the reaction container (flask), and the flask was heated to a prescribed temperature with a heater.

(b) Air was bubbled through the solution from a line connected to the flask, and the decomposed inorganic carbon was volatilized. At this time, we checked whether the line was properly connected by visually checking for the existence of air bubbles by opening and closing the line.

(c) Total oxidation of imperfect mineralization (CO etc.) of the volatilized carbon was carried out with a copper oxide catalyst, and the oxidized volatilized carbon was passed through a cold trap (cooled on ice) and was collected by an alkali trap (10 cm³ of 1 mol/dm³ NaOH solution).

(d) Alkali traps, a line washing liquid, and the reaction solution were collected, and their C-14 concentrations were measured with a liquid scintillator, mixing, and a liquid scintillation detector, respectively.

(e) The decomposition rate of C-14 was searched for using the following formula:
$$D = (A_T / A_I) \times 100 \qquad (1)$$
where $D$ is the decomposition rate, $A_T$ is the amount of C-14 collected in the alkaline solution, and $A_I$ is the amount of C-14 supplied to the experimental unit.

(2) Improving the catalyst used in the oxidation furnace

In order to measure C-14 gas from an immersed sample, the C-14 gas was aerated in an oxidation furnace and was collected in the form of $CO_2$ in an alkaline solution. Improving the performance of oxidation furnaces is an important issue. A CuO catalyst provides a $CO_2$ conversion ratio of only about 85.8% for $CH_4$, showing that $CH_4$ is not sufficiently decomposed to $CO_2$ using a CuO catalyst alone. We examined the decomposition performance when using a platinum catalyst in addition to CuO. We expected that the decomposition reaction would progress under the platinum+CuO catalyst. We checked the $CH_4$ decomposition performance to measure the $CO_2$ conversion ratio for $CH_4$.

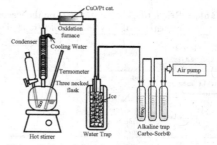

**Figure 2.** Experimental unit of wet oxidation.

(3) Examining the influence of the composition of the uptake liquid on LSC measurements

Figure 2 shows an outline of the experimental unit used in the wet oxidation of an organic compound. The organic compound added to the reaction vessel was decomposed into $CO_2$ by the oxidizing agent. An oxidation furnace decomposed C-14 gas into $CO_2$ to be absorbed in NaOH aqueous solution in an alkali trap. The uptake liquid was mixed with the liquid scintillation cocktail, and the amount of C-14 was measured by the LSC. However, the analytical value of the

144

amount of C-14 in the uptake liquid was only 80% of the amount of C-14 used in the experimental unit.

Then, the amounts of C-14 in the reaction liquid, the catalyst used in the oxidation furnace, the uptake liquid of the alkali trap, and the piping of the decomposition test system were measured in a wet oxidizing unit to investigate where the 20% undetectable C-14 was in the experimental unit. A controlled amount of C-14 labeled ethanol and an ethanol reagent were put into the wet oxidizing unit and were decomposed to $CO_2$, which was collected in NaOH aqueous solution. The amount of C-14 and total organic carbon TOC) were measured to investigate the mass balance of carbon before and after the reaction.

(4) Improving chemical species identification

The chemical species of an organic compound can be identified from a comparison of the retention value in HPLC; however, the chemical species of C-14 compounds cannot be determined by HPLC alone. We have to use nuclear magnetic resonance (NMR) and liquid chromatography–mass spectroscopy (LC-MS), which were developed in the 1980s for identification of chemical species [10][11]. We can separate typical organic compounds into each chemical species by HPLC now. By combining the latest analytical techniques, such as mass analysis and NMR, with HPLC, the conventional method of identifying carbon compounds released from irradiated metal can be developed. In this research, the newest analytical techniques for identifying carbon compounds were investigated.

## RESULTS AND DISCUSSION

(1) Evaluating the oxidizing agents used in the wet oxidation method

When $K_2S_2O_8$+$AgNO_3$ was used as an oxidizing agent, the decomposition rate of the carbon compound was 90%. With other oxidizing agents, the decomposition rate was 10% for $Ce(SO_4)_2$ and 16% for $KMnO_4$. With $K_2Cr_2O_7$, $KIO_3$, $H_2O_2$, and $KBrO_3$, organic compounds were not decomposed. As a result, among the oxidizing agents selected, $K_2S_2O_8$+$AgNO_3$ was found to be the most promising oxidizing agent for the wet oxidation technique. Although reaction time dependency of the decomposition rate was not observed in 0.5 to 2 hours, a high decomposition rate was obtained. It took 30 minutes to decompose an organic compound. The decomposing time for the wet oxidation of an

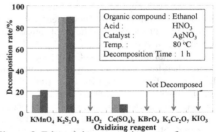

**Figure 3.** Ethanol decomposition rate of oxidizing reagents

organic compound was set to 60 minutes.

(2) Improving the catalyst used in the oxidation furnace

The oxygen concentration dependency of the decomposition rate is shown in Table VI. The decomposition rate was calculated by dividing the amount of recovered $CO_2$ by the amount of $CH_4$ used. At oxygen concentrations in the range 20% to 100%, the platinum catalyst functioned well in the decomposition of $CH_4$. On the other hand, at an oxygen concentration of 0%, the recovery rate was only 72.8%, and we confirmed that the effect by the oxygen supply from CuO is not expectable.

**Table VI** Decomposition performance of CH$_4$ in an oxidation furnace using a CuO+Pt catalyst.

| | Oxygen concentration /% | Used CH$_4$ /mol | Recovered CO$_2$ /mol | D /% |
|---|---|---|---|---|
| 1 | 0 | $8.2 \times 10^{-5}$ | $6.0 \times 10^{-5}$ | 72.8 |
| 2 | 20 | $8.2 \times 10^{-5}$ | $8.3 \times 10^{-5}$ | 101.5 |
| 3 | 100 | $4.1 \times 10^{-5}$ | $4.0 \times 10^{-5}$ | 98.4 |

(3) Examining the influence of the composition of the uptake liquid on LSC measurements

The results are shown in table VI. When the residual amount of C-14 in each part of the wet decomposition unit was measured, C-14 was not detected other than in the uptake liquid of the alkali trap. The amount of C-14 in the uptake liquid was 80% of the initial amount of C-14 used.

On the other hand, C-14 in the form of CO$_2$ was not detected in the case where an ethanol reagent was decomposed, other than in the alkali trap. The amount of CO$_2$ in the uptake liquid was consistent with the amount of CO$_2$ calculated from the amount of ethanol used.

By analyzing the amount of C-14 collected in the NaOH aqueous solution and comparing it with the mass balance of C-14 labeled compound and a nonradioactive ethanol reagent, it was found that only 80% of the available C-14 in the wet oxidation unit was detected. Then, when wet oxidation of the C-14 labeled compound was carried out by using Carbo-Sorb® as the uptake liquid, the consistency of the mass balance was confirmed.

By adding acid to C-14 in the form of CO$_2$ dissolved in the NaOH aqueous solution to collect it in Carbo-Sorb®, we found that the analytical value of the amount of C-14 was 120% of the controlled amount. When C-14 in the form of ethanol was distilled and collected an alkali trap, namely, both NaOH aqueous solution and Carbo-Sorb®, the analytical values were 100% of the whole quantity, confirming the consistency of the mass balance.

The above result showed that the whole amount of C-14 had been trapped in the NaOH alkali trap, and the analytical value obtained by the LSC was 80% of the amount of C-14 available in the experimental unit, when C-14 in the form of CO$_2$ was collected in the NaOH aqueous solution. Therefore, it is necessary to use Carbo-Sorb® for wet oxidation of organic C-14 to collect it in the form of CO$_2$ (shown in Table VII).

(4) Improving chemical species identification

The amount of C-14 released from 8 g of spent fuel cladding per year is $10^{-13}$ g per sample, and it is difficult to identify $10^{-13}$ g of organic C-14[4]. The cladding tube contains 250 ppm of C-12, which is a stable isotope[13]. In the case where C-12 is released as the cladding corrodes, the amount of the carbon compound released from a cladding tube in one year will be $1.2 \times 10^{-7}$ g. Mass Spectroscopy (MS) analysis is appropriate for identifying the chemical species of C-12. By combining HPLC and an ion exchange method, separation and analysis of typical organic compounds can be conducted. It is also possible to use MS analysis to determine the mass number of the organic compound that forms each peak. The combination of separation such as chromatography and mass spectrometer that can measure mass with high precision will allow identification of the chemical species (see Table VIII).

**Table VII** Recovery rates of a mixed sample containing five organic compounds ($CH_3COOH$, HCOOH, HCHO, $CH_3OH$, and $C_2H_5OH$) decomposed to $CO_2$ for collection in NaOH and Carbo-Sorb®.

| | RUN 1 | RUN 2 | RUN 3 |
|---|---|---|---|
| | C-14 tracer | Organic compound reagent | C-14 tracer |
| Organic compound | Five kinds of organic compound | Five kinds of organic compound | Five kinds of organic compound |
| Uptake solution | NaOH aqueous solution | NaOH aqueous solution | Carbo-Sorb® |
| Analytical method | LSC | TOC | LSC |
| Initial amount | 941 Bq | 12.8 mg | 941 Bq |
| Solution after reaction | N.D. | 0.26 mg | N.D. |
| Piping | N.D. | 0.05 mg | N.D. |
| Catalyst | N.D. | – | N.D. |
| Water trap | N.D. | 0.02 mg | N.D. |
| Alkaline trap | 747 Bq | 13.6 mg | 941 Bq |
| Total | 747 Bq | 13.9 mg | 941 Bq |
| Recovery rate | 79% | 108% | 100% |

**Table VIII** Influence of the composition of the uptake liquid on LSC measurements.

| No. | Initial | Initial soln. | Initial amount | Final form | Uptake soln. | Measured value | Chemical process |
|---|---|---|---|---|---|---|---|
| 1 | $CO_2$ | NaOH aq. | A1 | $CO_2$ | NaOH aq. | A1 × 100% | $CO_2$ recovery by acid |
| 2 | $CO_2$ | NaOH aq. | A2 | $CO_2$ | Carbo-Sorb® | A2 × 120% | $CO_2$ recovery by acid |
| 3 | $C_2H_5OH$ | NaOH aq. | A3 | $C_2H_5OH$ | NaOH aq. | A3 × 100% | Distillation |
| 4 | $C_2H_5OH$ | NaOH aq. | A4 | $CO_2$ | NaOH aq. | A4 × 80% | Oxidation |
| 5 | $C_2H_5OH$ | NaOH aq. | A5 | $CO_2$ | Carbo-Sorb® | A5 × 100% | Oxidation |
| 6 | $C_2H_5OH$ | NaOH aq. | A6 | $C_2H_5OH$ | Carbo-Sorb® | A6 ×1 00% | Distillation |

## CONCLUSIONS

(1) In wet oxidation, most kinds of organic compounds can be decomposed by using $K_2S_2O_8$-$AgNO_3$.

(2) By adding Pt to CuO as a catalyst in the oxidation furnace, the C-14 gas can be completely decomposed to $CO_2$ form.

(3) The analytical value of the amount of $CO_2$ labeled by C-14 collected in NaOH aqueous solution was 80% of a controlled amount in LSC measurements. In order to quantify C-14 in the form of $CO_2$, it is necessary to collect C-14 in Carbo-Sorb®.

(4) The improved analysis for C-14 has been applied to quantification of C-14 released from STEPIII cladding tube.

(5) An organic compound cannot be identified by ion exchange or HPLC alone, but needs to be identified with a mass spectrometer. By combining ion exchange, HPLC, and a mass spectrometer, it will become possible to identify the organic compound in the separated fraction.

## ACKNOWLEDGMENTS
This research is a part of "Research and development of processing and disposal technique for TRU waste containing I-129 and C-14" program funded by the Ministry of Economy, Trade and Industry (METI).

## REFERENCES
1. Japan Atomic Energy Agency and the Federation of Electric Power Companies: Second Progress Report on Research and Development for TRU Waste Disposal in Japan, JAEA-Review 2007-010, 2007.
2. Takashi Yamaguchi, Susumu Tanuma, Isamu Yasutomi, Tadashi Nakayama, Hiromi Tanabe, Kiyomichi Katsurai, Wataru Kawamura, Kazuto Maeda, Hideo Kitao, Moriyuki Saigusa: A Study on Chemical Forms and Migration Behavior of Radionuclides in HULL Waste, ICEM1999, September, Nagoya, Japan, 1999.
3. RWMC, Annual Reports on Research & Development for Leaching Behavior of C-14 from Irradiated Metals, under contract with the Agency for Natural Resources and Energy, Ministry of Economy, Trade and Industry, 2011.
4. RWMC, Annual Reports on Research & Development for Leaching Behavior of C-14 from Irradiated Metals, under contract with the Agency for Natural Resources and Energy, Ministry of Economy, Trade and Industry, 2012.
5. M. Sasoh, The Study for The Chemical Forms of C-14 Released from Activated Metal, Proceedings of Workshop on the Release and Transport of C-14 in repository environments., NAGRA NIB04-03 (2004).
6. Y. Miyauchi, Y. Yamashita, J. Sakurai and M. Sasoh, Nuclide release behavior from activated stainless and measurement of Kd-value, The Atomic Energy Society of Japan Autumn meeting, B23, 2011. (in Japanese)
7. M. Isobe, T. Yamamoto, R. Takahashi, M. Sasoh, Y. Nakane and H. Sakai, Chemical Form of Organic C-14 Leaching from Irradiated Graphite in Tokai Plant, The Atomic Energy Society of Japan Autumn meeting, L28, 2008. (in Japanese)
8. S. Satoru, et al., A Study on the Chemical Forms and Migration Behavior of Carbon-14 Leached from the Simulated Hull Waste in the Underground condition, 2002 MRS FALL MEETING, December 2-6, Boston, MA (2002).
9. M. Isobe, K. Numata, R. Takahashi, M. Sasoh, E. Hirose and J. Sakurai, Distribution Coeffient of C-14 released Irradiated Graphite in Tokai Plant, The Atomic Energy Society of Japan Autumn meeting, N57, 2007.
10. Y. Tanaka and Y. Okura, eds., *Bunsekikagaku I*, (Nankodo, Tokyo, 1982), p. 79–81.
11. The Chemical Society of Japan, *Zikkenkagakukouza Vol. 12, Kakuzikikyoumeikyushu I*,(Maruzen, Tokyo, 1967).
12. R. M. Holt, M. J. Newman, F. S. Pullen, D. S. Richards and A. G .Swanson, *J. Mass Spectrom.* **32**, p.64 (1997).
13. ASTM B349-93, Annual Book of ASTM Standards, Sec. 2, Volume 02, 04.
14. International Workshop on Mobile Fission and Activation Products in Nuclear Waste Disposal, L'Hermitage, La Baule, France, January 16-19, 2007.

Mater. Res. Soc. Symp. Proc. Vol. 1665 © 2014 Materials Research Society
DOI: 10.1557/opl.2014.640

# $K_d$ setting approach through semi-quantitative estimation procedures and thermodynamic sorption models: A case study for Horonobe URL conditions

Yukio Tachi[1], Michael Ochs[2], Tadahiro Suyama[1] and David Trudel[2]
[1] Japan Atomic Energy Agency, 4-33, Muramatsu, Tokai, Ibaraki, 319-1194, Japan
[2] BMG Engineering Ltd, Ifangstrasse 11, CH-8952 Schlieren-Zürich, Switzerland

## ABSTRACT

The use of generic sorption data in PA requires the transfer of the data to the PA-specific conditions. A site-specific $K_d$ setting approach for PA calculations was tested, comparing two data transfer procedures. First transfer of sorption data can be done through semi-quantitative estimation procedures, by considering differences between experimental and PA geochemical conditions (sorption capacity, radionuclide speciation, competitive reactions, etc.). On the other hand, thermodynamic sorption models allow to estimate $K_d$ variations directly, based on quasi-mechanistic understanding. The present paper focuses on illustrating example calculations regarding the derivation of $K_d$ values, and their uncertainties, of Cs, Ni, Am and Th, for the mineralogical and geochemical conditions of the mudstone system at the Horonobe URL. Clay minerals (illite and smectite) were considered as sorption-relevant minerals in all cases. The $K_d$ setting results were compared with $K_d$ measured for Horonobe mudstone by batch experiments. The results indicate that $K_d$ can be quantitatively evaluated from generic sorption data when adequate data and models are available. The careful evaluation and conjunctive use of calculated and measured $K_d$ values can enhance the reliability of $K_d$ setting and uncertainty assessments.

## INTRODUCTION

Sorption of radionuclides (RNs) on host rocks is a key process in the safe geological disposal of radioactive waste. For performance assessment (PA) calculations, the magnitude of sorption is normally expressed by a distribution coefficient ($K_d$). $K_d$ depends critically on relevant geochemical conditions. Therefore, the values to be used in PA calculations need to correspond to the specific PA conditions, and need to take into account geochemical variability or uncertainty. Since it is not feasible to measure $K_d$ values for all PA conditions, the use of existing sorption data obtained under generic experimental conditions and transferring such data to a range of PA-specific conditions is therefore a key challenge. Such data transfer can be done through expert judgment, semi-quantitative estimations and use of a thermodynamic sorption model, by considering the differences between experimental and PA-specific geochemical conditions, such as sorption capacity, RN speciation, competitive reactions, etc. [e.g., 1, 2, 3]. JAEA has adopted the site-specific $K_d$ setting approach for PA based on the simultaneous use of different transferring approaches [4, 5]. The present paper focuses on illustrating example calculations regarding the derivation of $K_d$ values and their uncertainties of Cs, Ni, Am and Th, for geochemical conditions at the Horonobe Underground Research Laboratory (URL).

## GEOCHEMICAL CONDITIONS AND $K_d$ SETTING APPROACHES

### Geochemical/experimental conditions

$K_d$ values recommended for PA calculations of RN sorption on host rocks are often associated with relatively large uncertainties. A significant part of the overall uncertainty may be related to uncertainties in the specified geochemical conditions. Possible variations of geochemical conditions are typically considered based on borehole investigations in the respective host formation. For the calculations shown in this paper, only one set of conditions was focused on in terms of comparative discussions with measured $K_d$ values. The mineralogical and groundwater compositions listed in Table I are seen as representative for the Wakkanai Formation (the HDB-6 borehole from a depth of ~500m) at Horonobe URL[5, 6]. Batch sorption experiments for Cs, Ni, Am and Th were conducted using crushed rock samples (<355 μm) and synthetic groundwater (Table I). Other experimental conditions are also shown in Table I. All tests were conducted in duplicate or triplicate in parallel with blank tests without rocks [4, 6].

**Table I.** Mineralogical/groundwater composition for $K_d$ setting and batch experimental condition.

| | |
|---|---|
| Mineralogical composition | Smectite: 16wt.%, Illite: 10wt.%, Opal-CT: 42wt.%, Quartz: 13wt.%, Albite: 6wt.%, K-feldspar: 5wt.%, Pyrite: 3wt.%, minors: 5wt.% |
| Groundwater composition | pH: 8.5, Ionic strength: 0.26, Na$^+$:0.23M, K$^+$:0.00021M, Mg$^{2+}$:0.0058M, NH$_4^+$:0.0078M, Cl$^-$:0.22M, HCO$_3^-$:0.03M |
| Batch sorption experimental condition | Liquid/solid ratio: 100mL/g, Reaction time: 14days, RN initial concentration: Cs 1×10$^{-5}$M / Ni 1×10$^{-6}$M / Am 5×10$^{-10}$M / Th 1×10$^{-6}$M, Atmosphere: aerobic-Cs, Ni / anaerobic-Am, Th, Filtration: 0.45μm membrane filter |

## Semi-quantitative estimation approaches

A semi-quantitative estimation procedure has been developed based on the pertinent approaches previously proposed [1, 2, 5]. The conversion flow and factors accounting for differences in sorption capacity and surface as well as solution chemistry in the experimental vs. application (PA) conditions are summarized in Figure 1 [see details in 5]. The working hypothesis for the present $K_d$ setting is that RN sorption on the Horonobe mudstone is determined by the respective content of clay minerals (smectite and illite). All available evidence suggests that these clay minerals are able to sorb RNs mainly through surface complexation and ion exchange [e.g., 3]. The $K_d$ values for the Horonobe mudstone system were derived using available generic sorption data for these clay minerals taken from the JAEA-SDB [7].

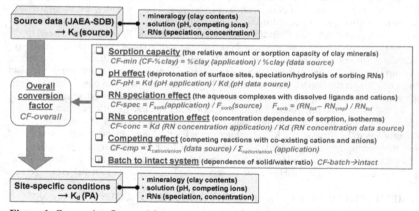

**Figure 1.** Conversion flow and factors used in the semi-quantitative estimation approach.

Scaling for differences in sorption capacity can be done on the basis of the direct use of the respective content of sorption-relevant minerals, i.e., clay minerals. The pH is the most critical parameter for the sorption of most RNs as it determines the speciation of the sorbing element itself, as well as the protonation/deprotonation of surface complexation sites, etc. Due to the complexity of pH effects, it was found preferable to select data sources at pH values corresponding to the application conditions. Where this is not possible, scaling to the application pH (CF-pH) can be done on the basis of separate sorption edges for the same RN. The treatment of differences in RN speciation (CF-spec), concentration (CF-conc) and competing ions (CF-cmp) is discussed for each RN below. In addition, the more general issue of using batch-type data (corresponding to disperse systems) to derive $K_d$ values for intact rock needs to be assessed. This can be difficult to assess in a fractured/ heterogeneous rock (e.g., granite), but no conversion factor (CF-batch→intact) is necessary to account in the case of homogeneous porous media such as bentonite and clay rocks [3]. Uncertainties for each $K_d$ setting were also calculated following [5] taking into account all uncertainties related to the scaling process.

## Thermodynamic sorption modeling (TSM) approaches

Quasi-mechanistic sorption models and a model parameter database have been developed for bentonite/smectite systems [4]. These models were developed for various RNs based on experimental data from the literature, which had originally been obtained in batch sorption experiments under simplified conditions. The models are generally based on a 1-site surface complexation model without electrostatic correction terms and a 1-site ion exchange model. All models are parameterized to be consistent with the JAEA-TDB [8]. TSM parameters of Ni, Am and Th for smectite in [4] are directly applied to the Horonobe rock system. For the present illustration, these models are assumed to be applicable to illite as well, as a first approximation. For Cs, the existing sorption models for smectite [9] and for illite [10] were directly used. All calculations were done using the PHREEQC code [11] and TSM parameters listed in Table II.

**Table II.** TSM parameters for Cs, Ni, Am and Th used in $K_d$ setting approaches.

| | *Ion exchange* | | | | *Surface complexation* | |
|---|---|---|---|---|---|---|
| | **Illite** | | **Smectite** | | **Smectite (Illite)** | |
| Site type | FES/Q | Type II/X | PS/Y | PS/Z | SOH | |
| Site capacity | 0.5 [meq/kg] | 40 [meq/kg] | 160 [meq/kg] | 1100 [meq/kg] | $4.31\times10^{-2}$ [mol/kg] | |
| Site-K | 2.4 | 2.1 | 1.1 | 0.42 | $SOH^{2+}$: 5.75 / $SO^-$: $-8.30$ | |
| Site-NH$_4$ | 3.5 | 2.8 | 0.46 | 0.46 | $SONi^+$ $-2.32$ | $SOAm^{2+}$ 0.30 |
| Site-Ca | – | – | 0.69 | 0.69 | $SONiOH$ $-12.16$ | $SOAmOH^+$ $-7.62$ |
| Site-Mg | – | – | 0.67 | 0.67 | $SONi(OH)_2^-$ $-21.02$ | $SOAm(OH)_2$ $-16.57$ |
| Site-Cs | 7 | 3.6 | 1.6 | 1.6 | $SOTh^{3+}$ 7.07 | $SOAm(OH)_3^-$ $-27.30$ |
| (Site)$_2$-Ni | – | – | 0.75 | 0.75 | $SOTh(OH)_3$ $-10.40$ | $SOAmCO_3$ 6.74 |
| (Site)$_3$-Am | – | – | 2.05 | 2.05 | $SOTh(OH)_4^-$ $-18.59$ | $SOAm(CO_3)_2^{2-}$ 10.42 |
| (Site)$_4$-Th | – | – | 2.92 | 2.92 | | |

## RESULTS AND DISCUSSION

### Cesium

Cesium exists only as $Cs^+$ ion in solution and sorbs on clay minerals mainly by ion exchange reactions. Figure 2(a) shows a compilation of $K_d$ datasets which were chosen in terms of good quality and coverage of a range of relevant conditions. As can be expected, the data show that Cs sorption is almost independent of pH, but strongly influenced by the concentration of competing cations and of Cs itself. Sorption on illite is higher than sorption on smectite at trace Cs concentration, mainly because of the frayed edge sites (FES) on the illite surface.

It is expected that sorption on Horonobe mudstone is influenced by both illite and smectite [6], therefore the respective data for these minerals were taken into account in the semi-quantitative $K_d$ evaluation. As shown in Figure 2(b), the $K_d$ source data close to the application conditions can be transferred using CF-min and CF-cmp factors (the competitive effect of all cations is taken to be equal, considering the high Cs concentration). CF-conc was not used considering no difference in Cs concentration ($1 \times 10^{-5}M$). The $K_d$ values derived from the various data sources (smectite; [12,13], illite; [12,14]) were averaged for illite and smectite, and their contributions were summed up according to the component additivity (CA) approach [3, 15]. There is reasonable quantitative agreement within a factor of two between $K_d$ values derived by the semi-quantitative estimation, the TSM (ion exchange) and the batch measurements (Figure 4(a)). However, at trace concentration of Cs, the estimation procedure would have a significant disadvantage. It is not clear under such conditions which correction factor takes competition by major cations including best into account, considering the importance of FES. It is therefore recommended to use TSM-derived values for supporting PA-specific measurements.

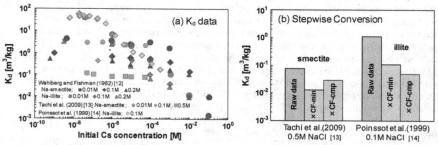

**Figure 2.** Cs sorption data on smectite and illite (a), and stepwise conversion results (b).

## Nickel

Several relevant $K_d$ datasets for Ni on smectite and illite shown in Figure 3(a) indicate that sorption is dominated by ion exchange at low ionic strength and low pH (up to pH $\approx$ 8). The further increase of sorption with pH can be explained by surface complexation at clay edges. The sorption edges seems to differ significantly between smectite and illite at lower pH, however, $K_d$ values under all conditions are relatively similar at the maximum of the pH edges at about pH 8-9. It is therefore assumed that sorption on Horonobe mudstone is influenced by illite and smectite, and the respective data for both minerals were considered in the semi-quantitative evaluation.

The $K_d$ source data close to the application conditions can be transferred using CF-min, CF-conc and CF-spec factors (Figure 3(b)). CF-pH was not needed, considering in the nearly identical pH between data source and the application conditions. Scaling to the appropriate RN concentration was done on the basis of isotherm data obtained under similar solution conditions. In the case of Ni, complexation with carbonate in the aqueous phase is viewed as fully competitive with respect to sorption, since there is no evidence for the existence of ternary carbonate complexes (dissolved or surface-bound) for Ni. The $K_d$ values derived from the various data sources (smectite; [16, 17], illite; [18]) were averaged for illite and smectite. Overall $K_d$ was again calculated based on the CA approach. As shown in Figure 4(b), it can be seen that both semi-quantitative estimation and TSM (smectite model) give nearly identical results, due to similar sorption values of illite and smectite under the application condition. Both approaches give slightly lower values than the measured data. A comparison of the various uncertainties can be used to supply reasonable and conservative $K_d$ values for PA.

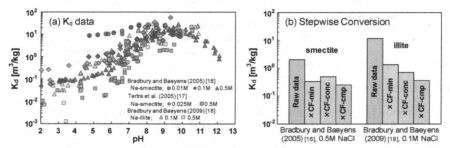

**Figure 3.** Ni sorption data on smectite and illite (a), and stepwise conversion results (b).

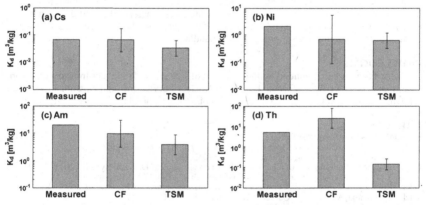

**Figure 4.** Comparison of $K_d$ values of (a) Cs, (b) Ni, (c) Am and (d) Th, obtained by batch sorption measurement, semi-quantitative estimation (CF) and sorption model (TSM).

### Americium/Thorium

As strongly hydrolyzing cations, +III- and +IV-valent actinides sorb on clay minerals mainly through inner-sphere surface complexation. Their complex chemistry makes the scaling of sorption data from experimental to application conditions more challenging. Significant uncertainties are in particular associated with corrections for differences in RN speciation. Several relevant $K_d$ datasets for Am and Eu (as a chemical analog) on smectite [19, 20] and illite [18, 19] are selected. The $K_d$ source data close to the application conditions can be transferred using CF-min and CF-spec factors. Based on the effect of dissolved carbonate on sorption shown in the datasets by [20], 50 % of the Am-carbonate complexes predicted by the aqueous speciation calculation are assumed to contribute to $K_d$ as ternary surface Am-carbonate complex (CF-spec). As shown in Figure 4(c), the $K_d$ value derived by the TSM for smectite including ternary complex is slightly lower than measured and scaled values, but the agreement is reasonably good as all values fall within one order of magnitude.

In the case of Th (Figure 4(d)), it appears that the experimental database is insufficient for both estimation methods used here. The semi-quantitative estimation overestimates $K_d$ because of large gap RN concentration, The TSM significantly underpredicts $K_d$ due to the lack of model parameters describing Th sorption in the presence of dissolved carbonate. The $K_d$ derivation for Th can only be improved by obtaining systematic experimental sorption data as a function of Th and carbonate concentrations.

## CONCLUSIONS

The present paper focuses on illustrating a range of example calculations regarding the derivation of $K_d$ values and their uncertainties for four key RNs, Cs, Ni, Am and Th, for the geochemical conditions of Horonobe mudstone system. This $K_d$ setting exercise allows to estimate the magnitude of sorption and the related uncertainty under the expected in-situ conditions when adequate data and models are available. On the other hand, some critical gaps in the existing experimental data and in process understanding are identified. The results indicate that the simultaneous use of different estimation procedures can increase the confidence in $K_d$ setting. However, a careful selection of results from the different methods is needed, according to the data available and the level of process understanding.

## ACKNOWLEDGMENTS
This study was partly funded by the Ministry of Economy, Trade and Industry of Japan. The experiments were partly performed by Tôkyo Nuclear Service Inc. and Mitsubishi Material Corporation. This study was supported by the Horonobe Underground Research Center, JAEA.

## REFERENCES
1. M.H. Bradbury, B. Baeyens, PSI Bericht Nr. 03-08 (2003).
2. M. Ochs, C. Talerico, P. Sellin, A. Hedin, *Phys. Chem. Earth,* **31**, 600-609 (2006).
3. OECD/NEA, NEA Sorption Project Phase III report (2012).
4. JAEA, Assessment Methodology of Chemical Effects on Geological Disposal System (2011, 2012).
5. M. Ochs, Y. Tachi, D. Trudel, T. Suyama, JAEA-Research 2012-044 (2013).
6. Y. Tachi, K. Yotsuji, Y. Seida, M. Yui, *Geochim. Cosmochim. Acta* **75**, 6742–6759 (2011).
7. Y. Tachi, T. Suyama, M. Ochs, C. Ganter, JAEA-Data/Code 2010-03 (2011).
8. A. Kitamura, K. Fujiwara, R. Doi, Y. Yoshida, M. Mihara, M. Terashima, M. Yui, JAEA-Data/Code 2009-024 (2010).
9. H. Wanner, Y. Albinsson, E. Wieland, *Fresenius J. Anal. Chem.* **354**, 763-769 (1996).
10. M.H. Bradbury, B. Baeyens, *J. Contam. Hydrol.* **42**, 141–163 (2000).
11. D.L. Parkhurst, C.A.J. Appelo, User's guide to PHREEQC (ver.2) (1999).
12. J.S. Wahlberg, M.J. Fishman, USGS Bull. 1140-A (1962).
13. Y. Tachi, K. Yotsuji, Y. Seida, M. Yui, *Mater. Res. Soc. Symp. Proc.* **1193**, 545–552 (2009).
14. C. Poinssot, B. Baeyens, M.H. Bradbury, PSI Bericht Nr. 99-06 (1999).
15. J.A. Davis, J.A. Coston, D.B. Kent, C.C. Fuller, *Environ. Sci. Technol.* **32**, 2820-2828 (1998).
16. M.H. Bradbury, B. Baeyens, *Geochim. Cosmochim. Acta,* **69**, 875-892 (2005).
17. E. Tertre, G. Berger, S. Castet, M. Loubet, E. Giffaut, *ibid,* **69**, 4937-4948 (2005).
18. M.H. Bradbury, B. Baeyens, *ibid,* **73**, 990-1013 (2009).
19. L. Gorgeon, L. Ph.D. thesis. Université Paris 6 (1994).

20. M. Marques Fernandes, B. Baeyens, M.H. Bradbury, *Radiochim. Acta* **96**, 691-697 (2008).

Mater. Res. Soc. Symp. Proc. Vol. 1665 © 2014 Materials Research Society
DOI: 10.1557/opl.2014.641

## Migration Behavior of Selenium in the Presence of Iron in Bentonite

Kazuya Idemitsu, Hikaru Kozaki, Daisuke Akiyama, Masanao Kishimoto, Masaru Yuhara,
Noriyuki Maeda, Yaohiro Inagaki and Tatsumi Arima
Dept. of Applied Quantum Physics and Nuclear Engineering, Kyushu Univ., Fukuoka, Japan

### ABSTRACT

Selenium (Se) is an important element for assessing the safety of high-level waste disposal. Se is redox-sensitive, and its oxidation state varies from -2 to 6 depending on the redox conditions and pH of the solution. Large quantities of ferrous ions formed in bentonite due to corrosion of carbon steel overpack after the closure of a repository are expected to maintain a reducing environment near the repository. Therefore, the migration behavior of Se in the presence of Fe in bentonite was investigated by electrochemical experiments. $Na_2SeO_3$ solution was used as tracer solution. Dry density range of bentonite was from 0.8 to $1.4 \times 10^3 \, kg/m^3$.

Results indicated that Se was strongly retained by the processes such as precipitation reaction with ferrous ions in bentonite. Se K-edge X-ray absorption near-edge structure (XANES) measurements were performed at the BL-11 beamline at SAGA Light Source, and the results revealed that the oxidation state of Se in the bentonite remained Se(IV).

### INTRODUCTION

Selenium-79 ($^{79}$Se ) has a long half-life of $2.95 \times 10^5$ years and is one of the key radionuclides for examining the safety of high-level waste disposals [1]. In aqueous solution, Se is present mainly in anionic form, therefore is expected to be less affected by sorption than cations in the natural and engineered barriers. Se is redox-sensitive, and its oxidation state varies from -2 to 6 depending on the redox conditions and the pH of the solution.

Carbon steel is a candidate overpack material for high-level waste disposal. In Japan, this material is expected to ensure complete containment of vitrified waste glass during an initial period of 1000 years [2]. Thus, large quantities of ferrous ions formed in bentonite by corrosion of carbon steel overpack after closure of a repository are expected to maintain a reducing environment in the vicinity of the repository. Therefore, the migration behavior of Se in the presence of Fe in bentonite was investigated by electrochemical experiments in this study.

### EXPERIMENTAL

#### Materials

A typical Japanese sodium bentonite, Kunipia-F, was used in this study. It contains approximately 99 wt % montmorillonite. The chemical composition of Kunipia-F is shown elsewhere [3]. The chemical formula of Kunipia-F is estimated to be $(Na_{0.3}Ca_{0.03}K_{0.004})(Al_{1.6}Mg_{0.3}Fe_{0.1})Si_4O_{10}(OH)_2$. Bentonite powder was compacted into cylinders with a diameter of 10 mm and a height of 10 mm with a dry density of around 0.8 to $1.4 \times 10^3$

kg/m$^3$. Each cylinder was inserted in an acrylic resin column and saturated with a 0.01 M NaCl solution for 30 days.

## Electrochemical and Diffusion Experiments

An iron coupon was assembled with bentonite that was saturated by immersion in a contact solution of 0.01 M NaCl in an apparatus for electrochemical experiments (Figure 1). Ten microliters of tracer solution containing $Na_2SeO_3$ (0.13 M) was spiked onto the interface between the iron coupon and the bentonite. An Ag/AgCl reference electrode and a platinum foil counter electrode were inserted in the upper part of the apparatus into a 0.01 M NaCl solution. The pH and Eh of the upper solution after contact with bentonite are 7 to 8 and ca. 700 mV respectively. The iron coupon was connected to a potentiostat to serve as a working electrode, and an electric potential of 0 mV vs. the Ag/AgCl electrode at 25 °C was applied. In all cases, the potential of the iron coupon as a working electrode was kept at ca. 1 V higher than the counter electrode. The iron corrosion products can migrate into bentonite as ferrous ion through the interlayer of montmorillonite replacing exchangeable sodium ions in the interlayer [4]. Electric potential was applied for 3 days (energization period) in these experiments. After energization period the pH and Eh of the upper solution changed to 9 to 10 and 0 to -20 mV respectively. The sample was optionally kept without an electric potential for a further period of up to 21 days to allow for further diffusion in grove box filled with argon containing 5% $H_2$ gas. After a set time, the bentonite specimens were extracted from the columns and cut into slices of 0.5 to 2 mm thick. Each slice was submerged in 1 N $HNO_3$ solution to extract Se and Fe, and the liquid phase was separated by centrifugation (2000 xg, 5 minutes). The supernatant was collected to measure the concentrations of Se by inductively coupled plasma-mass spectrometry (ICP-MS: Agilent; 7500C) and Fe by atomic absorption spectrometry (Shimazu; AA-6300).

Diffusion experiments were also carried out, where the experimental configuration was essentially the same as in the electrochemical experiments; however, an acrylic coupon was used instead of an iron coupon, and the coupon was not connected to the potentiostat.

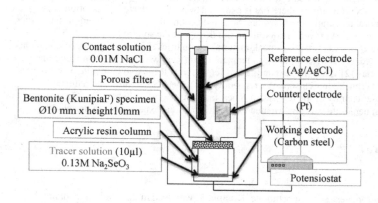

**Figure 1.** Experimental apparatus for electrochemical experiments.

## RESULTS AND DISCUSSION

### Apparent diffusion coefficients for Se in bentonite

The concentration profiles of Se infiltrated into bentonite specimens are shown in Figure 2 (open circles). The profiles obtained in the diffusion experiments were in close agreement with the instantaneous plane source model based on the following equation:

$$C(x,t) = \frac{M}{\sqrt{\pi D_a^* t}} \exp\left\{-\frac{x^2}{4 D_a^* t}\right\}, \quad (1)$$

where $D_a^*$ is the apparent diffusion coefficient and $M$ is the total amount of tracer. Fitting the measured profiles yields the apparent diffusion coefficient. The obtained values are plotted as a function of dry density in Figure 3. The $D_a^*$ values of selenite decreased with increased dry density. Similar tendencies have been reported for anionic species in bentonite by many researchers and are attributed to anion exclusion and complexity of the diffusion path [5].

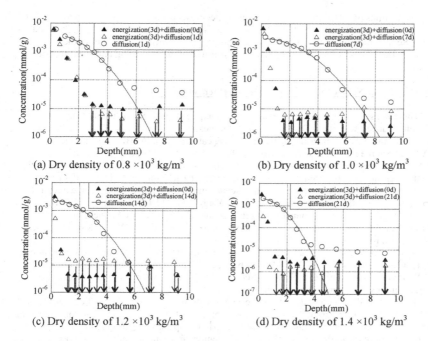

(a) Dry density of $0.8 \times 10^3$ kg/m$^3$

(b) Dry density of $1.0 \times 10^3$ kg/m$^3$

(c) Dry density of $1.2 \times 10^3$ kg/m$^3$

(d) Dry density of $1.4 \times 10^3$ kg/m$^3$

**Figure 2.** Selenium profiles in bentonite by both electrochemical and diffusion experiments.

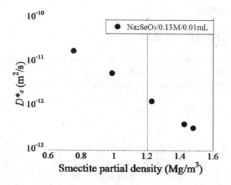

**Figure 3.** Obtained apparent diffusion coefficient $D*a$ plotted against dry density.

## Profiles of Se in bentonite specimens in electrochemical experiments

The profiles of Se infiltrated into bentonite specimens in electrochemical experiments are also shown in Figure 2 (triangles). In each case, there is little difference between the observed profile of Se just after application of electric potential and the profile obtained after the additional period allowed for diffusion. Se clearly diffused at the beginning and then stopped due to Fe diffusion.

## X-ray Absorption Near-edge Structure Measurements (XANES)

There are several possible mechanisms by which Fe stopped Se diffusion. One is that ferrous ions reduced selenite to Se(0) or to Se(-II) in FeSe.

Therefore, XANES measurements of Se were carried out at the BL-11 beamline of SAGA Light Source to investigate the chemical form of Se. A Si(111) double-crystal monochromator was used, and the beam was focused by using bent conical mirrors coated with Rh. At the measurement position, the beam had a width of 1 to 5 mm and a height of 1 mm.

XANES spectra were acquired in fluorescence mode by using a silicon drift detector (SDD). In this mode, the electrical signal of Se $K_\alpha$ X-rays (11,222.4 eV) from the SDD was selected by means of a single-channel analyzer at the K-edge. Data analysis, including background subtraction, normalization, and linear combination fitting of XANES spectra, was performed with REX2000 version 2.5 (Rigaku).

In order to determine the chemical form of Se in the presence of ferrous ions in bentonite, the first slices of the samples with dry density between 1.0 and $1.4 \times 10^3$ kg/m$^3$ were collected for XANES measurements. Se, FeSe, SeO$_2$, H$_2$SeO$_3$, and H$_2$SeO$_4$ powders were also prepared as reference samples. The samples were individually sealed in vinyl plastic and examined in air.

The results of the Se K-edge (12,652 eV) XANES measurements for Se, FeSe, SeO$_2$, H$_2$SeO$_3$, H$_2$SeO$_4$, and the samples with dry density between 1.0 and $1.4 \times 10^3$ kg/m$^3$ are shown in Figure 4. The background was subtracted from the original spectra by extrapolation of the linear absorption or curve (as defined by the Victoreen equation) from the pre-edge region. From Figure 4, the peaks of the first slice of the samples with dry density between 1.0 and $1.4 \times 10^3$

kg/m$^3$ are in close agreement with the peaks of SeO$_2$ and H$_2$SeO$_3$, meaning that in the presence of ferrous ions in bentonite, under the experimental conditions of our experiment, Se remains tetravalent.

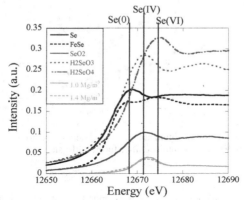

**Figure 4.** Se K-Edge XANES spectra of Se, FeSe, SeO$_2$, H$_2$SeO$_3$, H$_2$SeO$_4$, and the first slices of bentonite in the electrochemical experiments (dry density between 1.0 and 1.4 ×10$^3$ kg/m$^3$).

### Fixation of Se by ferrous ions

Some authors reported that selenite adsorption by ferric oxides is extensive and rapid, and that it decreases with pH between 3 and 8 [6-8]. Fe profiles are shown in Figure 5. In most cases, Fe diffused further during the optional diffusion period. It is possible that some of Fe precipitated as ferric oxide and adsorbed Se because the concentration of Fe was much higher than that of Se. However, it is unclear whether enough ferric oxide was produced to stop diffusion of all the spiked Se. It is also possible that ferrous ions react with selenite and precipitate as FeSeO$_3$ in the bentonite because the reported solubility product of FeSeO$_3$ is 10$^{-10}$ [9]. There is also another possibility, namely that selenite ions could be indirectly adsorbed onto the basal plane of montmorillonite via ferrous ions, for example, montmorillonite-Fe(II)-Se(IV) as an outer-sphere complex [10]. In any case, ferrous ions slow the diffusion of selenite ions in compacted bentonite, due to increase retention. However diffusion coefficients in bentonite with the presence of Fe could not obtained, because of little difference between profiles of Se just after energization and after additional period for further diffusion as shown in Figure 5.

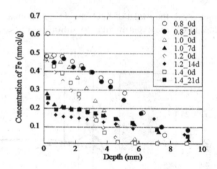

**Figure 5.** Fe profiles in bentonite obtained by electrochemical experiments. Closed symbols indicate data obtained after further diffusion of Fe.

## CONCLUSION

The apparent diffusion coefficient, $D^*_a$, of selenite without ferrous ions was found to be in the range of $10^{-11}$ to $10^{-13}$ m$^2$/s and decreased with increasing bentonite density. Furthermore, selenite ions were strongly retained by ferrous ions in bentonite. The oxidation state of Se remained tetravalent despite the reducing environment in bentonite in the presence of ferrous ions.

## ACKNOWLEDGMENTS

This research is partly financed by a Grant-in-Aid for Scientific Research (grant number S24226021). XANES measurements were performed with the approval of SAGA Light Source (1304023R).

## REFERENCES

1. National Nuclear Data Center, Brookhaven National Laboratory. Information extracted from the NuDat 2.6 database.
2. JNC, H12: *Project of Establish the Scientific and Technical Basis for HLW Disposal in JAPAN*, JNC, Tokai Japan (2000).
3. K. Idemitsu, S. Yano, X. Xia, Y. Kikuchi, Y. Inagaki, T. Arima, *Scientific Basis for Nuclear Waste Management XXVI*, edited by R. J. Finch and D. B. Bullen (Mater. Res. Soc. Proc. **757**, Pittsburgh, PA, 2003) pp.657-664.
4. K. Idemitsu, S. A. Nessa, S. Yamazaki, H. Ikeuchi, Y. Inagaki and T. Arima in *Scientific Basis for Nuclear Waste Management XXXI*, edited by W. E. Lee, J. W. Roberts, N. C. Hyatt and R. W. Grimes (Mater. Res. Soc. Proc. **1107**, 2008), pp.501-508.
5. T. E. Eriksen, M. Jansson, *Diffusion of $I^-$, $Cs^+$ and $Sr^{2+}$ in Compacted Bentonite – Anion Exclusion and Surface Diffusion*, SKB Technical Report 96-16, SKB ( 1996 ).
6. L. S. Balistrieri, T. T. Chao, Soil Science Society of America Journal **51**, 1145-1151(1987).
7. C. Su, D. L. Suarez, Soil Science Society of America Journal **64**, 101-111(2000).

8.  M. Duc, G. Lefevre, M. Fedoroff, J. Jeanjean, J. C. Rouchaud, F. Monteil-Rivera, J. Dumonceau, S. Milonjic, Journal of Environmental Radioactivity **70**, 61-72(2003).
9.  F. Seby, M. Potin-Gautier, E. Giffaut, G. Borge and O.F.X. Donard, Chemical Geology 171, 173-194(2001).
10. L. Charlet, A. C. Scheinost, C. Tournassat, J. M. Greneche, A. Gehin, A. Fernandez-Martinez, S. Coudert, D. Tisserand, J. Brendle, Geochemica and Cosmochemica Acta 71, 23, 5731-5749(2007).

Mater. Res. Soc. Symp. Proc. Vol. 1665 © 2014 Materials Research Society
DOI: 10.1557/opl.2014.642

## The Behavior of Radiocaesium Deposited in an Upland Reservoir After the Fukushima Nuclear Power Plant Accident

Hironori Funaki, Hiroki Hagiwara, and Tadahiko Tsuruta

Fukushima Environmental Safety Center, Headquarters of Fukushima Partnership Operations, Japan Atomic Energy Agency, Sahei Building, 1-29, Okitama-cho, Fukushima-shi, Fukushima, 960-8034, Japan

### ABSTRACT

In the autumn of 2012, the Japan Atomic Energy Agency (JAEA) launched a new research project named F-TRACE (Long-Term Assessment of Transport of Radioactive Contaminant in the Environment of Fukushima). The aims of this project are to develop a system for prediction of radiation exposure, taking into consideration the transport, deposition, and remobilization behavior of radiocaesium (RCs) from the highest contaminated mountain forests, down through the biosphere, before deposition in a number of different aquatic systems. Especially, it is important to understand balances of suspended and deposited particles and RCs inventory in inflow water, discharge water and bottom sediments of an upland reservoir. In this paper, we describe current research activities performed by JAEA at the Ogi Reservoir, Fukushima prefecture, Japan.

According to our analyses the specific sediment yield and the average rate of storage capacity loss at the Ogi Reservoir are 210 m$^3$ km$^{-2}$ year$^{-1}$ and 0.15 % year$^{-1}$, respectively. The vertical distribution of RCs exhibits clear peaks at several sites in the reservoir formed by deposition of eroded soil particles from the catchment that were contaminated by accident fallout. Above the depth of each of the RCs peaks, the distribution of RCs was found to be variable with depth, with concentrations ranging over five orders of magnitude for a single core. The peaks in the sedimentation profiles are probably formed from eroded soil particles entering the reservoir from the surrounding contaminated watershed (most probably during storm events). Results from grain size analyses suggest that contaminated fine sediment tends to be deposited thickly within deeper parts of the reservoir. In addition, above the depth of RCs peaks at these deeper sites, the concentration of RCs approximately increased or decreased as the proportion of fine sediments increased or decreased. However, some fine particles are possibly discharged downstream during operational releases from the dam.

### INTRODUCTION

The Fukushima Daiichi nuclear power plant (NPP) accident occurred as a consequence of the massive earthquake and associated tsunami on March 11, 2011. Large quantities of volatile radionuclides were released to the atmosphere from the NPP, and consequently caused widespread contamination of residential, agricultural, forested and aquatic areas of northeast Japan. The total activities of $^{131}$I and $^{137}$Cs discharged into the atmosphere from 10 Japanese Standard Time (JST) on March 12 to 0 JST on April 6 are estimated to be approximately $1.5 \times 10^{17}$ and $1.3 \times 10^{16}$ Bq, respectively [1]. After decay of shorter-lived nuclides, the major radionuclides of concern from the standpoint of long-term external and internal radiation exposure, are [137]Cs and [134]Cs (physical half-lives of 30 and 2 years, respectively). RCs has a tendency to be strongly sorbed onto fine soil particles, especially clays. In general, 80 % or more of the [137]Cs and [134]Cs inventory was present within about the top few cm (~ 5) of soil surface layers [2-4]. Decontamination work based on the distribution of the RCs concentration at residential and agricultural areas was initiated by both national and local governments [5]. However, there has been little decontamination within forested areas which cover about 70 % of the Fukushima prefecture.

In the autumn of 2012, JAEA launched a new research project named F-TRACE. The aims of this project are to develop a system for prediction of radiation exposure, taking into consideration the transport, deposition, and remobilization behavior of RCs from the highest contaminated mountain forests, down through the biosphere, before deposition in a number of different aquatic systems. In this paper, we describe current research activities performed by JAEA within an upland reservoir system.

## MATERIAL AND METHODS

### Site description

The area between the Pacific and the Abukuma Mountains is called the Hamadohri area (see figure 1) and is situated in the east of the Fukushima prefecture. The rivers of the Hamadohri area rise in the Abukuma Mountains and are generally short in length (~ 45 km). In addition, river bed gradients are steep in upstream areas and are more gently sloping in downstream areas. In many rivers of the Hamadohri area, dams were constructed on the middle stretches of rivers, based on the rivers geographical characteristics.

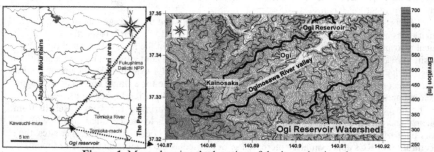

**Figure 1.** Maps showing the location of the investigation site.

The Ogi Reservoir watershed, a tributary of the Tomioka River catchment, is located in Kawauchi Village. The reservoir's primary purpose is to provide storage for supplemental irrigation to Tomioka Town. The Ogi Reservoir watershed covers an area of ~4.2 km$^2$ and mainly consists of mountainous land covered by vast forests. The watershed area also includes paddy fields (although, these have been out of crop since the NPP accident) and 15 residences, concentrated within two small communities (Ogi and Kainosaka) in the Oginosawa River valley. Elevation within the watershed ranges from approximately 350 m near the reservoir to 650 m on the western rim of the watershed. The bedrock in the watershed is predominantly Early

Cretaceous intrusive rocks [6] and soils are mainly Brown forest soil that correspond to Inceptisols [7]. In Kawauchi Village, data acquired by the Automated Meteorological Data Acquisition System (AMeDAS) of the Japan Meteorological Agency (37° 20.2' N, 140° 48.5' E, elevation 410 m), during 2001-2010, recorded mean annual precipitation as being 1300-1900 mm and mean annual temperature 10-12 °C. The Ogi Reservoir is an earthfill structure, with an original storage capacity of 568 000 m³, when it was completed in 1934. The reservoir level is regulated by three intakes.

## Methods

The bathymetry of the Ogi Reservoir was examined using a narrow multibeam echo-sounder in deeper water area and a singlebeam echo-sounder in very shallow water areas (2 m water depth). The narrow multibeam echo-sounder data were acquired using a hull-mounted RESON SEABAT 8125 echo-sounder, which operates at 455 kHz; it has a max swath angle of 120° whereas the along and across track beam widths are 1.0° and 0.5°, respectively. The singlebeam echo-sounder data were acquired with a hull-mounted TAMAYA TECNIC INC. TDM-9000, which operates at 200 kHz.

Bottom sediment samples were collected at five locations using gravity core samplers with an inner diameter of either 76 mm or 110 mm. Core samples were divided into either 1 or 2 cm-thick sections (depending on core diameter) and were dried at 80 °C for 24 hours for determination of dry weight and water content. Each sediment sample was packed into a polystyrene container (58 mm φ×35 mm height). The radioactivity concentrations of $^{137}$Cs (662 keV) and $^{134}$Cs (605 and 796 keV) in the sediment samples were determined by gamma-ray spectrometry. The gamma-ray spectrometry was carried out using high purity germanium detectors (GMX40P4-76, ORTEC®). The measurement time was 600 sec. The activities obtained were corrected for radioactive decay to the sampling date (January 18-25, 2013).

After measurement of radionuclides, sediment samples were passed through a 2 mm sieve to remove coarse organic debris and gravel. The grain size distributions of bottom sediment were analyzed using a laser diffraction particle size analyzer (SALD-3100, Shimadzu Co., Ltd., Kyoto, Japan). Following Wentworth's grain-size scale, coarse sands, fine sands, coarse silts, fine silts and clays are defined as 2000-500, 500-63, 63-16, 16-4 and <4 µm, respectively.

## DISCUSSION

### Sediment yield

The bathymetry of the Ogi Reservoir is displayed in figure 2. The minimum lake bottom elevation and surcharge water levels are 346 and 362 m, respectively. Within the first few months of the accident in March 2011, the upper-intake was opened by the operator of Tomioka town in case there had been any damage to the earthfill and its appurtenant structure during the earthquake. At the time of this survey, the water level was at an elevation of approximately 358 m and the lake bottom in the upstream part of the reservoir was exposed.

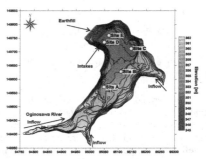

**Figure 2.** Bathymetric map and bottom sediment sampling sites at the Ogi Reservoir.

The estimated effective storage capacity of the reservoir in this survey was approximately 500 000 m³. The storage capacity loss due to sedimentation since 1934 to the present day is 68 000 m³. The specific sediment yield (from the entire catchment) and the average rate of storage capacity loss are 210 m³ km⁻² year⁻¹ and ~ 0.15 % year⁻¹, respectively. These are relatively low values in Japan [8, 9], and indicates that the majority of RCs will remain in the surrounding forest soils for longer than many other forested watersheds. The future challenge therefore is to compare findings with other nearby contaminated watersheds in the Fukushima prefecture.

## Depth distribution of RCs

The distribution of $^{137}$Cs, $^{134}$Cs and total RCs in five sediment cores is shown figure 3. The measured $^{134}$Cs concentration for all cores was lower than the $^{137}$Cs concentration as would be expected due to the faster decay of $^{134}$Cs, but although lower, the concentration profiles mirrored each other in increases and decreases in concentration throughout the cores. The ratio of $^{137}$Cs to $^{134}$Cs concentration was on average 1.8. The total RCs concentration for the extracted cores at each of the reservoir sampling locations will now be discussed.

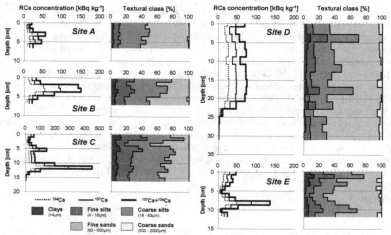

**Figure 3.** Vertical distribution of RCs and grain size fractions in sediment cores. Note, that the specific activity of RCs in Core C is plotted on a larger scale than for the 4 other cores.

The depth distribution of RCs at site C shows a clear peak in concentration at a depth of around 11-12 cm. This was the highest concentration measured within any section of the five sampled cores (445 kBq kg⁻¹), and approximately three times the peak concentration within cores B and E (149 and 136 kBq kg⁻¹, respectively). The peak in RCs concentration within core E is as well defined as the peak within core C, but this is not the case for cores A and D, with core D showing a relatively constant distribution in RCs with depth, from the surface down to a depth of ~ 20 cm. The relatively constant distribution of RCs within core D could be due to bioturbation, but the proximity of this site to the reservoir intakes suggests that dam operations could be a significant factor in sediment disturbance at this location. Below the depth at which each of the

168

peaks in RCs appears, there is a sharp decrease in the RCs concentration to the base of the core (with the exception of core E which shows a small increase within the bottom section). Cores C and E were extracted from the deepest parts of the reservoir and their RCs profiles indicate that they are relatively undisturbed in comparison to those at the other three locations. Therefore, the peaks in RCs concentration within these cores could indicate addition of RCs from the catchment during a storm event such as a typhoon (see below for further discussion).

The peak RCs concentration in paddy field soils (0 to 1 cm depth) obtained from the watershed in December of 2011 varied from 47 to 60 kBq kg$^{-1}$ [5]. The peak RCs concentration in the reservoir bottom sediments at sites B, C and E are between twice to nine times higher than the peak concentration found in the paddy field soils. These results strongly indicate that RCs present in the reservoir bottom sediments cannot solely be from aerial deposition of accident fallout but that there is a significant contribution from eroded soil particles coming from the surrounding contaminated watershed.

## Grain size characteristics

Results from grain size analyses for the five sample cores are shown in figure 3. The proportion of fine sediments (below 63μm) at sites A and D ranges from 37% to 49% and from 12% to 69%, respectively. In contrast, sediments at sites B and C are mainly composed of silt- and clay-size particles. At site D, sediments are dominated by sand-size particles; having > 60% sand sized particles on average. At site A, the sand sized particles make up ~ 50-55% of the profile with the distribution of sediment particle sizes being relatively uniform with depth. The fine sediment proportions at sites B and C varies from between 45% to 78% and between 50% to 93%, respectively. For, clay-size particles (below 4μm) the content at site C was the highest value of the five sampling sites, making up on average 10% of the grain size distribution. The sediment core extracted from site E showed quite a variable grain size distribution with depth. Two silty layers were observed within this core, one at the surface (1-2 cm) and another at ~10 cm depth. The coarsest layer, composed of silty sand, was observed at 6-7cm depth, with the fine sand size fraction making up ~ 70% of this depth increment.

Above the depth of RCs peaks at both sites C and E, the concentration of RCs approximately increased or decreased as the proportion of fine sediments increased or decreased (but the correlation between proportion of fine particles and RCs concentration is weak). In contrast, the peak concentration of RCs at site A occurs where the dominant grain size fraction is composed of sand-particles. Both these observations indicate that the majority of RCs within the cores is likely to be deposited during storm events, (either as bed load or suspended particles that settled in upstream parts of the reservoir) which will mobilize both larger sized particles in addition to the fine silt and clay particles. The larger size particles, on entering the reservoir will settle out first, whilst the silt- and clay-size particles remain in suspension for longer and travel a greater distance into the reservoir from the point of entry. Additionally, these fine particles may be resuspended by underflow and vertical circulation flow and be trapped in the deeper parts of the reservoir. These phenomena are probable explanations as to why the fine sediment proportion at site A is lower than sites B, C and E where the fine sediments tend to be deposited more thickly. At site D, fine contaminated particles are likely to be partially discharged downstream during dam operations due to the proximity of this location the reservoir intakes (as this routine procedure is performed in order to prevent sediment blocking the intakes).

## CONCLUSIONS

Analysis of the vertical distribution of RCs shows clear peaks at several sites in the reservoir, formed by deposition of eroded soil particles from the catchment that were contaminated by accident fallout. Results from grain size analyses reveal that contaminated fine sediment tends to be (but not always) deposited thickly at deeper parts of the reservoir. In addition, above the depth of RCs peaks at two sites (C and E), the concentration of RCs approximately increased or decreased as the proportion of fine sediments increased or decreased. However, the low % of fine clay particles at one deeper site (D), is likely due to the resuspension of fine sediment caused during the opening of intakes within the dam (see figure 4). This resuspended fine sediment, will then be partially discharged downstream along with reservoir water. The volume of contaminated sediment deposited in a reservoir can be estimated from the difference between fluvial sediment inflow and discharge. Future work on the distribution of RCs, will investigate, in addition to bottom sediments, suspended material captured at both the inflows and discharge waters of the reservoir. In addition, measurement of RCs within the individual grain size fractions (as opposed to bulk sediment measurement) will give a clearer indication of RCs particle association within bottom sediments.

**Figure 4.** Characteristics of sediment deposited at the Ogi Reservoir.

## ACKNOWLEDGMENTS

The authors wish to thank Dr. S.M.L. Hardie for her technical review comments and all team members of F-TRACE for assisting with field work and measurements.

## REFERENCES

1. M. Chino, H. Nakayama, H. Nagai, H. Terada, G. Katata and H. Yamazawa, J. Nuc. Sci. & Tec., **48** (2011).
2. T. Fujiwara, T. Saito, Y. Muroya, H. Sawahata, Y. Yamashita, S. Nagasaki, K. Okamoto, H. Takahashi, M. Uesaka, Y. Katsumura and S. Tanaka, J. Env. Rad., **113** (2012)
3. H. Kato, Y. Onda and M. Teramage, J. Env. Rad., **111** (2012).
4. H. Sato, T. Niizato, K. Amano, S. Tanaka and K. Aoki, Mater. Res. Soc. Symp. Proc. **1518** (2012).
5. JAEA, *Decontamination pilot projects mainly for evacuated zones contaminated with radioactive materials discharged from the Fukushima Daiichi nuclear plant* (2012) [in Japanese].
6. K. Kubo, Y. Yanagisawa, T. Yoshioka and Y. Takahashi, Quadrangle Series Scale 1:50,000 Niigata (7) No.46-47, Geology of the Namie and Iwaki-Tomioka District, 104p.
7. USDA Soil Conservation Service, Keys to Soil Taxonomy. 6th ed. USDA, Washington, DC. (1994).
8. K. Ashida and T. Okumura, Annuals of Disas. Prev. Res. Inst., Kyoto Univ., No.17B (1974) [in Japanese].
9. K. Hasegawa, K. Wakamatsu and M. Matsuoka, J. JSNDS 24-3 (2005) [in Japanese].

Mater. Res. Soc. Symp. Proc. Vol. 1665 © 2014 Materials Research Society
DOI: 10.1557/opl.2014.643

# Development of simplified biofilm sorption and diffusion experiment method using *Bacillus* sp. isolated from Horonobe Underground Research Laboratory.

Kotaro Ise, Tomofumi Sato, Yoshito Sasaki and Hideki Yoshikawa
Radionuclide Migration Research Group, Geological Isolation, Research and Development
Directorate, Japan Atomic Energy Agency (JAEA), 4-33 Muramatsu Tokai-Mura, Naka-gun,
Ibaraki 319-1194, Japan

## ABSTRACT

We developed a simplified biofilm sorption and diffusion experiment method. The biofilms of the *Bacillus cereus* were incubated on cellulose acetate membrane filters (pore size 0.2 μm, diameter 47 mm) placed on thick NB broth agar medium (thickness was about 30 mm) to support sufficient biofilm growth of the *Bacillus cereus*. The thickness of the formed biofilms was about 1 mm. The formed biofilms were applied to through-diffusion method, which has been used to measure diffusion coefficient of crystalline and sedimentary rocks and clay minerals. The obtained copper sorption coefficient by batch experiments was about 100 ml/g (wet weight) at the case of the concentration of cupper ion was over 0.074mmol/L. And diffusion coefficients by through diffusion experiment was De=1.1 x $10^{-10}$ ($m^2$/s). From these results, this simplified biofilm sorption and diffusion experiment may make possible to obtain these parameters with ease.

## INTRODUCTION

Recently, many studies have been conducted to develop a sorption model of radionuclides to predict transport phenomena after geological disposal [1]. Most of these studies treat only sedimentary rocks, crystalline rocks and clay minerals like bentonite [2]. However, in reality it is thought that these solid surfaces are covered with natural organic matter and most of these organic matters are biofilms [3]. It is thought that these affect aqueous phase concentration and sorption capacity of rocks [4]. Anderson et al. reported that *Gallionella ferruginea* biofilms concentrate trace metals up to 1000 fold higher than found within the host rock [5]. From these results, it is thought that the existence of biofilms obviously affects transport of radionuclides. Bacterial cell walls contain acid functional group and if these are deprotonated, that adsorb much cations. Therefore, the existence of bacteria affects groundwater chemistry [6]. A lot of studies about bacterial adsorption have been done, but there are quite few studies about biofilm adsorption. From these results, there are no consensus whether biofilm retard or stimulate radionuclide transport. To solve these problems, we should quantify sorption capacity and diffusion coefficient to predict the effect of biofilm covering. So far, there are almost no quantitative data about sorption capacity for radionuclides. This may due to the difficulties of sample preparation; biofilms are too fragile to treat quantitative analysis.

In this study, we conducted sorption and diffusion experiment to establish simplified biofilm sorption experiment using *Bacillus cereus* related bacteria which were isolated from the Horonobe Underground Research Laboratory groundwater.

## EXPERIMENT

### Strain source, isolation method polymerase chain reaction (PCR) and sequence analysis

*Bacillus* sp. was isolated from groundwater sample collected from 07-V140-M01 borehole excavated in the horizontal gallery at the depths of -140 m in Horonobe Underground Research

Laboratory which is located in the city of Horonobe, Hokkaido, Japan. 100 µl of the collected groundwater was spread on a plate containing of Nutrient Broth (Nissui) and agar (Wako) and then cultured aerobically at 25 °C. Then, a shiny colony was picked up and spread on a new plate, and cultured again to purify bacterial strain. The colony was picked up and applied as template for colonyPCR to amplify 16S rRNA gene. The primers used for the PCR reactions were Eu10F and Eu1500R [7], these primers were known as universal primer for eubacterium. The PCR product was purified and sequenced with the ABI Big Dye Terminater Cycle Sequencing kit (Applied Biosystems) and an ABI genetic analyzer (Model 3130/3130xl Applied Biosystems). The primers used for the sequencing were also Eu10F and Eu1500R. The obtained sequence was deposited in the DNA Data Bank of Japan (DDBJ). The phylogenetic trees were constructed by MEGA5 [8].

## Biofilm formation method

N B medium(30g/L) and agar (10g/L) containing medium was sterilized by autoclaving. And 150ml of the medium was introduced to each sterilized tall type glass petri dish (diameter 85mm, height 45mm). After cooled down to room temperature, sterilized membrane filter (diameter 47mm, cellulose acetate, Advantec) was placed on the solidified agar medium. After that, preincubated *Bacillus* sp. medium 10 µl was dropped on the center of the membrane filter. The petri dishes were sealed with parafilm (Pechiney Plastic Packaging Company) to avoid drying. The dishes were incubated at 25 °C for about three weeks.

## Confocal laser scanning microscope analysis of incubated biofilm

Biofilms were fluorescently stained with SYBR® Green I (TAKARA Bio). SYBR® Green I was diluted to 1/10 concentration. 50 µl of the solution was transferred to the biofilm surface, and placed for about 20 minutes in the dark. The stained biofilms were transferred to slide glass and covered with cover glass. The biofilm was visualized using Leica DM5500B. Violet Diode laser (405 nm) was used as excitation and 520-540nm fluorescent image was taken. Differential interference contrast microscopic image was captured by Nikon DS-Fi1.

## Biofilm sorption experiment of $Cu^{2+}$ with ion selective electrode

Along with membrane filters, incubated biofilms were picked up with tweezer and immersed in 30 ml saline solution to rinse out NB medium for 30 minutes. This procedure was repeated three times. After rinsed out medium, the films were transport to 50 ml centrifuge tube, and mess up with saline to 30 ml, and maintained for about 30 minutes to remove the membrane filters from the biofilms. After the biofilm was fallout from the membrane filter, the membrane filter was removed from the centrifuge tube with tweezer. Copper concentration was measured with copper ion selective electrode (CU-2021,TOA DKK) and multi water quality meter (MM-60R, TOA DKK). The weight of biofilm used in sorption experiment was about 1.7g. Ideally, sorption experiment should be conducted in neutral pH condition to simulate underground water condition. But because, copper ion selective electrode can detect only $Cu^{2+}$ ion and the objective of this study was to develop biofilm sample preparation method for sorption and diffusion experiment, sorption experiment was conducted in pH range of 3 − 5 (copper in the solution of over pH 6, major speciation is $CuOH^+$ and or $Cu(OH)_2$) (Figure 5). The solution pH was adjusted with 1 N HCl and NaOH solution.

## Biofilm diffusion experiment of Cu²⁺ with through diffusion experiment

The formed biofilms were applied to through-diffusion method, which has been used to measure diffusion coefficient of crystalline and sedimentary rocks and clay minerals [9]. Diffusion experiments were conducted in aerobic and room temperature condition. The biofilm samples were striped out from the membrane filters and filled in the hall (diameter 24 mm, depth 5 mm) of sample holder (Figure 1). And the hall was sandwiched with membrane filters. Inlet and outlet reservoirs (50 cm³) were filled with saline solution. After pre- equilibration for about 1 hour, 500 µl of copper standard solution (1000 mg/L) was added in inlet reservoir to adjust 10 mg/L. Copper concentrations in both inlet and outlet reservoirs were monitored with ion selective electrode as described in sorption experiments. The pH of the solution displayed gradual increase in the diffusion experiments, routinely 1 N HCl was added to maintain pH value at 4.0.

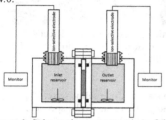

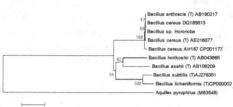

**Figure 1.** Schematic representation of the through-diffusion cell used in this study

**Figure 2.** Phylogenetic tree based on the 16S rRNA gene. Scale bar = 2% nucleotide substitution. Phylogenetic trees were produced using the neighbor-joining algorithm.

## DISCUSSION

### Characterizations of biofilm cultivated on membrane filter

#### Isolation and phylogenetic analysis

The biofilm forming bacteria was selected from 44 isolates from the groundwater collected from borehole 07-V140-M01 in Horonobe Underground Research Center by biofilm forming ability evaluated with the method of Toole et al. [10]. Phylogenetic analysis was conducted to determine bacterial strain by genetic analyzer, figure 2 showed that this strain was related to *Bacillus cereus* and the similarity of 16S rRNA gene was 99.9% (1449/1451). *Bacillus cereus* is known as biofilm forming bacteria [11].

#### Confocal microscopic image of biofilm

In order to examine the biofilm formation, confocal laser scanning microscopic analysis was conducted on the biofilm, and epifluorescence microscopic analysis was conducted on planktonic phase cells. Biofilm was build up mainly with chainformed *Bacillus* sp. which twisted to form film structure, and the voids between these chains were filled with EPS. On the other hand, planktonic phase cells were aggregated randomly [12]. The difference showed biofilm on the membrane filters was not only aggregate of planktonic cell and was organized to bacterial community. In biofilm forming process in *Bacillus* sp., chain formation is key step and EPS synthesis gene was expressed in this condition [13].

(A)　　　　　　　　　　　(B)　　　　　　　　　　　(C)

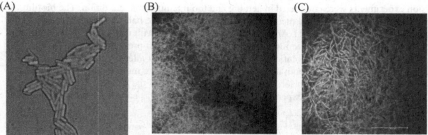

**Figure 3.** Microscopic images (A) differential interference microscopy of planktonic cells aggregates (B) Confocal laser scanning fluorescence microscopy of *Bacillus* sp. biofilm (x 400) (C) Confocal laser scanning fluorescence microscopy of Bacillus sp. biofilm (x 630)

## Cu$^{2+}$ sorption experiment

Prior to Cu$^{2+}$ sorption experiment, to confirm the existing state of Cu in saline water, the relationship of Cu and pH was plotted in Figure 4. And Figure 5 shows the Cu(II) speciation calculated with the geochemical code PHREEQC. Added Cu standard existed as Cu$^{2+}$ under pH 5 condition. The decline of Cu$^{2+}$ over pH 5 condition might be due to Cu(OH)$^+$, Cu(OH)$_2$ and sorbed to centrifuge tube and ion selective electrode. From these results, Cu$^{2+}$ sorption experiments of biofilms were conducted in pH 5 condition. Figure 6 shows the obtained Adsorption isotherm, and fitted with Langmuir equation and Freundlich equation.

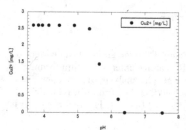

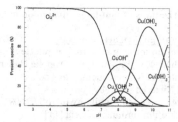

**Figure 4.** Cu$^{2+}$ concentration detected by ion selective electrode at different pH condition

**Figure 5.** Cu( II ) speciation in saline solution which was calculated with the geochemicalcode PHREEQC.

174

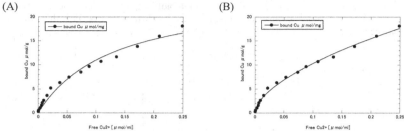

**Figure 6.** (A) Sorption isotherms of $Cu^{2+}$ to *Bacillus* biofilm, simultaneous fitting of Langmuir equation (B) simultaneous fitting of Freundlich equation

Figure 7 (A) shows that the relationship of $K_d$ and Cu concentration in the solution. In the low concentration condition, $K_d$ values were ranged from 200-600 ml/g, on the other hand, in the case of over 0.074 mmol/L, $K_d$ values were about 100 ml/g. From these results, biofilms tend to show high $K_d$ values at lower concentration condition. The water content ratio of biofilms are thought as 90-99%, therefore $K_d$ value per dry weight might be 1000- 10000 ml/g (dry weight). This value was higher value than the value of rocks. Figure 7 (B) shows the relationship of $K_d$ and pH. Even at pH 3, 75 % of Cu was sorbed to biofilm, it indicate that biofilm had many strong sorption sites. And from about pH 4.5, $K_d$ values were sharply increased, this infer that carboxyl group may contribute these Cu sorption. EPS produced by *Bacillus* spp. are composed with polyglutamic acid, these glutamic acid may contribute many copper ion sorption.

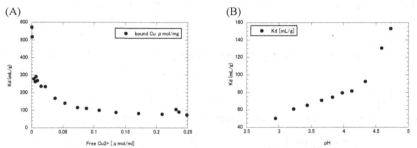

**Figure 7.** (A) $K_d$ values at different concentration of $Cu^{2+}$ (B) $Cu^{2+}$ $K_d$ values at different pH condition

### $Cu^{2+}$ diffusion experiment

Figure 8 shows the breakthrough curves and tracer-depletion curves from diffusion experiments. The calculated effective diffusivity from breakthrough curve was 1.1 x $10^{-10}$ $m^2$ $s^{-1}$, this value was one tenth lower than 1.2 x $10^{-9}$ $m^2$ $s^{-1}$ [14], the effective diffusivity of $Cu^{2+}$ in free water. And calculated distribution coefficient from breakthrough curve was 420 ml/g, this value was similar to the value obtained from the sorption experiment. Stewart [15] reviewed effective diffusiviti es of various organic compounds and inorganic compound in biofilms, in that report the ratio of De/Daq was ranged from 0.05-1.1. In this study, relative effective diffusivity was 0.09, this low diffusivity may be due to high cell concentration in biofilm.

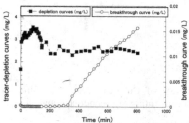

**Figure 8.** Tracer-depletion curves obtained by Cu²⁺ diffusion experiments

## CONCLUSIONS

From diffusion experiment, the biofilm of *Bacillus* sp. isolated from Horonobe URL showed about one tenth lower than diffusion coefficient in water. From sorption experiment, $K_d$ of this biofilms were about 100-600(ml/g). From these results, it is thought that if biofilm cover the rock surface, radionuclide may diffuse into biofilm and sorbed high concentration, and then these radionuclide may diffuse to rock matrix.

So far, there were few quantitative researches about biofilm effect on radionuclide migration, especially acquisition of biofilm diffusion coefficient of radionuclide was difficult. In this study, we developed simplified diffusion and sorption experiment method. This simplified biofilm sorption and diffusion experiment may make possible to obtain these parameters with ease.

## ACKNOWLEDGMENTS

This work was supported by the Ministry of Economy, Trade and Industry (METI) of Japan.

## REFERENCES

1. Bradbury MH, Baeyens B. J. Contam. Hydrol. 27:223–248 (1997)
2. Tachi Y, Nakazawa T, Ochs M, Yotsuji K, Suyama T, Seida Y, Yamada N, Yui M. Radiochimica Acta 98:711–718 (2010)
3. Pedersen K. (SKB) IPR-05-05:85 (2005)
4. Anderson C, Pedersen K, Jakobsson A-M. Geomicrobiol J 23:15–29 (2006)
5. Anderson CR, Pedersen K. Geobiology 1:169–178 (2003)
6. Fein J, Daughney C, Yee N, Davis T. Geochim. Cosmochim. Acta 61:3319–3328 (1997)
7. Takami H, Inoue A, Fuji F, Horikoshi K. FEMS Microbiol. Lett. 152:279–285 (1997)
8. Tamura K, Peterson D, Peterson N, Stecher G, Nei M, Kumar S. Molecular biology and evolution 28:2731–2739 (2011)
9. Tachi Y, Yotsuji K, Seida Y, Yui M. Geochim. Cosmochim. Acta 75:6742–6759 (2011)

10. O'Toole GA, Pratt LA, Watnick PI, Newman DK, Weaver VB, Kolter R. Methods Enzymol. 310:91–109 (1999)
11. Wijman JGE, de Leeuw PPL a, Moezelaar R, Zwietering MH, Abee T. Appl. Environ. Microbiol. 73:1481–8 (2007)
12. Branda SS, González-Pastor JE, Ben-Yehuda S, Losick R, Kolter R. Pnas 98:11621–6 (2001)
13. Branda S, Dervyn E. J. Bacteriol. 186:3970–3979 (2004)
14. Ribeiro ACF, Esteso M a., Lobo VMM, Valente AJM, Simões SMN, Sobral AJFN, Burrows HD. Journal of Chemical & Engineering Data 50:1986–1990 (2005)
15. Stewart PS. Biotechnol. Bioeng. 59:261–72 (1998)
16. Kersting A, Efurd D, Finnegan D, Rokop D, Smith D, Thompson J. Nature 397:56–59 (1999)

Mater. Res. Soc. Symp. Proc. Vol. 1665 © 2014 Materials Research Society
DOI: 10.1557/opl.2014.644

# The effect of alkaline alteration on sorption properties of sedimentary rock

Satoko Shimoda[1], Toshiyuki Nakazawa[1], Hiroyasu Kato[1], Yukio Tachi[2], Yoshimi Seida[2]
[1]Mitsubishi Materials Corporation, 1002-14, Mukoyama, Naka, Ibaraki, 311-0102, Japan
[2]Japan Atomic Energy Agency, 4-33, Muramatsu, Tokai, Ibaraki, 319-1194, Japan

## ABSTRACT

The potential effect of high pH plume caused by cementitious materials must be evaluated in the performance assessment for HLW geological disposal. Alkaline plume would lead to change sorption properties of host rock by primary mineral dissolution, secondary mineral precipitation and sequential change of pore water chemistry. In this study, the effect of alkaline alteration on sorption of Cs, Ni and Th was investigated using rock samples from the Horonobe Underground Research Laboratory. Crushed rock samples were reacted in high pH alkaline solution at 90 °C for 45 days, 95 days and 1,383 days, respectively. As a result of sample analysis, it was supposed that zeolitic mineral was precipitated as secondary mineral. The cation exchange capacity slightly increased in comparison with the unaltered sample. Distribution coefficients ($K_d$) of Cs, Ni and Th on unaltered and altered rock sample were measured by batch sorption experiment in synthetic groundwater. $K_d$ of Cs increased with the alteration period. These results show that secondary minerals contribute to the increase in Cs sorption. By contrast, $K_d$ of Ni and Th decreased with the alteration period. This change might be caused by dissolution of clay minerals and amorphous silicates controlling Ni and Th sorption by surface complexation. These results imply that effects of alkaline alteration on $K_d$ of rocks depend on the dissolution/precipitation of minerals, their surface properties and sorption mechanisms.

## INTRODUCTION

Sorption of radionuclides in host rock is one of the key processes in the performance assessment for HLW geological disposal [1]. The sorption properties of rocks have been investigated using key radionuclides under a wide variety of geochemical conditions, and distribution coefficients ($K_d$) setting approaches have been developed [2]. Cementitious materials will be used as tunnel support in the HLW disposal facility. High pH plume arising from dissolution of these materials will spread into a near field host rock. Alkaline plume would lead to mineral dissolution in the rock and secondary mineral precipitation, and cause to change in migration properties of the radionuclides. Therefore, it is important to understand the rock alteration and its effect on radionuclide sorption for the long-term performance assessment [1, 3, 4]. In this study, the mineral properties of the rock altered by alkaline solution and their sorption properties of Cs, Ni and Th were investigated using sedimentary rock samples from the Horonobe Underground Research Laboratory (URL).

## EXPERIMENTAL

### Rock sample preparation

Sedimentary rock (siliceous mudstone) was sampled from about 510 m depth (Wakkanai formation) of the borehole HDB-6 at the Horonobe URL. The dominant minerals are opal-CT, quartz and plagioclase [5]. Clay minerals (smectite, illite and chlorite) are also present in small percentage, they represent a 5% or less of the whole. K-feldspars, siderite and pyrite are also present. The rock sample was crushed into fine grain with particle sizes of under 250 μm. The crushed rock sample was washed with deionized water and ethanol, and oven-dried at 60 °C for a day.

Altered rock samples were prepared by immersing the above rock samples in Na-K-Ca mixed alkaline solution (~ 0.05 mol/l NaOH, ~0.05 mol/l KOH and ~0.005 mol/l $Ca(OH)_2$) with a liquid/solid (L/S) ratio of 0.01 $m^3$/kg. The alteration treatment was performed in flororesin (PFA)-container under a $N_2$ gas purged atmosphere (glove box). Containers were put into an oven at 90 °C for three different periods. After the aging time, the altered rock samples were separated from the alkaline solution by 0.45 μm membrane filtration, and the altered rock samples were washed in ethanol, and oven-dried at 60 °C for a day.

## Sample analysis

The unaltered and altered rock samples were characterized by X-ray powder diffraction (XRD, MAC SCIENCE MXP3), X-ray fluorescence analysis (XRF, JEOL 3SX-3100R II), Brunauer-Emmett-Teller (BET) specific surface area analysis (YUASA-IONICS QUANTASORB QS-18) and measurement of the cation exchange capacity (CEC) by Schollenberger method [6]. The pH and Eh of the reacted alkaline solutions were measured by pH/ORP meter (Yokogawa PH82). Inorganic carbon (IC) and Total Organic Carbon (TOC) was measured by TOC analysis (SHIMADZU TOC-5000A), and Na, K, Ca, Si and Al were measured using Inductively Coupled Plasma-Atomic Emission Spectrometry (ICP-AES, SII SPS 3100).

## $K_d$ measurement

$K_d$s of Cs, Ni and Th on the unaltered and altered rock samples were determined by batch sorption experiments using a synthetic groundwater (SGW). The SGW was prepared by reagents based on a representative groundwater composition of borehole HDB-6 at the Horonobe URL. The pH, ionic strength (IS) and chemical composition of the SGW is shown in Table I [2].

**Table I.** pH, IS and chemical composition of the SGW

| pH | IS | Na (mol/l) | K (mol/l) | Ca (mol/l) | Mg (mol/l) | IC (mol/l) | Cl (mol/l) |
|---|---|---|---|---|---|---|---|
| 8.5 | $2.6 \times 10^{-1}$ | $2.3 \times 10^{-1}$ | $2.1 \times 10^{-3}$ | $2.1 \times 10^{-3}$ | $5.8 \times 10^{-3}$ | $3.0 \times 10^{-2}$ | $2.2 \times 10^{-1}$ |

The L/S ratios on sorption experiments for Cs, Ni and Th were 0.1025 $m^3$/kg, 0.1025 $m^3$/kg and 1.01 $m^3$/kg, respectively. The unaltered and altered rock samples were contacted with the synthetic groundwater in polypropylene bottles. After a day, tracer solutions (carrier-free Cs-137, Ni-63, Th-232) were added. Initial concentrations of Cs-137, Ni-63 and Th-232 were $2 \times 10^{-10}$ mol/l (90 Bq/ml), $8 \times 10^{-10}$ mol/l (110 Bq/ml) and $1 \times 10^{-6}$ mol/l (230 μg/l), respectively. The bottles were shaken by hand once per day during seven days. These sorption experiments were carried out under Ar gas purged atmosphere (glove box; $O_2 < 1$ppm), at room temperature. After the reaction time of 7 days, the supernatants in each sample solutions were sampled and filtrated through 0.45 μm membrane filter (DISMIC 25HP045AN). The concentrations (radioactivities) of Cs-137 and Ni-63 in the filtrate solutions were analyzed by Liquid Scintillation Counter (LSC, PACKARD TRI-CARB2750TR/LL). Th concentration in the filtrate solution was measured by

Inductively Coupled Plasma Mass Spectrometry (ICP-MS, Yokogawa HP4500 and Agilent 7700x).

The distribution coefficient ($K_d$) values were calculated as the following equation:

$$K_d = \frac{C_0 - C}{C} \times \frac{V}{M}$$

where $C_0$ is the initial concentration of traces determined from blank solutions (Bq/ml or µg/l), $C_i$ is the equilibrium concentration of traces in test solutions (Bq/ml or µg/l), $V$ is the volume of liquid phase (m³), and $M$ is the mass of solid phase (kg).

## RESULTS AND DISCUSSION

### Properties of unaltered and altered rock samples

XRD patterns of the unaltered and altered rock samples are shown in Figure 1. Opal-CT, quartz, plagioclase, illite and K-feldspar were identified in these rock samples by XRD. These minerals were original minerals in the unaltered rock [5]. However, plagioclase has the possibility of a secondary mineral in the altered rock sample [7]. Although not detected explicitly by the XRD analysis, there is a possibility of existence of smectite, chlorite, pyrite and siderite as primary minerals [5]. In the unaltered rock sample, a lot of amorphous minerals were contained, which have the broad peak. However, for altered rock samples, peaks of crystalline minerals were comparatively clear, because the amorphous had dissolved in the alkaline solution. Although a certain kind of zeolite, illite, chlorite, phillipsite and CSH-gel precipitated by reactions of rocks and alkaline solutions in previous studies [3, 4, 7, 8], these minerals were not identified in this study. The concentration of the alkaline solution in this study was slightly low as compared with alkaline solutions in previous studies. Therefore, it was presumed that the secondary minerals had little precipitated.

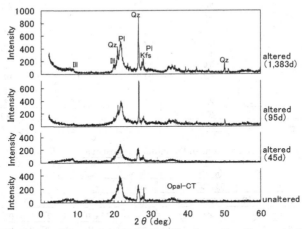

**Figure 1.** XRD patterns of the unaltered and altered rock samples. Qz: quartz; Pl: plagioclase; Ill: illite; Kfs: K-feldspar (Cu-Kα1).

181

The results of XRF analysis for the rock samples are summarized in Table II. While amounts of CaO and $K_2O$ increased with the increase of alteration period, amounts of $SiO_2$ and $Al_2O_3$ were not substantially changed. From these results, it was predicted that secondary minerals were precipitated by reactions of K, Ca, Si and Al in alkaline solution. The measurement result of post-reaction alkaline solutions (Table III) showed a similar tendency. In early period of alteration, primary minerals (Opal-CT, smectite etc.) dissolved, and Si and Al concentration in alkaline solution increased. The alteration period became longer, some secondary mineral containing Ca, K, Si and Al precipitated, and Ca, K, Si and Al concentration in alkaline solution decreased. Mineral saturation indices (SI) in post-reaction alkaline solutions were calculated using the PHREEQC code [9]. Zeolitic minerals (analcime, laumontite) SI values were changed from positive values to negative values with the increase of alteration period from 95 days to 1,383 days. Therefore, it was predicted that zeolitic minerals precipitated. Precipitation of zeolitic minerals was showed in previous studies [4, 8]. Moreover, the concentration of TOC also increased with the alteration period by elution of organic carbon in the rock sample.

The result of the CEC and the specific surface area measurements are shown in Table II. The CEC slightly increased with the increase of alteration period. On the other hand, the specific surface area decreased in early period of alteration. This was caused by dissolution of some minerals, especially the amorphous and smectite. Moreover, as the alteration period became longer, the specific surface area increased. This would be because the secondary minerals with large specific surface area precipitated. Precipitation of zeolitic minerals was estimated from the above results.

**Table II.** XRF analysis, CEC and specific surface area of the unaltered and altered rock samples.

|  | Unaltered | Altered (45d) | Altered (95d) | Altered (1,383d) |
|---|---|---|---|---|
| XRF analysis (wt%) |  |  |  |  |
| $SiO_2$ | 80.7 | 77.9 | 77.9 | 78.4 |
| $TiO_2$ | 0.5 | 0.6 | 0.5 | 0.7 |
| $Al_2O_3$ | 8.8 | 9.2 | 9.3 | 7.3 |
| $Fe_2O_3$ | 3.7 | 4.1 | 4.0 | 5.6 |
| MgO | 1.5 | 1.2 | 1.5 | 1.1 |
| CaO | 0.6 | 1.2 | 1.1 | 1.8 |
| $Na_2O$ | 0.6 | 1.2 | 1.0 | 0.0 |
| $K_2O$ | 1.7 | 3.5 | 3.4 | 4.7 |
| $SO_3$ | 2.0 | 1.1 | 1.2 | 0.3 |
| CEC (meq/100g) | 26.1 | 26.8 | 27.4 | 28.5 |
| Specific surface area ($m^2/g$) | 66.1 | 49.1 | 47.9 | 61.8 |

**Table III.** Chemical compositions of the initial and post-reaction alkaline solutions.

|  | pH | Eh (mV) | Na (mol/l) | K (mol/l) | Ca (mol/l) | Si (mol/l) | Al (mol/l) | IC (mol/l) | TOC (ppm) |
|---|---|---|---|---|---|---|---|---|---|
| Alkaline solution | 12.0 | 86 | $5.2 \times 10^{-2}$ | $4.7 \times 10^{-2}$ | $4.6 \times 10^{-3}$ | $-^*$ | $-^*$ | $-^*$ | $-^*$ |
| Altered (45d) | 11.4 | 150 | $4.2 \times 10^{-2}$ | $2.0 \times 10^{-2}$ | $7.8 \times 10^{-5}$ | $5.0 \times 10^{-2}$ | $-^*$ | $3.7 \times 10^{-3}$ | 401 |
| Altered (95d) | 11.1 | 167 | $4.1 \times 10^{-2}$ | $2.0 \times 10^{-2}$ | $7.2 \times 10^{-5}$ | $3.8 \times 10^{-2}$ | $1.0 \times 10^{-5}$ | $7.5 \times 10^{-3}$ | 360 |
| Altered (1,383d) | 9.6 | 244 | $8.9 \times 10^{-2}$ | $3.0 \times 10^{-2}$ | $4.8 \times 10^{-5}$ | $8.2 \times 10^{-5}$ | $<7.4 \times 10^{-6}$ | $3.2 \times 10^{-2}$ | 876 |

* unmeasured

## The effect of alteration on $K_d$ value

The averages of calculated $K_d$ values are shown in Table IV and are plotted as function of alteration period and the CEC. The $K_d$ of Cs increased from 0.6 m³/kg to 1.6 m³/kg with the alteration period (Figure 2(a)). The CEC increased with the alteration period because of possible precipitation of secondary minerals. Since the sorption takes place by ion exchange [10] increased with the CEC, $K_d$ of Cs increased (Figure 2(b)). Gaboreau et al. [11] reported that the $K_d$ of an assemblage of minerals representative of the concrete/Callovian-Oxfordian clayrock interface (CSH, zeolites, ettringite and saponite mixture) were lower than the $K_d$ of host clayrock. In this study, the $K_d$ of Cs increased with the alteration period. The cause of this difference might be that the CSH-gel was not precipitated in this experiment. Thus, $K_d$ of Cs increased because the mineral such as the zeolite having large CEC precipitated in this experiment.

**Table IV.** $K_d$ values of Cs, Ni and Th obtained by batch sorption experiments.

| Alteration period (d) | Cs pH_final | Cs Kd(m³/kg) | Cs σ | Ni pH_final | Ni Kd(m³/kg) | Ni σ | Th pH_final | Th Kd(m³/kg) | Th σ |
|---|---|---|---|---|---|---|---|---|---|
| 0 | 8.3 | 0.5 | 0.03 | 8.3 | 0.7 | 0.01 | 8.6 | 6.1 | 0.79 |
| 0 | 9.2 | 0.7 | 0.01 | 9.4 | 1.4 | 0.04 | 9.3 | 4.0 | 0.60 |
| 0 | 8.3 | 0.5 | 0.01 | 8.2 | 0.5 | 0.01 | 8.4 | 5.4 | 3.62 |
| 45 | 8.4 | 1.0 | 0.06 | 8.4 | 0.8 | 0.13 | 8.6 | 6.7 | 1.30 |
| 95 | 9.4 | 1.1 | 0.01 | 9.4 | 0.6 | 0.09 | 9.3 | 8.5 | 1.94 |
| 1383 | 8.4 | 1.6 | 0.07 | 8.4 | 0.2 | 0.01 | 8.4 | 1.1 | 0.15 |

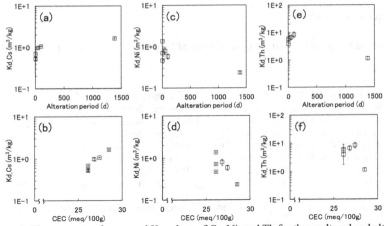

**Figure 2.** The averages of measured $K_d$ values of Cs, Ni, and Th for the unaltered and altered rock samples, as functions of alteration period (a, c, e), CEC (b, d, f). The error bars were shown one standard deviations.

The $K_d$ values of Ni and Th decreased with the alteration period. The $K_d$ of Ni decreased from about 1.0 m³/kg to 0.2 m³/kg (Figure 2(c)). The $K_d$ of Ni was independent of CEC (Figure

2(d)). The Ni sorption is dominated by surface complexation at clay edges at pH 8-10 [12]. The reduction of $K_d$ was suggested to be caused mainly by the dissolution of clay minerals (smectite, illite and chlorite) contributing the surface complexation at clay edges. The quartz (having small $K_d$ of Ni [13]) increased relatively by the dissolution of clay minerals.

The $K_d$ of Th decreased from about 6.0 $m^3$/kg to 1.1 $m^3$/kg (Figure 2(e)). The $K_d$ of Th was independent of CEC (Figure 2(f)). Sorption of Th takes place by surface complexation [2], the $K_d$ of Th was estimated to be caused mainly by the dissolution of clay minerals contributing to the surface complexation at clay edges as same as Ni case.

## CONCLUSIONS

The sedimentary rock from the Horonobe URL was altered by alkaline solution, and the altered samples were analyzed and sorption of Cs, Ni and Th on the samples were investigated. It was presumed that clay minerals (smectite, illite and chlorite) dissolved and some zeolitic minerals with large CEC precipitated at altered samples. The CEC were increased with the alteration period, which resulted in the increase of $K_d$ of Cs with the rock sample alteration period. On the other hand, the $K_d$ of Ni and Th decreased with the alteration period. Such decrease in $K_d$ values were supposed to be caused mainly by the dissolution of clay minerals (smectite, illite and chlorite) which contributes to the surface complexation at clay edges.

## ACKNOWLEDGMENTS

This study was partly funded by the Ministry of Economy, Trade and Industry of Japan.

## REFERENCES

1. JNC (present JAEA), JNC technical report, TN1410 2000-004 (2000).
2. M. Ochs, Y. Tachi, D. Trudel, T. Suyama, JAEA-Research 2012-044 (2013).
3. U. Vuorinen, J. Lehikoinen, A. Luukkonen, H. Ervanne, POSIVA 2006-01 (2006).
4. E. C. Gaucher, P. Blanc, *Waste Management* **26**, 776–788 (2006).
5. K. Takahashi, JNC technical report, TN5400 2005-010 (2005) (Japanese).
6. C. J. Schollenberger, R. H. Simon, *Soil Sci.* **59**, 13–24 (1945).
7. L. Wang, D. Jacques, P. D. Cannière, SCK-CEN-ER-28 (2010).
8. S. Ramirez, P. Vieillard, A. Bouchet, A. Cassagnabere, A. Meunier, E. Jacquot, *Applied Geochemistry* **20**, 89-99 (2005).
9. D. L. Parkhurst, C.A.J. Appelo, USGS Water-Resources Investigations Report 99-4259 (1999).
10. M. H. Bradbury, B. Baeyens, *J. Contam. Hydrology* **42**, 141-163 (2000).
11. S. Gaboreau, F. Claret, C. Crouzet, E. Giffaut, Ch. Tournassat, *Applied Geochemistry* **27**, 1194–1201 (2012).
12. M. Bradbury, B. Baeyens, PSI Bericht Nr. 03-08 (2003).
13. K. V. Ticknor, *Radiochim. Acta* **66/67**, 341–348 (1994).

**Corrosion studies of zircaloy, container and carbon steel**

Mater. Res. Soc. Symp. Proc. Vol. 1665 © 2014 Materials Research Society
DOI: 10.1557/opl.2014.645

# C-14 Release Behavior and Chemical Species from Irradiated Hull Waste under Geological Disposal Conditions

Yu Yamashita[1], Hiromi Tanabe[2], Tomofumi Sakuragi[2], Ryota Takahashi[1] and Michitaka Sasoh[1]

[1]Toshiba Corporation, 4-1 Ukishima-cho, Kawasaki-ku, Kawasaki 210-0862, Japan
[2]Radioactive Waste Management Funding and Research Center, Pacific Marks Tsukishima, 1-15-7 Tsukishima, Chuo-ku, Tokyo, 104-0052, Japan

## ABSTRACT

C-14 contained in Hull waste is one of the most important radionuclides in the safety assessment of transuranic (TRU) waste disposal. For more realistic safety assessment, it is important to clarify the release mechanism and chemical species of C-14 from Hull waste. In this research, leaching tests were conducted using an irradiated Zry cladding tube from a boiling-water reactor (BWR) to obtain leaching data and to investigate the relationship between Zry metal corrosion and C-14 release behavior. Both organic and inorganic C-14 compounds existed in the the liquid phase, and some C-14 moved to the gaseous phase. The release rate of C-14 obtained from the BWR cladding tube after two-year leaching tests was lower than the release rate from a pressurize water reactor (PWR) cladding tube. It is considered that the BWR cladding tube used in this test did not easily corrode since it used a comparatively new material. The release rate of C-14 was slightly lower as compared with the corrosion rate of unirradiated Zry. This is thought to be the result of improved corrosion resistance conferred by neutron irradiation, which encouraged the dissolution of grain boundary precipitation elements, such as Fe, Cr, and Ni, into the crystal grains. The leaching tests will be continued for 10 years.

## INTRODUCTION

C-14 contained in Hull and End-pieces is an important radio nuclide in the safety assessment of transuranic (TRU) waste. Various studies have been conducted on the C-14 release behavior from such radioactive metal waste. However, there are insufficient long-term leaching data for evaluating C-14 release and migration behavior in the geological environment, C-14 chemical species and their adsorption action, and the relationship between C-14 release and metal corrosion. For realistic safety assessment, it is necessary to clarify the C-14 inventory of radioactive waste, the release mechanism of C-14, and the C-14 chemical form. Regarding the C-14 inventory, Sakuragi et.al. have reported the total amount and concentration of C-14 in various kinds of metal waste [1]. Although it is assumed that the C-14 release behavior is related with metal corrosion, there are few studies that considered the relationship over a long period of time. There is also a report that the chemical species of released C-14 has a superior organic form [2]. Adsorption of C-14 in an organic form into cement, which is the main material used to construct disposal facilities, cannot be expected, and this leads to too conservative input condition for safety assessment.

Against this background, in the research described here, we examined the C-14 release behavior and compare with a metal corrosion behavior in a simulated disposal environment for the purpose of conducting a more realistic safety assessment. We report the results of leaching tests and C-14 release behavior using irradiated Zry claddings from a boiling-water reactor (BWR). A corrosion test using unirradiated Zry for comparison was carried out independently [3], and the effects of neutron irradiation were considered.

## EXPERIMENT

### Specimen

Irradiated Zry-2 claddings from a BWR were used for the tests. The features of the specimens are shown in table I. The claddings were cut to heights of 2 cm for the leaching tests. The pretreatment procedure is shown in figure 1. Reprocessing steps were imitated and fuel dissolution processing was carried out with 4 N boiling nitric acid for 4 hours. After fuel dissolution processing, the specimens were cleaned ultrasonically with pure water. Underwater polishing with a diamond file was carried out in the order of #150 and #600 grits to remove oxide films on the inside and outside surfaces and expose the base material. The polished specimens were kept in a vacuum desiccator until the start of the leaching tests.

Inventory analysis was conducted as shown in figure 2 using a specimen that was subjected to the same pretreatment as described above. Almost all C-14 was released in the dissolving process, and the C-14 content in the gas from the wet oxidation was about 2%.

**Table I.** Feature of specimens

| Fuel Type | Materials | Burnup (GWd/t) | Cycles | Inventory of Zr metal ($\times 10^4$ Bq/g) | | | |
|---|---|---|---|---|---|---|---|
| | | | | $^{14}C$ | $^{60}Co$ | $^{125}Sb$ | $^{137}Cs$ |
| STEP3 | Zry-2 | 39.7 | 3 | 1.74 | 117 | 1040 | 495 |

The inventory values were calculated on December 22$^{nd}$, 2010.

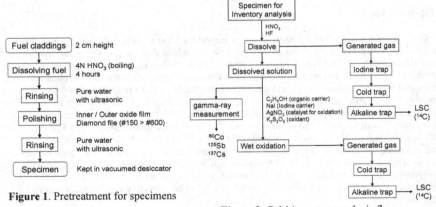

**Figure 1.** Pretreatment for specimens

**Figure 2.** C-14 inventory analysis flow.

## Procedure

A NaOH solution adjusted to pH 12.5 was used for the test solution. Although it is possible for a saturated solution of $Ca(OH)_2$ to exist in an actual disposal environment due to the presence of cement material, only the pH was imitated in order to avoid the generation of $CaCO_3$ precipitates. $N_2$ and $H_2$ gases were bubbled in the test solution just before the specimen was immersed, which reduced the oxidation/reduction potential to less than -250 mV. Immersion was carried out within a simple glove box containing $N_2$ so that the inside of the immersion container was an inert atmosphere. ·

Since it is necessary to maintain the atmosphere for a long period of time, the container shown in figure 3 was used. In order to avoid contact of the specimen with different metals, a glass vial was used for immersion as an inner container. Furthermore, in the event of generation of gas containing C-14, an outer container made from stainless steel was prepared. A gas sampling line was not installed in the outer container in order to maintain prolonged sealing performance. Therefore, at the time of sampling, the inner and outer containers were opened within a simple glove box or a glove bag for collecting C-14 in the gaseous phase. The outer container was sealed with bolts using a silver-plated stainless steel gasket. A preliminary leak test using helium was carried out in advance, and after confirming that the stainless steel container was properly sealed, it was used for the leaching test.

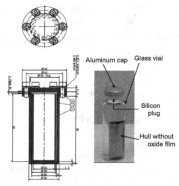

**Figure 3.** Schematic diagram and photograph of test container.

Figure 4 shows a flowchart of the analysis procedure performed in the leaching test. The method was based on the one reported by Sasoh [4]. The outer and inner containers were opened inside the glove box or glove bag, and C-14 contained in the gaseous phase in the inner and outer containers were collected in the alkali traps (1.0 mol dm$^{-3}$ NaOH solution) through an oxidization furnace. C-14 collected in the alkali traps was moved to the gaseous phase by the

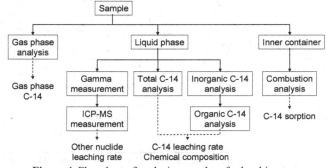

**Figure 4.** Flowchart of analysis procedure for leaching test.

189

addition of concentrated sulfuric acid, was re-collected in Carbosorb®, and was measured by using a liquid scintillation counter (LSC). Part of the collected liquid phase was used to measure the gamma-emitting radionuclides with a germanium semiconductor detector, and another part of the liquid phase was used to measure nonradioactive elements by inductively coupled plasma mass spectroscopy. As target nonradioactive elements, Zr, which is the main constituent of Zry, and the additional elements Sn, Ni, and Cr were chosen. Total C-14 analysis and inorganic/organic C-14 analysis were conducted using part of the immersion solution, and the amount of C-14 contained in the liquid phase and the ratio of organic and inorganic forms were measured. Finally, the presence of C-14 remaining in the inner glass container was checked by conducting combustion analysis of the inner container at 900 °C.

The release fraction, R, of the radioactive nuclides released was calculated by

$$R = (A_G + A_L + A_C) / I$$

where $A_G$ is the activity of the gas phase, $A_L$ is the activity of the liquid phase, $A_C$ is the activity of nuclides remaining in the inner container, and I is the radioactivity inventory of nuclides.

## RESULTS AND DISCUSSION

### Release behavior

The results of the leaching tests are shown in table II. Total C-14 analysis and inorganic/organic C-14 analysis were carried out two times respectively only with the liquid phase for two-year immersion. C-14 was detected in the gaseous phase at 0.5 and 0.75 years. Remaining C-14 in the container could not be detected. In analysis of the liquid phase, the results of the samples immersed for two years showed a wide variation. This tendency will be checked using longer-term test data.

**Table II.** Results of leaching test

| Sample No | Hull weight /g | Test period /year | Gas | Liquid | Container | Sb-125 | Co-60 | Cs-137 | C-14 analysis method |
|---|---|---|---|---|---|---|---|---|---|
| 1 | 5.8763 | 0.5 | $1.62 \times 10^{-1}$ | $1.96 \times 10^{-1}$ | ND | $3.10 \times 10^1$ | $2.12 \times 10^0$ | $1.45 \times 10^2$ | Total C-14 |
| | | | | $1.29 \times 10^{-1}$ (IC 39%, OC 61%) | | | | | IC / OC |
| 2 | 5.8060 | 0.75 | $1.76 \times 10^{-1}$ | $1.82 \times 10^{-1}$ | ND | $2.91 \times 10^1$ | $1.97 \times 10^1$ | $6.58 \times 10^1$ | Total C-14 |
| | | | | $1.51 \times 10^{-1}$ (IC 36%, OC 64%) | | | | | IC / OC |
| 3 | 5.8415 | 1.0 | ND | $1.36 \times 10^{-1}$ | ND | $5.06 \times 10^1$ | $1.63 \times 10^1$ | $7.37 \times 10^2$ | Total C-14 |
| | | | | $1.26 \times 10^{-1}$ (IC 72%, OC 28%) | | | | | IC / OC |
| 4 | 5.7924 | 2.0 | ND | $2.35 \times 10^0$ | ND | $9.30 \times 10^0$ | $7.19 \times 10^0$ | $2.93 \times 10^2$ | Total C-14 |
| | | | | $3.41 \times 10^0$ | | | | | |
| | | | | $7.12 \times 10^{-1}$ (IC 19%, OC 81%) | | | | | IC / OC |
| | | | | $6.64 \times 10^{-1}$ (IC 20%, OC 80%) | | | | | |

Header note: Quantity of nuclide leaching / Bq (C-14: Gas, Liquid, Container)

IC: Inorganic carbon; OC: Organic carbon; ND: not detectable

The test results were compared with the results obtained with a Zry-4 cladding from a pressurized water reactor (PWR; figure 5) [2]. Up to one-year immersion, the BWR samples showed a lower C-14 release fraction than the PWR samples. The PWR cladding used for the leaching test had a thick oxide film compared with the BWR cladding, and corrosion may have been accelerated under the influence of hydrogen absorption.

Figure 6 shows a comparison with the results of a corrosion test using unirradiated Zry metal [3]. The BWR samples showed a lower C-14 release rate than the unirradiated Zry samples. The largest difference between these samples was whether or not they had been subjected to neutron irradiation. Neutron irradiation has various effects on the corrosion resistance of materials [5-8], including both improvement and deterioration of corrosion resistance, and it is thought that the corrosion resistance was greatly improved by neutron irradiation in our tests. According to Etoh [5], the corrosion resistance of Zry is improved with a small dose of fast neutrons because precipitates that exist in Zry, such as Fe, Cr, and Ni, are dissolved into the Zr grains by neutron irradiation. However, the C-14 release rates of the samples immersed for two years showed a large variation, and the average value was almost equivalent to the results of the unirradiated Zry corrosion test. To explain this behavior, further investigation based on longer-term test results will be needed.

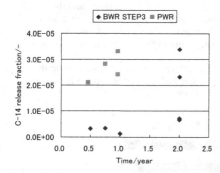 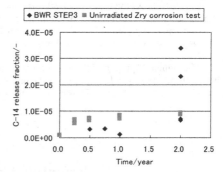

**Figure 5.** Comparison between BWR and PWR claddings.　**Figure 6.** Comparison between BWR cladding and unirradiated Zry.

The left side of figure 7 shows a comparison of the release fractions of C-14 and gamma-emitting radionuclides. When compared with the gamma nuclides, until one year, the C-14 release rate was comparable to those of Sb-125 and Co-60, which are activated products, but at two years, the C-14 release rate was comparable to that of Cs-137, which is a fission product. It seems that Cs-137 is not directly related with the release of C-14 because Cs-137 is a knock-on atom produced by nuclear fission or contamination at the time of sample pretreatment. It will be necessary to examine this more detail in future work. The right side of figure 7 shows a comparison of the release fractions of C-14 and nonradioactive elements. Compared with C-14, the release fractions of Zr and Sn, which are basic components of Zry, were small, whereas those of Ni and Cr were comparable. The C-14 release behavior may be related to the forms in which these elements exist in the oxide film.

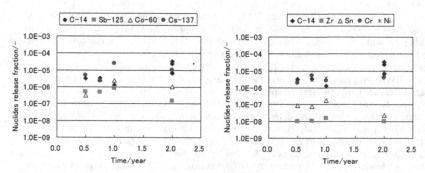

**Figure 7.** Comparison with other nuclides (Left: gamma nuclides, Right: nonradioactive elements).

## C-14 speciation

Figure 8 shows the C-14 chemical form analysis results of the gaseous phase and the liquid phase. Although 50% of C-14 existed in the gaseous phase until 9 months, it was not detected after 9 months. In the liquid phase, both organic and inorganic C-14 co-existed, and the amount of organic C-14 was higher than that of inorganic C-14, except in the samples immersed for 1 year.

Yamaguchi [2] reported that C-14 released from a PWR sample was detected only in the liquid phase and most of the C-14 was the organic form. For both the BWR and PWR claddings, the release rate of organic C-14 was higher than that of inorganic C-14.

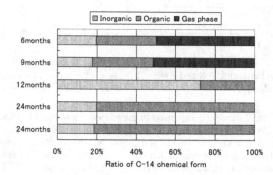

**Figure 8.** C-14 chemical form.

## FUTURE WORK

Two more samples identical to the samples used in these tests remain, and these are scheduled to be opened after three years and five years from the start of the tests, respectively. On the other hand, data for five to ten year old samples will be acquired using STEPI samples which were first immersed in 2008, and the results of those tests will be used to evaluate long-term C-14 release behavior.

## CONCLUSIONS

- In order to perform a realistic safety assessment, leaching tests using irradiated Zry claddings from a BWR were performed.
- From the leaching tests, the C-14 release rate of a BWR Zry cladding was smaller than that of a PWR Zry cladding.
- The C-14 release rate was equivalent to or less than the corrosion rate of unirradiated Zry.
- The release rate of C-14 was not consistent with that of gamma-emitting radionuclides; however, it was similar to those of Cr and Ni.
- C-14 was detected in the gaseous phase in the initial period of the test.
- In the liquid phase, both organic and inorganic C-14 co-existed in the irradiated Zry from a BWR, and the amount of organic C-14 was higher than that of inorganic C-14, except in the samples immersed for 1 year.
- The leaching tests will be continued for 10 years.

## ACKNOWLEDGMENTS

This research is a part of "Research and development of processing and disposal technique for TRU waste containing I-129 and C-14 (FY2012)" being financed by the Agency of Natural Resources and Energy of the Ministry of Economy, Trade and Industry of Japan.

## REFERENCES

1. T. Sakuragi, et al., Estimation of Carbon 14 Inventory in Hulls and Endpieces Wastes from Japanese Commercial Reprocessing Plant, ICEM2013, September, Brussels, Belgium (2013)
2. T. Yamaguchi, et al., A Study on Chemical forms and Migration Behavior of Radionuclides in HULL Waste, ICEM1999, September, Nagoya, Japan (1999).
3. O. Kato, et al., Corrosion Tests of Zircaloy Hull Waste to confirm applicability of corrosion model and to evaluate influence factors on corrosion rate under Geological Disposal Conditions, Res. Soc. Symp. Proc. Scientific Basis for Nuclear Waste Management XXXVII (2013).
4. M. Sasoh, et al., Improvement of C-14 Measurements for Inventory and Leaching Rate for Hull Waste, and Separation of the Organic Compound for Chemical Species Identification, Res. Soc. Symp. Proc. Scientific Basis for Nuclear Waste Management XXXVII (2013).

5. Y. Etoh, S. Shimada, and K. Kikuchi, Irradiation Effects on Corrosion Resistance and Microstructure of Zircaloy-4, Journal of Nuclear Science and Technology, 29, pp.1173–1183 (1992).
6. B. Cheng, et al., Corrosion behavior of irradiated Zircloy, 10$^{th}$ Int. Symp., ASTM-STP-1245, American Society for Testing and Materials, W. Conshohocken. PA, pp. 400–418 (1994).
7. T. Kido, et al., PWR Zircaloy cladding corrosion behavior: quantitative analyses, Journal of Nuclear Materials 248, pp.281–287 (1997).
8. V. Bouineau, et al., A New Model to Predict the Oxidation Kinetics of Zirconium Alloys in Pressurized Water Reactor, Zirconium in the Nuclear Industry; 15$^{th}$ International Symposium, pp. 405–427 (2009).
9. H. Tanabe, et al., Long Term Corrosion of Zircaloy Hull Waste under Geological Disposal Conditions – Corrosion Correlations, Factors Influencing Corrosion, Corrosion Test Data, and a Preliminary Evaluation, Res. Soc. Symp. Proc. Scientific Basis for Nuclear Waste Management XXXVII (2013).
10. FEPC and JAEA, Second Progress Report on Research and Development for TRU Waste Disposal in Japan (2007).
11. H. Tanabe, et al., Characterization of Hull Waste in Underground Condition, Proceedings of the International Workshop on Mobile Fission and Activation Products in Nuclear Waste Disposal, L'Hermitage, La Baule, France, January 16–19 (2007).
12. The Japan Society of Mechanical Engineers, Zirconium Alloy Handbook (1997).
13. RWMC, Research and development of processing and disposal technique for TRU waste containing I-129 and C-14 (FY2012) (2013).

Mater. Res. Soc. Symp. Proc. Vol. 1665 © 2014 Materials Research Society
DOI: 10.1557/opl.2014.646

# Corrosion Tests of Zircaloy Hull Waste to Confirm Applicability of Corrosion Model and to Evaluate Influence Factors on Corrosion Rate under Geological Disposal Conditions

Osamu Kato[1], Hiromi Tanabe[2], Tomofumi Sakuragi[2], Tsutomu Nishimura[1], and Tsuyoshi Tateishi[3]

[1]Kobe Steel, LTD., JAPAN
[2]Radioactive Waste Management Funding and Research Center, JAPAN
[3]Kobelco Research Institute Inc., JAPAN

## ABSTRACT

Corrosion behavior is a key issue in the assessment of disposal performance for activated waste such as spent fuel assemblies (i.e., hulls and end-pieces) because corrosion is expected to initiate radionuclide (e.g., C-14) leaching from such waste. Because the anticipated corrosion rate is extremely low, understanding and modeling Zircaloy (Zry) corrosion behavior under geological disposal conditions is important in predicting very long-term corrosion. Corrosion models applicable in the higher temperature ranges of nuclear reactors have been proposed based on considerable testing in the 523–633 K temperature range.

In this study, corrosion tests were carried out to confirm the applicability of such existing models to the low-temperature range of geological disposal, and to examine the influence of material, environmental, and other factors on corrosion rates under geological disposal conditions. A characterization analysis of the generated oxide film was also performed.

To confirm applicability, the corrosion rate of Zry-4 in pure water with a temperature change from 303 K to 433 K was obtained using a hydrogen measuring technique, giving a corrosion rate for 180 days of $8 \times 10^{-3}$ μm/y at 303 K.

To investigate the influence of various factors, corrosion tests were carried out. The corrosion rates for Zry-2 and Zry-4 were almost same, and increased with a temperature increase from 303 K to 353 K. The influence of pH (12.5) compared with pure water was about 1.4 at 180 days at 303 K.

## INTRODUCTION

The development of a corrosion model to evaluate long-term Zry corrosion under disposal conditions has been discussed in [1], which proposed a corrosion model using corrosion correlations obtained in the higher temperature range of 523 to 633 K. In this study, to confirm the applicability of high-temperature corrosion models to the lower temperature range of geological disposal, and to examine the influence of various factors on corrosion rates, corrosion tests were carried out to evaluate the following issues:

1) Applicability of high-temperature corrosion equations to low-temperature corrosion
   a) Zry corrosion rate in pure water and changes over time in lower temperature range
   b) Influence of differences in corrosion rate measurement methods (hydrogen measurement technique at low temperatures, weight gain measurement technique at high temperatures)
   c) Influence of annealing condition differences in actual Zry tube and test specimen
   d) Influence of composition differences in actual Zry tube and test specimen

e) Properties of the oxide film forming at low temperatures
2) Key parameters possibly affecting corrosion rate
  2-1) Material factor
     a) Type of Zry (Zry-4, Zry-2, etc.)
  2-2) Environmental factors
     a) Temperature,   b) pH,   c) Cations (Na, Ca),
     d) Chemical composition of groundwater (cement-equilibrium, seawater-derived)

## EXPERIMENT

### Specimen

The Zry specimen was prepared using the same method as in the previous study [2]. To produce a specimen of 0.1 mm$^t$ (t stands for thickness, the Zry underwent two sets of cold-rolling followed each time by vacuum annealing (< $10^{-4}$ Pa at 1023 K for 10 h, and < $10^{-2}$ Pa at 873 K for 1 h) to remove the hydrogen absorbed during cold-rolling. The 0.1 mm$^t$ specimen was ground to produce a specimen of 0.05 mm$^t$ for absorbed hydrogen measurement. Hydrogen was measured at various steps with an inert gas fusion method system coupled with gas chromatography (LECO RH-404). The empirical error for measuring Zry hydrogen content is ±0.3 ppm. Analysis of the oxide film was conducted with TEM-EDX (JEM-2010F) and electron diffraction. A thin oxide film of approximately 5 nm had already been observed. Before being supplied for corrosion tests, the specimens were washed with acetone and stored in a vacuumed desiccator. The Zry-4 and Zry-2 compositions after pretreatment are shown in Table I.

**Table I.** Composition of Zry used in corrosion tests (wt%, measured after pretreatment).

|  |  | Sn | Fe | Cr | Ni | O | H | N |
|---|---|---|---|---|---|---|---|---|
| Zry-4 | Specification | 1.20 - 1.70 | 0.18 - 0.24 | 0.07 – 0.13 | <0.0070 | - | <0.0025 | <0.008 |
|  | Measurements | 1.24 | 0.18 | 0.10 | <0.006 | 0.15 | 0.0010(0.1mm$^t$) 0.0009(0.05mm$^t$) | 0.03 |
| Zry-2 | Specification | 1.20 - 1.70 | 0.07 - 0.20 | 0.05 – 0.15 | 0.03 – 0.08 | - | <0.0025 | <0.008 |
|  | Measurements | 1.23 | 0.13 | 0.10 | 0.04 | 0.13 | 0.0007(0.1mm$^t$) 0.0006(0.05mm$^t$) | 0.02 |

### Procedure

As the corrosion rate of Zry is very low at the 303–353 K expected in geological disposal conditions, weight gain measurements were difficult. Therefore, the hydrogen measurement technique was applied to measure the amount of hydrogen generated by the anaerobic corrosion reaction between Zry and the solution. The hydrogen volume was then converted to the equivalent corrosion rate. Corrosion tests were carried out using the ampoule method for batch tests described by Honda [3]. ,The initial steps were carried out in a glove box that had been purged by nitrogen gas with oxygen concentration below 0.1 ppm. A Zry specimen (0.1 mm$^t$, surface area of $1.13 \times 10^{-2}$ m$^2$; and 0.05 mm$^t$, surface area of $0.06 \times 10^{-2}$ m$^2$ for the evaluation of absorbed hydrogen) was placed in a glass ampoule, and a stop-cock was attached. The ampoule was then filled with solution and the stop-cock closed. The solutions used were pure water, and alkaline solutions and simulated groundwater with pH 12.5. The alkaline solutions were adjusted by NaOH or Ca(OH)$_2$. The simulated groundwater (SGW) that simulates the sea-water derived groundwater interacted with cement was prepared using the same method as in the previous study [4].

The ampoule was moved outside the glove box and sealed by heating. After a predetermined period at a constant temperature, the ampoule was set on the ampoule opening apparatus, a vacuum gas collecting system connected with a gas chromatograph (Shimadzu GC1400). The ampoule was opened and the hydrogen gas inside was measured. Zry is known as hydrogen absorption metal; thus, the hydrogen absorbed into the specimen was measured with an inert gas melting system coupled with a gas chromatograph (LECO RH-404)

The weight of the specimen was measured before and after the corrosion test at 433 K. The corrosion rate was evaluated from the weight gain and compared with the equivalent corrosion rate using the hydrogen measuring method.

After hydrogen gas measurement, the oxide layer was analyzed with TEM-EDX (JEM-2010F) and electron diffraction.

## Analysis Method

Zry weight gain was measured using a hydrogen measurement technique to measure the hydrogen generated by the anaerobic corrosion reaction between Zry and the solution as shown in equation (1). The part of hydrogen generated by corrosion was released as gas and the rest was absorbed in Zry.

$$Zr + 2H_2O \rightarrow ZrO_2 + 4H \quad (4H \rightarrow \text{gas released } (2H_2) \text{ or absorbed in Zry}) \tag{1}$$

The equivalent corrosion rate, $R_{total}$ ($\mu$m/y), which represents Zry corrosion by oxidation and hydrogenation according to the stoichiometry in equation (1), can be obtained from the following formula. The equivalent corrosion rates by released hydrogen gas, $R_{gas}$ ($\mu$m/y), and by absorbed hydrogen, $R_{abs}$ ($\mu$m/y), were calculated using equations (3) and (4) respectively, and the equivalent corrosion rate, $R_{total}$, is the sum of these two.

$$R_{total} = R_{gas} + R_{abs} \tag{2}$$

$$R_{gas} = \frac{V \times 10^{-3} \times M \times 365}{2 \times Vm \times S \times \rho \times T} \qquad R_{abs} = \frac{C \times 10^{-6} \times W \times M \times 365}{4 \times S \times \rho \times T} \tag{3),(4}$$

where,
$R_{gas}$:Equivalent corrosion rate by released hydrogen, $R_{abs}$:Equivalent corrosion rate by absorbed hydrogen, $V$:Volume of released hydrogen gas (cm$^3$), $C$:Hydrogen density absorbed into the specimen (mass ppm), $M$:Atomic weight of zirconium (= 91.22), $Vm$: Molar volume(=22.4 dm$^3$/mol), $W$: Weight of Zry specimen(g), $S$:Specimen surface area (m$^2$), $\rho$: Density of Zry (=6.5g/cm$^3$), $T$:Test time (day)

The relation between equivalent weight gain and Zry metal corrosion is to be 14.9 mg/dm$^2$ of weight gain = 0.66 $\mu$m of metal consumed.

## RESULTS AND DISCUSSION

The results for the amount of hydrogen generated by Zry corrosion, hydrogen pick-up ratio, equivalent corrosion rate and equivalent weight gain are shown in Table II. The hydrogen pick-up ratio is calculated from the ratio of absorbed hydrogen and total amount (released + absorbed) of hydrogen.

197

**Table II**. Corrosion test results.

| Test No. | Material | Solution | pH | Temp. /K | Period /day | Amount of hydrogen /H-mol·m⁻² Released (a) | Absorbed (b) | Hydrogen pick-up ratio/% (b)/[(a)+(b)] | Equivalent corrosion rate /$\mu m \cdot y^{-1}$ | Equivalent weight gain /$mg \cdot dm^{-2}$ |
|---|---|---|---|---|---|---|---|---|---|---|
| 1-1-1 |  |  |  | 303 | 90 | $6.8 \times 10^{-5}$ | $6.5 \times 10^{-4}$ | 90.5 | $1.0 \times 10^{-2}$ | $5.6 \times 10^{-2}$ |
| 1-1-2 |  |  | — | 303 | 90 | $8.2 \times 10^{-5}$ | $9.9 \times 10^{-4}$ | 92.2 | $1.5 \times 10^{-2}$ | $8.4 \times 10^{-2}$ |
| 1-1-3 |  |  |  |  |  | $5.2 \times 10^{-5}$ | $8.9 \times 10^{-4}$ | 94.4 | $1.4 \times 10^{-2}$ | $7.8 \times 10^{-2}$ |
| 1-2-1 |  |  |  | 303 | 180 | $7.4 \times 10^{-5}$ | $1.1 \times 10^{-3}$ | 93.7 | $8.5 \times 10^{-3}$ | $9.5 \times 10^{-2}$ |
| 1-2-2 |  |  | — | 303 | 180 | $8.6 \times 10^{-5}$ | $9.1 \times 10^{-4}$ | 91.3 | $7.2 \times 10^{-3}$ | $8.0 \times 10^{-2}$ |
| 1-2-3 |  |  |  |  |  | $7.9 \times 10^{-5}$ | $1.1 \times 10^{-3}$ | 93.3 | $8.5 \times 10^{-3}$ | $9.5 \times 10^{-2}$ |
| 1-3-1 |  |  | — | 323 | 90 | $1.9 \times 10^{-4}$ | $1.3 \times 10^{-3}$ | 87.4 | $2.2 \times 10^{-2}$ | $1.2 \times 10^{-1}$ |
| 1-3-2 |  |  |  |  |  | $1.7 \times 10^{-4}$ | $1.2 \times 10^{-3}$ | 87.7 | $2.0 \times 10^{-2}$ | $1.1 \times 10^{-1}$ |
| 1-4-1 | Zry-4 | Pure water | — | 323 | 180 | $2.2 \times 10^{-4}$ | $1.7 \times 10^{-3}$ | 88.7 | $1.4 \times 10^{-2}$ | $1.6 \times 10^{-1}$ |
| 1-4-2 |  |  |  |  |  | $2.4 \times 10^{-4}$ | $1.9 \times 10^{-3}$ | 89.1 | $1.6 \times 10^{-2}$ | $1.8 \times 10^{-1}$ |
| 1-5-1 |  |  | — | 353 | 90 | $4.5 \times 10^{-4}$ | $1.8 \times 10^{-3}$ | 79.8 | $3.2 \times 10^{-2}$ | $1.8 \times 10^{-1}$ |
| 1-5-2 |  |  |  |  |  | $4.0 \times 10^{-4}$ | $2.1 \times 10^{-3}$ | 83.6 | $3.5 \times 10^{-2}$ | $2.0 \times 10^{-1}$ |
| 1-6-1 |  |  | — | 353 | 180 | $4.9 \times 10^{-4}$ | $2.5 \times 10^{-3}$ | 83.4 | $2.2 \times 10^{-2}$ | $2.5 \times 10^{-1}$ |
| 1-6-2 |  |  |  |  |  | $4.6 \times 10^{-4}$ | $2.8 \times 10^{-3}$ | 85.8 | $2.3 \times 10^{-2}$ | $2.6 \times 10^{-1}$ |
| 1-7-1 |  |  |  | 433 | 150 | $1.1 \times 10^{-2}$ | $5.5 \times 10^{-3}$ | 33.9 | $1.4 \times 10^{-1}$ | 1.3 |
| 1-7-2 |  |  | — | 433 | 150 | $1.2 \times 10^{-2}$ | $6.6 \times 10^{-3}$ | 34.9 | $1.6 \times 10^{-1}$ | 1.5 |
| 1-7-3 |  |  |  |  |  | $1.1 \times 10^{-2}$ | $6.5 \times 10^{-3}$ | 36.2 | $1.6 \times 10^{-1}$ | 1.5 |
| 2-1 |  |  | 12.5 | 303 | 90 | $9.2 \times 10^{-5}$ | $1.1 \times 10^{-3}$ | 92.3 | $1.7 \times 10^{-2}$ | $9.5 \times 10^{-2}$ |
| 2-2-1 |  |  | 12.5 | 303 | 180 | $1.1 \times 10^{-4}$ | $1.5 \times 10^{-3}$ | 93.5 | $1.2 \times 10^{-2}$ | $1.3 \times 10^{-1}$ |
| 2-2-2 |  |  |  |  |  | $1.2 \times 10^{-4}$ | $1.3 \times 10^{-3}$ | 91.7 | $1.1 \times 10^{-2}$ | $1.2 \times 10^{-1}$ |
| 2-3-1 |  |  | 12.5 | 303 | 365 | $1.5 \times 10^{-4}$ | $1.7 \times 10^{-3}$ | 91.9 | $6.7 \times 10^{-3}$ | $1.5 \times 10^{-1}$ |
| 2-3-2 |  |  |  |  |  | $1.3 \times 10^{-4}$ | $1.5 \times 10^{-3}$ | 92.1 | $5.7 \times 10^{-3}$ | $1.3 \times 10^{-1}$ |
| 2-4-1 |  |  | 12.5 | 303 | 730 | $2.0 \times 10^{-4}$ | $1.8 \times 10^{-3}$ | 90.1 | $3.5 \times 10^{-3}$ | $1.6 \times 10^{-1}$ |
| 2-4-2 |  |  |  |  |  | $1.7 \times 10^{-4}$ | $1.9 \times 10^{-3}$ | 92.1 | $3.7 \times 10^{-3}$ | $1.7 \times 10^{-1}$ |
| 2-5-1 |  |  | 12.5 | 323 | 30 | $1.8 \times 10^{-4}$ | $1.0 \times 10^{-3}$ | 84.7 | $5.2 \times 10^{-2}$ | $9.7 \times 10^{-2}$ |
| 2-5-2 |  |  |  |  |  | $2.0 \times 10^{-4}$ | $9.9 \times 10^{-4}$ | 83.2 | $5.1 \times 10^{-2}$ | $9.5 \times 10^{-2}$ |
| 2-6 |  |  | 12.5 | 323 | 90 | $2.9 \times 10^{-4}$ | $1.7 \times 10^{-3}$ | 87.1 | $2.9 \times 10^{-2}$ | $1.6 \times 10^{-1}$ |
| 2-7-1 |  |  | 12.5 | 323 | 180 | $2.5 \times 10^{-4}$ | $2.3 \times 10^{-3}$ | 88.4 | $1.8 \times 10^{-2}$ | $2.0 \times 10^{-1}$ |
| 2-7-2 |  |  |  |  |  | $3.0 \times 10^{-4}$ | $2.0 \times 10^{-3}$ | 87.9 | $1.6 \times 10^{-2}$ | $1.8 \times 10^{-1}$ |
| 2-8-1 | Zry-4 | NaOH | 12.5 | 323 | 365 | $2.7 \times 10^{-4}$ | $2.7 \times 10^{-3}$ | 87.7 | $1.1 \times 10^{-2}$ | $2.5 \times 10^{-1}$ |
| 2-8-2 |  |  |  |  |  | $3.8 \times 10^{-4}$ | $2.1 \times 10^{-3}$ | 86.3 | $8.4 \times 10^{-3}$ | $1.9 \times 10^{-1}$ |
| 2-9-1 |  |  | 12.5 | 323 | 730 | $3.2 \times 10^{-4}$ | $2.6 \times 10^{-3}$ | 85.4 | $5.4 \times 10^{-3}$ | $2.4 \times 10^{-1}$ |
| 2-9-2 |  |  |  |  |  | $4.4 \times 10^{-4}$ | $3.1 \times 10^{-3}$ | 88.8 | $6.2 \times 10^{-3}$ | $2.8 \times 10^{-1}$ |
| 2-10-1 |  |  | 12.5 | 323 | 1825 | $3.8 \times 10^{-4}$ | $4.1 \times 10^{-3}$ | 88.3 | $3.3 \times 10^{-3}$ | $3.7 \times 10^{-1}$ |
| 2-10-2 |  |  |  |  |  | $5.4 \times 10^{-4}$ | $3.5 \times 10^{-3}$ | 85.1 | $2.9 \times 10^{-3}$ | $3.3 \times 10^{-1}$ |
| 2-11-1 |  |  | 12.5 | 353 | 30 | $4.8 \times 10^{-4}$ | $1.6 \times 10^{-3}$ | 77.1 | $9.2 \times 10^{-2}$ | $1.7 \times 10^{-1}$ |
| 2-11-2 |  |  |  |  |  | $5.3 \times 10^{-4}$ | $1.7 \times 10^{-3}$ | 76.2 | $9.8 \times 10^{-2}$ | $1.8 \times 10^{-1}$ |
| 2-12 |  |  | 12.5 | 353 | 90 | $7.0 \times 10^{-4}$ | $2.7 \times 10^{-3}$ | 82.5 | $4.7 \times 10^{-2}$ | $2.6 \times 10^{-1}$ |
| 2-13-1 |  |  | 12.5 | 353 | 180 | $5.7 \times 10^{-4}$ | $3.3 \times 10^{-3}$ | 81.8 | $2.8 \times 10^{-2}$ | $3.1 \times 10^{-1}$ |
| 2-13-2 |  |  |  |  |  | $7.2 \times 10^{-4}$ | $3.5 \times 10^{-3}$ | 85.2 | $2.9 \times 10^{-2}$ | $3.2 \times 10^{-1}$ |
| 2-14-1 |  |  | 12.5 | 353 | 365 | $5.9 \times 10^{-4}$ | $4.3 \times 10^{-3}$ | 85.2 | $1.8 \times 10^{-2}$ | $4.1 \times 10^{-1}$ |
| 2-14-2 |  |  |  |  |  | $7.4 \times 10^{-4}$ | $4.0 \times 10^{-3}$ | 82.4 | $1.7 \times 10^{-2}$ | $3.8 \times 10^{-1}$ |
| 2-15-1 |  |  | 12.5 | 353 | 730 | $8.4 \times 10^{-4}$ | $4.7 \times 10^{-3}$ | 84.2 | $1.0 \times 10^{-2}$ | $4.5 \times 10^{-1}$ |
| 2-15-2 |  |  |  |  |  | $8.9 \times 10^{-4}$ | $4.3 \times 10^{-3}$ | 79.8 | $9.7 \times 10^{-3}$ | $4.4 \times 10^{-1}$ |
| 2-16-1 |  |  | 12.5 | 303 | 90 | $6.9 \times 10^{-5}$ | $1.4 \times 10^{-3}$ | 94.8 | $1.9 \times 10^{-2}$ | $1.1 \times 10^{-1}$ |
| 2-16-2 |  |  |  |  |  | $8.3 \times 10^{-5}$ | $1.3 \times 10^{-3}$ | 94.3 | $2.1 \times 10^{-2}$ | $1.2 \times 10^{-1}$ |
| 2-17-1 |  |  | 12.5 | 303 | 365 | $1.5 \times 10^{-4}$ | $1.8 \times 10^{-3}$ | 92.1 | $6.9 \times 10^{-3}$ | $1.6 \times 10^{-1}$ |
| 2-17-2 |  |  |  |  |  | $1.6 \times 10^{-4}$ | $2.0 \times 10^{-3}$ | 92.3 | $7.6 \times 10^{-3}$ | $1.7 \times 10^{-1}$ |
| 2-18-1 |  |  | 12.5 | 303 | 730 | $1.7 \times 10^{-4}$ | $2.3 \times 10^{-3}$ | 93.1 | $4.5 \times 10^{-3}$ | $2.0 \times 10^{-1}$ |
| 2-18-2 |  |  |  |  |  | $1.9 \times 10^{-4}$ | $2.0 \times 10^{-3}$ | 91.2 | $3.8 \times 10^{-3}$ | $1.7 \times 10^{-1}$ |
| 2-19-1 |  |  | 12.5 | 323 | 90 | $2.2 \times 10^{-4}$ | $2.0 \times 10^{-3}$ | 90.0 | $3.1 \times 10^{-2}$ | $1.7 \times 10^{-1}$ |
| 2-19-2 |  |  |  |  |  | $1.7 \times 10^{-4}$ | $2.5 \times 10^{-3}$ | 93.3 | $3.7 \times 10^{-2}$ | $2.1 \times 10^{-1}$ |
| 2-20-1 | Zry-2 | NaOH | 12.5 | 323 | 365 | $2.7 \times 10^{-4}$ | $2.7 \times 10^{-3}$ | 90.8 | $1.1 \times 10^{-2}$ | $2.5 \times 10^{-1}$ |
| 2-20-2 |  |  |  |  |  | $2.5 \times 10^{-4}$ | $3.0 \times 10^{-3}$ | 92.2 | $1.2 \times 10^{-2}$ | $2.7 \times 10^{-1}$ |
| 2-21-1 |  |  | 12.5 | 323 | 730 | $2.8 \times 10^{-4}$ | $3.3 \times 10^{-2}$ | 92.2 | $6.3 \times 10^{-3}$ | $2.8 \times 10^{-1}$ |
| 2-21-2 |  |  |  |  |  | $3.6 \times 10^{-4}$ | $4.4 \times 10^{-2}$ | 90.3 | $6.6 \times 10^{-3}$ | $3.0 \times 10^{-1}$ |
| 2-22-1 |  |  | 12.5 | 353 | 90 | $5.5 \times 10^{-4}$ | $2.5 \times 10^{-3}$ | 82.0 | $4.5 \times 10^{-2}$ | $2.5 \times 10^{-1}$ |
| 2-22-2 |  |  |  |  |  | $6.1 \times 10^{-4}$ | $3.0 \times 10^{-3}$ | 83.0 | $5.2 \times 10^{-2}$ | $2.9 \times 10^{-1}$ |
| 2-23-1 |  |  | 12.5 | 353 | 365 | $6.8 \times 10^{-4}$ | $4.2 \times 10^{-3}$ | 86.0 | $1.8 \times 10^{-2}$ | $4.1 \times 10^{-1}$ |
| 2-23-2 |  |  |  |  |  | $7.4 \times 10^{-4}$ | $4.6 \times 10^{-3}$ | 86.2 | $1.9 \times 10^{-2}$ | $4.3 \times 10^{-1}$ |
| 2-24-1 |  |  | 12.5 | 353 | 730 | $7.4 \times 10^{-4}$ | $4.8 \times 10^{-3}$ | 86.6 | $9.8 \times 10^{-3}$ | $4.4 \times 10^{-1}$ |
| 2-24-2 |  |  |  |  |  | $8.7 \times 10^{-4}$ | $5.2 \times 10^{-3}$ | 85.8 | $1.1 \times 10^{-2}$ | $5.0 \times 10^{-1}$ |
| 2-25 |  |  | 12.5 | 323 | 30 | $1.3 \times 10^{-4}$ | $1.2 \times 10^{-3}$ | 89.8 | $5.6 \times 10^{-2}$ | $1.0 \times 10^{-1}$ |
| 2-26 | Zry-4 | Ca(OH)₂ | 12.5 | 323 | 90 | $1.7 \times 10^{-4}$ | $1.9 \times 10^{-3}$ | 91.8 | $2.9 \times 10^{-2}$ | $1.6 \times 10^{-1}$ |
| 2-27 |  |  | 12.5 | 323 | 365 | $2.8 \times 10^{-4}$ | $2.0 \times 10^{-3}$ | 87.5 | $8.1 \times 10^{-3}$ | $1.8 \times 10^{-1}$ |
| 2-28 |  |  | 12.5 | 303 | 90 | $1.2 \times 10^{-4}$ | $1.5 \times 10^{-3}$ | 92.6 | $2.4 \times 10^{-2}$ | $1.3 \times 10^{-1}$ |
| 2-29 | Zry-4 | SGW | 12.5 | 303 | 180 | $1.4 \times 10^{-4}$ | $1.8 \times 10^{-3}$ | 93.0 | $1.4 \times 10^{-2}$ | $1.6 \times 10^{-1}$ |
| 2-30 |  |  | 12.5 | 303 | 270 | $1.5 \times 10^{-4}$ | $2.1 \times 10^{-3}$ | 93.2 | $1.1 \times 10^{-2}$ | $1.8 \times 10^{-1}$ |
| 2-31 |  |  | 12.5 | 303 | 365 | $1.7 \times 10^{-4}$ | $2.2 \times 10^{-3}$ | 92.8 | $8.3 \times 10^{-3}$ | $1.9 \times 10^{-1}$ |

## 1) Confirmation of applicability of high-temperature corrosion equation to low-temperature corrosion

### 1-a) Zry corrosion rate in pure water and changes over time at lower temperature ranges

The equivalent corrosion rates calculated by released and absorbed hydrogen in pure water are given in Figure 1 The equivalent corrosion rate for Zry-4 in pure water at 180 days immersion was $8 \times 10^{-3}$ μm/y at 303 K. Equivalent corrosion rates increased with an increase in temperature from 303 K to 353 K.

The hydrogen pick-up ratio of Zry-4 is shown in Figure 2. The pick-up ratio was about 80–95% and decreased with an increase in temperature from 303 K to 353 K. The pick-up ratio was about 35% at 433 K.

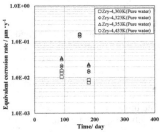

**Figure 1.** Equivalent corrosion rates with Zry-4 in pure water.

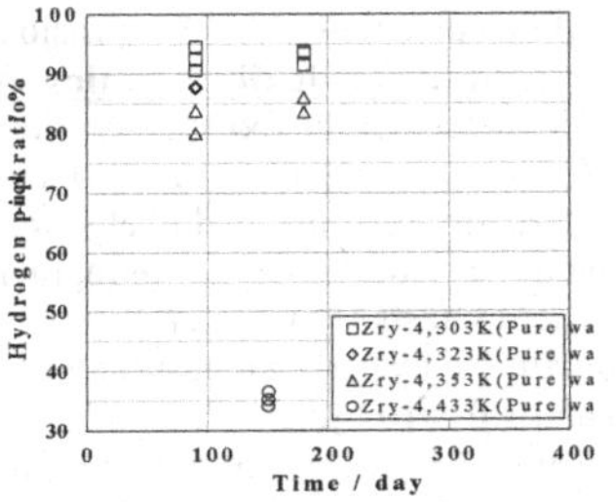

**Figure 2.** Hydrogen pick-up ratios with Zry-4 in pure water.

It has previously been established for high-temperature corrosion that the Zry corrosion rate in the pre-transition region [5] obeys the square rate law ($\Delta W = k \times t^{1/2}$) in the initial period of the high-temperature test and gradually shifts to the cubic rate law ($\Delta W = k \times t^{1/3}$), where $\Delta W$ is weight gain, $k$ is the rate constant, and $t$ is time [6]. To evaluate time-dependent changes, data fitting was performed for the data at 303 K and 353 K in alkaline water using the power law ($\Delta W = a \times t^b$), and multiplier changes were checked for the test period.

The equivalent corrosion rate during the initial period of the low-temperature test was higher than the approximation line of the cubic rate law ($\Delta W = k \times t^{1/3}$), and the data was close to the approximation line of the square rate law ($\Delta W = k \times t^{1/2}$). The data then appears to gradually approach the line of the cubic rate law.

### 1-b) Influence of corrosion rate measurement method differences

To evaluate the influence of the measurement method, weight gain was measured at 433 K with the amount of hydrogen using the hydrogen measurement method. The equivalent corrosion rate evaluated from weight gain measurement techniques were $1.3 \times 10^{-1}$ μm/y, $1.5 \times 10^{-1}$ μm/y, and $1.4 \times 10^{-1}$ μm/y. The equivalent corrosion rate evaluated from weight gain measurement and the hydrogen measurement method were in agreement within a 10% range of error.

### 1-c) Influence of annealing condition differences in actual Zry tube and test specimen

In this test, the specimen was annealed at 1023 K. The accumulated annealing parameter, $\Sigma A_i$ [7], of the specimen was found to be $10^{-16}$. This parameter is approximately 10 times larger than that of the actual Zry tube product [7] used in high temperature tests.

F. Garzarolli et al. [8] showed that there is no significant difference in corrosion resistance in the range of $\Sigma A_i = 10^{-17}$–$10^{-16}$. Thus, it was assumed that there is no significant difference in the corrosion resistance of the specimen in this test and the actual Zry tube. The particle diameter of intermetallic precipitates increased with $\Sigma A_i$ increases and affected Zry corrosion rates.

The particle diameter of the intermetallic precipitates in the specimen used for this corrosion test was identified through SEM. The mean particle diameter of the intermetallic precipitates was 0.25 μm. F. Garzarolli showed that the mean particle diameter was 0.18–0.32 μm at $\Sigma A_i = 10^{-16}$. The results for this test specimen were in agreement with this finding.

### 1-d) Influence of composition differences in actual Zry tube and test specimen

The composition of the actual Zry tube differs slightly from that of the corrosion test specimen, in that the Sn concentration in the older Zry tube is slightly greater. But no significant difference in the corrosion rate at 593–633 K was observed in [9]; thus, it was assumed that the influence of this difference on the corrosion rate at low temperatures was also insignificant.

### 1-e) Property of the oxide film forming at low temperatures

The thicknesses of the oxide films were about 15 nm (323 K, 1825 days, pH 12.5) and 20 nm (353 K, 730 days, pH 12.5) in TEM analysis. Oxygen atoms were observed in EDX analysis and crystallization was observed in electron diffraction analysis. The crystal structure was determined to be orthorhombic or tetragonal, and the monoclinic system observed in high-temperature tests [5] was not observed in either specimen. The equivalent corrosion rates measured with the hydrogen measurement technique were $7.5 \times 10^{-3}$ μm/y (323 K, 1825 days, pH 12.5) and $1.0 \times 10^{-2}$ μm/y (353 K, 730 days, pH 12.5). The thicknesses of the oxide films calculated from the equivalent corrosion rate were approximately 15 nm and 20 nm respectively, based on an assumed $ZrO_2$ density of 5.5 g/cm$^3$.

### 2) Influence of key parameters possibly affecting the corrosion rate

Corrosion tests were carried out to evaluate the influence of material, temperature, pH, cation, and chemical composition in solution, and the results were evaluated as acceleration factors calculated by comparing the results with the corrosion rate in pure water.

### 2-1) Material factor a) Influence of Zry type

The equivalent corrosion rates and hydrogen pick-up ratios with Zry-4 and Zry-2 in alkaline water (pH 12.5) are shown in Table III. The equivalent corrosion rate of Zry-4 was within approximately 15% agreement with that of Zry-2. The difference between the two equivalent corrosion rates over the 730 days of immersion was not significant.

### 2-2) Environmental factor a) Influence of temperature

The equivalent corrosion rate and the hydrogen pick-up ratio of Zry-4 in pure water are shown in Table III. The equivalent corrosion rate increased with a temperature increase from 303 K to 353 K. The corrosion rate at 353 K was about 2.6 times higher at 90 days and about 2.8 times higher at 180 days than that at 303 K.

The hydrogen pick-up ratio in the specimens was approximately 85–95%. The hydrogen pick-up ratio decreased with a temperature increase from 303 K to 353 K.

### 2-2) Environmental factor b) Influence of pH

The equivalent corrosion rate and the hydrogen pick-up ratio of Zry-4 in pure and alkaline water are presented in Figures 4 and 5. The equivalent corrosion rate in pure water and alkaline water (pH 12.5) at 90 days were $1.3 \times 10^{-2}$ μm/y and $1.7 \times 10^{-2}$ μm/y respectively at 303 K, and at 180 days were $8.1 \times 10^{-3}$ μm/y and $1.1 \times 10^{-2}$ μm/y respectively at 303 K. The results showed the acceleration factor of pH was about 1.3 at 90 days and about 1.4 at 180 days.

The hydrogen pick-up ratio in the specimens in alkaline water was about 77–92%, essentially the same as in pure water. The hydrogen pick-up ratio decreased with a temperature increase from 303 K to 353 K. This tendency was the same as in pure water.

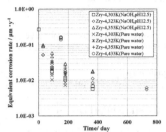

Figure 4. Equivalent corrosion rate with Zry-4 in pure water and alkaline water.

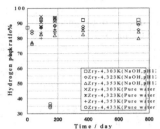

Figure 5. Hydrogen pick-up ratios with Zry-4 in pure water and alkaline water.

### 2-2) Environmental factor c) Influence of cation species

The equivalent corrosion rates and the hydrogen pick-up ratios of Zry-4 in NaOH alkaline water and $CaOH_2$ alkaline water of pH 12.5 are shown in Table III. The equivalent corrosion rates in NaOH alkaline water and $CaOH_2$ alkaline water at 90 days were $2.9 \times 10^{-2}$ μm/y and $2.9 \times 10^{-2}$ μm/y respectively and at 365 days were $9.7 \times 10^{-3}$ μm/y and $8.1 \times 10^{-3}$ μm/y respectively at 323 K. The results showed that the acceleration factor of Ca to Na was about 1.0 at 90 days and about 0.8 at 365 days.

The hydrogen pick-up ratios in the specimens were approximately 90–93% in $Ca(OH)_2$ alkaline water and 87–89% in NaOH alkaline water. The hydrogen pick-up ratios decreased with a temperature increase from 303 K to 353 K. These tendencies were the same as in pure water.

### 2-2) Environmental factor d) Influence of groundwater chemical composition

The equivalent corrosion rates and hydrogen pick-up ratios of Zry-4 in pure water, NaOH alkaline water of pH12.5, and SGW of pH 12.5 are shown in Table III. The equivalent corrosion rates in pure water, NaOH alkaline water of pH 12.5, and SGW of pH 12.5 at 90 days were $1.3 \times 10^{-2}$ μm/y, $1.7 \times 10^{-2}$ μm/y, and $2.4 \times 10^{-2}$ μm/y respectively and at 180 days were $8.1 \times 10^{-3}$ μm/y, $1.1 \times 10^{-2}$ μm/y, and $1.4 \times 10^{-2}$ μm/y respectively at 303 K. The results showed the acceleration factor of SGW compared with pure water was approximately 1.8 at 90 days and approximately 1.7 at 180 days. The acceleration factor of SGW over NaOH alkaline water was approximately 1.4 at 90 days and approximately 1.3 at 180 days, indicating that an element of SGW other than NaOH causes acceleration effect.

The hydrogen pick-up ratio in the specimens in SGW was approximately 92–95%. The hydrogen pick-up ratio decreased with a temperature increase from 303 K to 353 K. This tendency was the same as in pure water.

## CONCLUSIONS

### 1) Confirmation of applicability of high-temperature corrosion equation to low-temperature corrosion

1. The equivalent corrosion rate of Zry-4 in pure water was acquired.
2. The corrosion rate constants derived from low-temperature corrosion data gradually deviated from those estimated from a high-temperature corrosion equation and the rates were higher at

lower temperature as evaluated in [1].

3. To evaluate the possible reasons of the deviation, influence of the measurement method and annealing conditions and composition of the specimen was examined. But those influences were not significant.

4. The crystal structure of the generated oxide film was determined to be orthorhombic or tetragonal, and the monoclinic system observed in high-temperature testing was not observed.

5. Further acquisition of long-term corrosion data in pure water to evaluate the applicability of high-temperature corrosion models will be pursued.

**2) Influence of key parameters possibly affecting the corrosion rate**

1. The acceleration factor of temperature, pH, cation species and groundwater chemical composition was evaluated.

2. The effect of temperature was relatively large compared to the effect of pH and groundwater chemical composition.

3. The effect was unstable yet. Further acquisition of long-term corrosion data to evaluate the acceleration factor will be pursued.

## ACKNOWLEDGEMENT

This research is a part of the research and development of processing and disposal techniques for TRU waste containing I-129 and C-14 (FY2012) program funded by the Agency of Natural Resources and Energy, Ministry of Economy, Trade and Industry of Japan.

## REFERENCES

1.  H. Tanabe et al., 2013, "Long Term Corrosion of Zircaloy Hull Waste under Geological Disposal Conditions: Corrosion Correlations, Factors Influencing Corrosion, Corrosion Test Data, and Preliminary Evaluation" Mat. Res. Soc. Symp. Proc. Scientific Basis for Nuclear Waste Management XXXVII..

2.  T. Sakuragi et al., 2012, "Corrosion Rates of Zircaloy-4 by Hydrogen Measurement under High pH, Low Oxygen, and Low Temperature Conditions," Mater. Res. Soc. Symp. Proc. 1475, p. 311.

3.  A. Honda et al., 1999, Japan Patent 2912365.

4.  T. Yamaguchi et al., 1999, "Study on Chemical Forms and Migration Behavior of Radionuclides in Hull Waste", Proceedings of ICEM 99, September, Nagoya, Japan.

5.  IAEA, 1998, "Waterside corrosion of zirconium alloys in nuclear power plants", IAEA-TECDOC-, p996

6.  S. Hagi et al., 2005, "Introduction to Nuclear Fuel Engineering; Focused on LWR Fuel - (8) LWR Fuel Fabrication, Nuclear and Thermal-Hydronic Design", J. Nucl. Sci. Technol. Vol. 47, No. 1.

7.  F. Garzarolli et al, 1989, "Microstructure and Corrosion Studies for Optimized PWR and BWR Zircaloy Cladding", Zirconium in the Nuclear Industry, ASTM STP 1023.

8.  Japan Nuclear Energy Safety Organization Report, 2012.

9.  Takeda et al, 1996, "Effect of Oxide Film Structure on Resistance to Uniform Corrosion of Zircaloy-4", J. Japan Inst. Met. Mater. vol. 60, No. 9.

Mater. Res. Soc. Symp. Proc. Vol. 1665 © 2014 Materials Research Society
DOI: 10.1557/opl.2014.647

# Long-Term Corrosion of Zircaloy Hull Waste under Geological Disposal Conditions: Corrosion Correlations, Factors Influencing Corrosion, Corrosion Test Data, and Preliminary Evaluation

Hiromi Tanabe, Tomofumi Sakuragi, Hideaki Miyakawa, and Ryota Takahashi

Radioactive Waste Management Funding and Research Center (RWMC),
Pacific Marks Tsukishima, 1-15-7 Tsukishima, Chuo-ku, Tokyo, 104-0052, Japan
Tel.: 81-3-3534-4533; E-mail: tanabe.hiromi@rwmc.or.jp

## ABSTRACT

The carbon-14 generated in Zircaloy (Zry) hull waste is considered an important radionuclide in the TRU waste geological disposal concept in Japan. Given that the metal Zry is highly corrosion-resistant in the anaerobic and low-temperature conditions of the repository, and that the C-14 release rate is assumed to be controlled by the corrosion rate, a variety of corrosion and leaching tests have been performed. However, since the Zry corrosion rate is extremely slow, it is not possible to predict long-term corrosion behavior through low-temperature corrosion tests conducted in a reasonable time period. A vast amount of testing has been conducted in the higher-temperature range of 523 to 633 K, and corrosion correlations have been obtained from these tests. Corrosion correlations have been used to predict the corrosion rate of Zry in a tuff repository. Long-term Zry autoclave corrosion data have been analyzed to develop new corrosion correlations. Extrapolating these correlations to a lower temperature range requires verification that the mechanisms do not change over the range of testing and extrapolation. Factors that influence corrosion rates under geological disposal conditions, such as material and environmental factors, should also be examined. Corrosion correlations, factors influencing corrosion rates, the results of corrosion and leaching tests, and a preliminary evaluation are discussed.

## INTRODUCTION

A safe disposal concept of Transuranic (TRU) wastes generated from reprocessing and MOX fuel fabrication plants has been studied in Japan by the Federation of Electric Power Companies of Japan, Japan Atomic Energy Agency, and Japan Nuclear Fuel Limited with the support of research organizations [1]. This report is named as the second TRU report. Based on the results of the TRU waste disposal safety assessment, the I-129 in spent absorbents and C-14 in hull waste are considered nuclides which contribute substantially to radiation dose rate. To develop the safety concept, hot tests using simulated hulls generated from PWR spent fuel reprocessing have been pursued to evaluate C-14 inventories in Zry metal and Zry oxide film, C-14 release rates, chemical forms of C-14 released into simulated groundwater, and distribution coefficients of C-14 in cement [2]. Similar tests using BWR spent fuel and non-irradiated Zry metals were also conducted later [3, 4, 5, 6, 7]. However, it is not possible to predict long-term Zry corrosion behavior and the associated C-14 release rate solely through low-temperature corrosion tests conducted in a reasonable time period because of the extremely slow corrosion rate of Zry. This paper discusses how to develop a corrosion model using corrosion correlations obtained in the higher-temperature range of 523 to 633 K [6]. Preliminary evaluation was conducted by extrapolating these correlations to a lower temperature range taking into consideration low-temperature corrosion test data. The effect of the corrosion models on radiation dose rate was also evaluated preliminary.

In this report, high-temperature refers to temperature greater than 523K, low-temperature refers to geological disposal conditions below 373K , and medium-temperature refers to the temperature between the high and low temperature cut-offs. The relation used for the Zry oxide and metal is 14.9 mg/dm$^2$ of oxide weight gain = 1 μm of oxide thickness = 0.66 μm of metal consumed.

## CORROSION MODEL DEVELOPMENT

### Method to evaluate low-temperature corrosion based on knowledge of high-temperature corrosion

The method to evaluate low-temperature corrosion in a geological repository (GR) is shown in Figure 1. A vast amount of testing has been conducted in high-temperature conditions, and several correlations and factors influencing corrosion rate have been obtained and evaluated. Rothman [8] had attempted to predict the Zry corrosion rate in a tuff repository by extrapolating correlations obtained in high-temperature corrosion tests in 1984. Later, Hillner et al. [9] analyzed long-term Zry autoclave corrosion data to develop new correlations. Extrapolating these to a lower-temperature range requires verification that the mechanisms do not change over the range of testing and extrapolation. Factors influencing corrosion rates under geological disposal conditions, such as material and environmental factors, should also be examined.

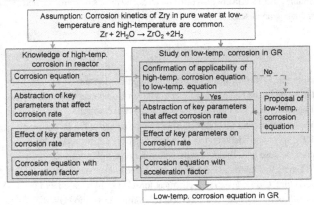

**Figure 1.** Method to evaluate low-temperature corrosion based on knowledge of high-temperature corrosion.

## Conclusions from high-temperature corrosion tests

There is a significant body of literature on high-temperature corrosion. Conclusions obtained are summarized in Table I [11].

**Table I.** High-temperature corrosion conclusions [11].

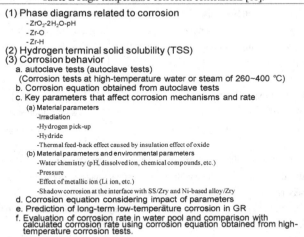

(1) Phase diagrams related to corrosion
- $ZrO_2$-$2H_2O$-pH
- Zr-O
- Zr-H

(2) Hydrogen terminal solid solubility (TSS)

(3) Corrosion behavior
  a. autoclave tests (autoclave tests)
    (Corrosion tests at high-temperature water or steam of 260~400 °C)
  b. Corrosion equation obtained from autoclave tests
  c. Key parameters that affect corrosion mechanisms and rate
    (a) Material parameters
      -Irradiation
      -Hydrogen pick-up
      -Hydride
      -Thermal feed-back effect caused by insulation effect of oxide
    (b) Material parameters and environmental parameters
      -Water chemistry (pH, dissolved ion, chemical compounds, etc.)
      -Pressure
      -Effect of metallic ion (Li ion, etc.)
      -Shadow corrosion at the interface with SS/Zry and Ni-based alloy/Zry
  d. Corrosion equation considering impact of parameters
  e. Prediction of long-term low-temperature corrosion in GR
  f. Evaluation of corrosion rate in water pool and comparison with calculated corrosion rate using corrosion equation obtained from high-temperature corrosion tests.

Many researchers have proposed corrosion equations based on corrosion test results obtained from autoclave tests in pure water. Most proposed the equations for the pre- and post-transition regions. The corrosion rate in the pre-transition region is considered to be in accordance with the cubic rate law, while that of post-transition region is in accordance with a linear rate law (Table II [10]).

**Table II.** Various constants of proposed corrosion equations.

| Model | $K_{co}$ | $Q_{co}$ | $K_{LO}$ | $Q_{LO}$ | Assumption |
|---|---|---|---|---|---|
| Hillner | 6.36E+11 | 27095 | 1.12E+08 | 24895 | |
| Van der Linde | 1.99E+13 | 31116 | 2.30E+09 | 28613 | |
| Dyce | 5.07E+13 | 32289 | 6.53E+09 | 29915 | |
| Dalgaard | - | - | 1.84E+07 | 22220 | Linear law |
| Stehle | - | - | 2.21E+09 | 28220 | Linear law |
| Garzarolli et al. | 5.07E+13 | 32289 | 2.21E+09 | 28200 | |

Proposed corrosion equations:

$$\Delta W^3 = Kco \times \exp(-Qco/RT)\,t \qquad \text{(1): Pre-transition}$$

$$\Delta W = K_{LO} \times \exp(-Q_{LO}/RT)\,t \qquad \text{(2): Post-transition}$$

$\Delta W$: Weight gain, mg/dm$^2$
$K_{LO}$: Rate constant for post-transition, mg/dm$^2 \cdot$d
$Q_{LO}$: Activation energy for post-transition, cal/mol
T : Temperature, K

$K_{co}$: Rate constant for pre-transition, mg$^3$/dm$^6 \cdot$d
$Q_{co}$: Activation energy for pre-transition, cal/mol
R : Gas constant=1.987, cal/mol·K
t : Exposure time, days

Hillner et al. [9] analyzed long-term Zry autoclave corrosion data to develop new Zry corrosion equations. Approximately 14500 data points were generated. The maximum exposure time was 10507 days (approx. 29 y) in a test at 589 K, and the maximum weight gain was 1665 mg/dm$^2$ (approx. 114 μm of oxide) in a test at 611 K. The post-transition rate constant is best described by two successive equations from stage 1 to stage 2, with the change occurring at approximately 400 mg/dm$^2$ oxide thickness (approx. 18 μm of metal). In this work, the following equations have been used to evaluate low-temperature corrosion in a geological repository. However, taking a conservative viewpoint, the change is assumed to occur at the minimum weight gain of 154 mg/dm$^2$ oxide thickness (approx. 6.8 μm of metal) shown in [9].

$$\Delta W^3 = 6.36 \times 1011 \exp(-13636/T)t \qquad \text{(3): Pre-transition [12]}$$
$$\Delta W(1) = 2.47 \times 10^8 \exp(-12880/RT)t \qquad \text{(4): Post-transition stage 1 [9]}$$
$$\Delta W(2) = 3.47 \times 10^7 \exp(-11452/RT)t + C_2 \qquad \text{(5): Post-transition stage 2 [9]}$$

Hillner commented how to apply the new corrosion equations to a geological repository.
- Since it is assumed Zry cladding will be in the post-transition kinetic region upon disposal, the use of eq. (5) is conservative.
- Extrapolation beyond 114 μm of oxide (75 μm of metal) may risk another transition to a region with faster corrosion.

## Key parameters possibly affecting corrosion mechanisms and kinetics in geological disposal

The differences in Zry corrosion conditions in a light water reactor (LWR) and a geological repository are shown in Table III. These figures should be considered typical examples.

**Table III.** Differences in LWR and GR corrosion conditions [1].

| Conditions | LWR | | Geological Repository |
|---|---|---|---|
| | PWR | BWR | |
| Temperature, K | 289 (Inlet) 325 (Outlet) | 278 (Inlet) 287 (Outlet) | Decrease gradually from 353 K (initial) to 303 K (ground temp.) |
| pH | 6.9-7.3 | 5.5 | Decrease from 13.2 (initial cement equilibrium ground water) to 8.5 (ground water) |
| Water chemistry | Additives (Li, B, H$_2$) | No additives | Transition from cement equilibrium ground water to underground water |
| Oxygen, ppb | <1 | 200 | Transition from atmosphere (initial) to reduction condition |
| Eh, mV | Potential at metal/oxide interface is -1.1~1.2V | | Ground water Eh is -281mV Potential at metal/oxide interface is -1.1~1.2V |
| Pressure, MPa | 15.5 | 7.17 | 5 (-500m) |
| Neutron irradiation, n/m$^2$ | High | ~10$^{25}$ | Little self-irradiation |
| Irradiation of water | High | High | γ : 4x10$^4$ Gy/y and n: 5x10$^{-1}$ Gy/y at the surface of waste package |
| Hydrogen conc., ppm | NA | NA | Initial H$_2$ conc. is same as LWR |
| Corrosion equation | Zr + 2H$_2$O → ZrO$_2$ +2H$_2$ | | Zr + 2H$_2$O → ZrO$_2$ +2H$_2$ |
| Corrosion rate, μm/y | A few | A few | < 0.02 (initial) |

Based on the conclusions of high-temperature corrosion tests in Table I, the key parameters of the material and environmental factors that might affect the Zry corrosion rate in a geological repository were selected [11]. The influence of these key parameters on corrosion rates was then qualitatively evaluated to be high, medium, or low. The result is shown in Table IV, which lists high influence factors.

**Table IV.** Key parameters possibly affecting corrosion rate. (summary).

| Key parameters | | Influence | Evidence | Method/Experiment plan | Remaining issues |
|---|---|---|---|---|---|
| Material factor | Irradiation effect | High | This influence is though to be large based on results of high temp. out-of-pile corrosion test. | Corrosion test using irradiated sample. | - |
| Material factor | Amount of hydrogen absorption | High | This influence is though to be large based on results of high temp. out-of-pile corrosion test. | Corrosion test using sample which absorbed various amount of hydrogen. | - |
| Material factor | Distribution of hydrides (uniform or surface) | High | This influence is though to be large based on results of high temp. out-of-pile corrosion test. | | - |
| Material factor | Long-term stability of hydrides | High | It is necessary in securing the long-term performance of the disposal site. | | Experimental methods, conditions, etc. need further consideration. |
| Environmental factor | Temperature | High | | Corrosion test at various temperatures. | - |
| Environmental factor | pH | High | | Corrosion test at various pHs. | - |
| Environmental factor | Influence of cation ions | High | | Corrosion test with Na or Ca ion. | |
| Environmental factor | Chemical composition of Groundwater (sea water derived groundwater) | High | | Corrosion test with simulated cement-equilibrium sea-water derived groundwater. | - |

## EXPERIMENTAL DATA OBTAINED, PRELIMINARY RESULTS, AND DISCUSSION

### Confirmation of applicability of high-temperature corrosion equations to low-temperature corrosion

Corrosion rate constants were calculated from previously obtained corrosion data in pure water [5, 6, 7, 11] by fitting these data to eq. (1) irrespective of actual time dependence. As the corrosion volumes are less than 1 μm, the data and rate constants are considered to be in a pre-transition region and were compared with the data and pre-transition rate constant in [12]. The results are shown in Figure 2. The pre-transition corrosion rate constant, $Kco$ in eq. (1), derived from low-temperature corrosion data gradually deviated from the estimate from the high-temperature corrosion equation and increased as temperature decreased. Possible reasons were considered and are given in Table V. Some have been already examined in [5], demonstrating that the difference in measurement method and annealing conditions will not cause sufficient deviation between the corrosion rate constant derived from the results of the low-temperature corrosion tests and Hillner's equation. However, the initial corrosion rate gradient from the low-temperature corrosion tests seems to be larger than it is in eq. (3). For example, the gradient of the initial phase with a NaOH solution of pH 12.5 of and 353 K is more than 0.4, decreasing to less than 1/3 after one year. The gradient of the initial phase for NaOH solution of pH 12.5 and 303 K is also more than 0.4 and decreases after that, but is more than 1/3 even after two years. This could be one of the reasons for the deviation, and the corrosion tests will be continued, in addition to analyzing the oxide properties to confirm the effect.

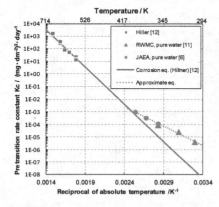

**Figure 2.**
Pre-transition rate constant, $Kc$ vs. temperature.

**Table V.** Possible reasons for corrosion rate constant deviation between high and low temperatures.

| Reason | Importance /Possibility | Present situation | Experimental plan | Confirmation method | Notes |
|---|---|---|---|---|---|
| It is said the corrosion of Zircaloy before the transition is $W=kt^{0.5}$ initially and it gradually becomes $W=kt^{1/3}$. Therefore, as for the corrosion test at low temperature, the apparent corrosion rate calculated by $W=kt^{1/3}$ eq. may become larger. | High | The tritium technique is under investigation for use in the experiment. | Corrosion test using sample with oxide | Evaluate assuming that the relation between time and oxide-thickness (corrosion equation) is correct. | If the hypothesis is correct, this is the same as the temperature-dependent property (activation energy) demonstrated with the high-temperature corrosion equation. |
| | | The O-18 technique is under investigation for use in the experiment. | | | |
| The property of the oxide formed at low temperature may differ from that formed at higher temperatures. | High | The applicability of the analysis method must be confirmed for very thin oxides (less than 30-50nm). | Corrosion test on low-temp (303-353 K) and mid-range temp. (approximately 453 K) | Analyze the oxide. | This year, the oxide formed in corrosion tests under NaOH conditions for two years will be analyzed. The results will then be compared with the oxide formed in corrosion tests in pure water. |
| Measurement method differences may influence it. | High | Experiment is underway. | Corrosion test at mid-range temp. (approximately 453 K) | Compare result of hydrogen measurement with the metal consumption. (amount of generated-hydrogen and weight gain) | |
| The solubility of hydrogen is temperature dependent, and there is a larger amount of hydrides at lower temperatures. Thus, there may be acceleration due to the effect of the hydrides. | Mid | Experiment is underway. | Corrosion test using samples with various amounts of hydrogen absorbed. | Conduct corrosion tests using samples with various amount of hydrogen absorbed at various temperatures. | |
| The annealing condition differences in the actual Zry tube and the test specimen may have an effect. | Mid | The test sample annealing parameter is approximately one digit larger than that of a fuel cladding. | | ·Analyze sample precipitate size. ·Conduct corrosion test at 633 K, and confirm whether corrosion is equal to that of a fuel cladding. | There are data on the bulk diffusivity of oxygen by metal particle size. |

## Effect of key parameters on corrosion rate and tentative acceleration factors

Corrosion data had been previously obtained in pure water, high pH, and other conditions [5, 11]. Leaching tests using irradiated PWR Zry-4 and BWR Zry-2 have been also carried out [2, 4, 11]. These corrosion data are shown in Figure 3, with corrosion volumes estimated from high-temperature corrosion eq. (3) [12]. A preliminary evaluation of acceleration factors was carried out using those data, the results of which are listed in Table VI. The acceleration factor for the irradiation and hydriding effect was evaluated using PWR Zry-4 and found to be around 2.4. This is almost agrees with the knowledge of high-temperature corrosion obtained by Hillner [9] that a factor of 2 could be applied to treat any prior irradiation effects in high-temperature corrosion. Kido et al. [14] also investigated the effect of hydriding at high-temperatures and demonstrated an obvious effect after the transition point. Jeong et al. [13] investigated the effect of high pH using various aqueous solutions, including 32.5 mmol NaOH and water at 623 K for 300 days, and concluded that corrosion rates are similar before transition and increase slightly in the NaOH solution after transition. Corrosion rates in pure water are slightly lower than in the NaOH solution, which suggests there are other factors cause unknown deviation accelerating the corrosion rate shown in Figure 3. Corrosion rates for BWR Zry-2 [4] were lower than for PWR Zry-4 during the first year. The reason could be an irradiation effects improving nodular corrosion resistance as Etoh et al. discussed in [15]. However, the corrosion data after two years seems to increase. Investigation of the effect of the parameters listed in Table IV to identify the reason will continue.

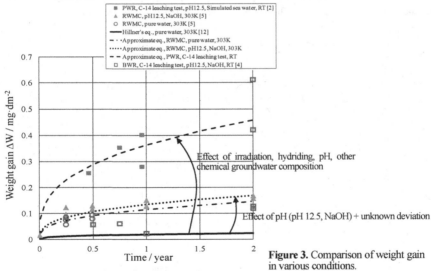

**Figure 3.** Comparison of weight gain in various conditions.

**Table VI.** Tentative acceleration factors for key parameters at low-temperatures.

| Key parameters | Temperature/K (Acceleration factor) | 353 ($a_1$) | 323 ($a_2$) | 303 ($a_3$) |
|---|---|---|---|---|
| pH 12.5/ Pure water (eq. (3))* | | 2.5 | 5.5 | 7.4 |
| Irradiation + hydride | | 2.4** | 2.4** | 2.4 |
| Surface increase | | 2*** | 2*** | 2*** |
| Sum of acceleration factors | | 12 | 30 | 40 |

\* Unknown deviation is included.
\*\* Assumed to be the same figure at 303 K.
\*\*\* Assumption based on previous experience [2].

## Estimation of long-term Zry corrosion

The lifetime of Zry was preliminary evaluated taking into consideration the discussion above and by multiplying high-temperature equations (3), (4), and (5) by the acceleration factors in Table VI. Several cases shown in Table VII were established. Case R0 was set in the second TRU report [1]. From the point of view of lifetime, case R2 or R3 is considered to be more conservative, as Hillner commented [9]. The temperature condition is assumed to be 353 K initially, the upper temperature limit of cement, and it is assumed to gradually decrease up to 323 K at 1,000 years after disposal. The calculation results are shown in Figure 4. The lifetime of Zry varied from 4.11E+6 when applying post-transition eq. (5) from the beginning to 3.59E+8 years when applying pre-transition eq. (3), post-transition eq. (4) and (5) sequentially, but these are both longer than the lifetime of 1.14E+4 years in the second TRU report.

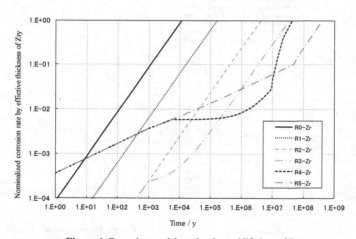

**Figure 4.** Corrosion models and estimated lifetime of Zry.

## Table VII. Corrosion Models and Assumptions

| Case | Corrosion model | Corrosion equations applied | Period to be applied | Calculated life time [y] |
|---|---|---|---|---|
| R0-Zr | Leaching rate 0 of the second TRU report<br>*Most conservative* | $Y = 0.02 \times t$    Y:Corrosion volume[μm]   t:Time[y]<br>The corrosion rate was determined based on the corrosion data at pH=12.4, $Cl^-$conc.=3,200ppm, 308 K, 600days.<br>The effective half thickness of Zry is the same as the 2nd TRU report and is assumed to be 228μm. | All period | 1.14E+04 |
| R1-Zr | Leaching rate 1<br>*Conservative* | $Y = 1.44 \times 10^{-3} \times t$    Y:Corrosion volume[μm]   t:Time[y]<br>The corrosion rate during 10 years is assumed. | All period | 1.58E+05<br>ibid. |
| R2-Zr | Leaching rate 2 at 353 K<br>*Probably concervative realistic* | $Y = 4.63 \times 10^{-6} \times t \times a_1$    Y:Corrosion volume[μm]   t:Time[y]   $a_1$:Acceleration factor (=12) | All period | 4.11E+06<br>ibid. |
| R3-Zr | Leaching rate 3 at 353 to 323 K<br>*Probably concervative realistic* | $Y(353\ K) = 4.63 \times 10^{-6} \times t \times a_1$<br>$Y(323\ K) = 2.28 \times 10^{-7} \times t \times a_2$<br>Y:Corrosion volume[μm]   t:Time[y]   $a_1$:Acceleration factor (=12)   $a_2$:Acceleration factor (=30) | Y(353 K):0 - 1000 y<br>Y(323 K):1000y - | 3.33E+07<br>ibid. |
| R4-Zr | Leaching rate 4 at 353 to 323 K<br>*Probably realistic* | Pre:$Y(353\ K) = (9.39 \times 10^{-10} \times t)^{1/3} \times a_1$<br>Pre:$Y(323\ K) = (2.60 \times 10^{-11} \times t)^{1/3} \times a_2$<br>Post1:$Y = 5.36 \times 10^{-11} \times t \times a_2$<br>Post2:$Y = 6.25 \times 10^{-10} \times t \times a_2$<br>Y:Corrosion volume[μm]   t:Time[y]   $a_1$:Acceleration factor (=12)   $a_2$:Acceleration factor (=30)<br>Weight gain at the first transition point from pre to post1 is assumed to be 30mg/dm² (1.33μm of metal).<br>The second transition point from post1 to post2 is assumed to be 154mg/dm²(6.82μm of metal). | Pre:Y(353 K):0 - 1000y<br>Pre:Y(323 K):1000 - 5.53x10³y<br>Post1:Y :5.53x103 - 9.37x10⁶ y<br>Post2:Y :9.37x10⁶ y - | 4.17E+07<br>ibid. |
| R5-Zr | Leaching rate 5 at 353 to 323 K<br>*Probably realistic* | Pre:$Y(353\ K) = (9.39 \times 10^{-10} \times t)^{1/3} \times a_1$<br>Pre:$Y(323\ K) = (2.60 \times 10^{-11} \times t)^{1/3} \times a_2$<br>Post1:$Y = 5.36 \times 10^{-11} \times t \times a_2$<br>Post2:$Y = 6.25 \times 10^{-10} \times t \times a_2$<br>Y:Corrosion volume[μm]   t:Time[y]   $a_1$:Acceleration factor (=12)   $a_2$:Acceleration factor (=30)<br>Weight gain at the first transition point from pre to post1 is assumed to be the intersection point of pre and post1.<br>The second transition poin from post1 to post2 is assumed to be 154mg/dm² (6.82μm of metal) multiplied by $a_2$, i.e. 204.6μm of metal. | Pre:Y(353 K):0 - 1000 y<br>Pre:Y(323 K):1000 - 4.46x10⁷ y<br>Post1:Y :4.46x10⁷ - 3.56x10⁸ y<br>Post2:Y :3.56x10⁸ y - | 3.59E+08<br>ibid. |

## Effect of corrosion model on radiation dose rate evaluation

The dose rates caused by C-14 released from Zry were also evaluated based on the estimates for long-term Zry corrosion. C-14 is assumed to be released congruently with Zry corrosion based on the model and assumptions of the second TRU report [1]. The dose rate in the reference scenario was reduced by 2 to 3 orders compared with that of the second TRU report. Zry lifetime in case R2 or R3 was shorter when applying the post-transition equation from the beginning than when applying the pre-transition equation then post-transition equations (case R4 or R5), but the C-14 dose rate was larger in the latter case. This is because the half-life of C-14 is 5,730 years. The dose rate of C-14 is dominated by the radioactivity released during the first few tens of thousands of years, and radioactivity during this period is larger in the latter case as shown in Figure 5. The effect of corrosion rate for radionuclides with longer half-lives such as Cl-36 should be considered separately.

211

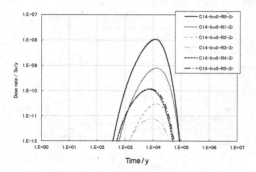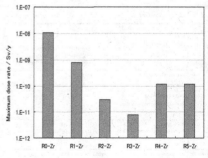

**Figure 5.** Influence of corrosion model on max. dose rate. Dose rate vs. time (left) and max. dose rate (right).

## CONCLUSIONS

1. A method to evaluate low-tempe..u..re corrosion in a GR based on knowledge of high-temperature corrosion was proposed.

2. Low-temperature corrosion tests have been pursued with a hydrogen measuring technique using thin plates of Zry-2 and Zry-4, and with C-14 leaching tests using irradiated Zry-2 and Zry-4.

3. Based on these data, a preliminary evaluation was conducted on the applicability of high-temperature corrosion equations to GR conditions, but the pre-transition corrosion rate constant, Kco, derived from the low-temperature corrosion data gradually deviated from those estimated from the high-temperature corrosion equation and increased as temperature decreased. The reason for this requires further investigation.

4. Preliminary evaluations based on these data were conducted on the influence of parameters such as temperature, pH, irradiation, and hydriding, and acceleration factors were determined. By multiplying high-temperature equations (3), (4), and (5) by these factors, the Zry lifetime was evaluated. This lifetime varied from 4.11E+6 when applying post-transition eq. (5) from the beginning to 3.59E+8 years when applying pre-transition eq. (3), and post-transition eq. (4) and (5) sequentially, but these lifetimes are longer than that in the second TRU report, 1.14E+4 years. Further testing has been performed to evaluate the influence of other parameters.

5. The dose rate caused by C-14 released from Zry was also evaluated using high-temperature equations multiplied by acceleration factors, and based on the model and assumptions in the second TRU report.

The dose rate for the reference scenario was reduced by 2 to 3 orders compared with that in the second TRU report. The Zry lifetime was shorter when applying the post-transition equation from the beginning than when applying the pre-transition equation followed by the post-transition equations, but the dose rate of C-14 was larger in the latter case. This is because the half-life of C-14 is 5730 years. The C-14 dose rate is dominated by the radioactivity released during the first few tens of thousands of years, and radioactivity during the period is larger in the latter case.

## ACKNOWLEDGMENTS

The authors wish to acknowledge the direction supplied by, and the many helpful discussions held with, Prof. Emeritus S. Ishino (Tokyo University), Prof. K. Idemitsu, Prof. T. Tanabe (Kyushu University), Prof. S. Yamanaka, Assistant Prof. H. Muta (Osaka University), Dr. T. Kido (NDC), Dr. K. Takeda (NSSMC), Dr. Y. Etoh (NFD), and Dr. K. Abe (Kobelco).

This research is part of the research and development of processing and disposal techniques for TRU waste containing I-129 and C-14 (FY2012) financed by the Agency of Natural Resources and Energy of the Ministry of Economy, Trade and Industry of Japan.

## REFERENCES

1.  FEPC and JAEA, *Second Progress Report on Research and Development for TRU Waste Disposal in Japan* (2007).
2.  T. Yamaguchi, S. Tanuma, I. Yasutomi, T. Nakayama, H. Tanabe, K. Katsurai, W. Kawamura, K. Maeda, H. Kitao and M. Saigusa, A Study on Chemical forms and Migration Behavior of Radionuclides in HULL Waste, ICEM1999, September, Nagoya, Japan (1999).
3.  H. Tanabe, T. Nishimura, M. Kaneko, T. Sakuragi, Y. Nasu and H. ASANO, Characterization of Hull Waste in Underground Condition, Proceedings of the International Workshop on Mobile Fission and Activation Products in Nuclear Waste Disposal, L'Hermitage, La Baule, France, January 16-19 (2007).
4.  Y. Yamashita, H. Tanabe, T. Sakuragi, R. Takahashi and M. Sasoh, C-14 Release Behavior and Chemical Species from Irradiated Hull Waste under Geological Disposal Conditions, Mat. Res. Soc. Symp. Proc. Scientific Basis for Nuclear Waste Management XXXVII (2013).
5.  O. Kato, H. Tanabe, T. Sakuragi, T. Nishimura, and T. Tateishi, Corrosion Tests of Zircaloy Hull Waste to confirm applicability of corrosion model and to evaluate influence factors on corrosion rate under Geological Disposal Conditions, Res. Soc. Symp. Proc. Scientific Basis for Nuclear Waste Management XXXVII (2013).
6.  T. Maeda, N. Chiba, T. Tateishi and T. Yamaguchi, Corrosion rate of Zircaloy-4 in deoxidized deionized water at 80 - 120 °C, Trans. At. Energy Soc. Japan (2013). (In Japanese)

7. Japan Atomic Energy Agency, Regulatory support research on safety evaluation methods for the safety licensing review of geological disposal of HLW & TRU waste (2013).

8. A. J. Rothman, Potential Corrosion and Degradation Mechanisms of Zircaloy Cladding on Spent Nuclear Fuel in a Tuff Repository, Lawrence Livermore National Laboratory Report UCID-20172 (1984).

9. E. Hillner, D. G. Franklin and J. D. Smee, Long-term Corrosion of Zircaloy before and after Irradiation, *Journal of Nuclear Materials* **278**, 334 - 345 (2000).

10. The Japan Society of Mechanical Engineers, Zirconium Alloy Handbook (1997).

11. RWMC, Research and development of processing and disposal technique for TRU waste containing I-129 and C-14 (FY2012) (2013).

12. E. Hillner, Proceedings, Corrosion of Zirconium-Base Alloy-An Overview, Zirconium in the nuclear industry, ASTM STP 633 (1977).

13. Y. H. Jeong, J. H. Baek, S. J. Kim, H. G. Kim and H. Ruhmann, Corrosion characteristics and oxide microstructures of Zircaloy-4 in aqueous alkali hydroxide solutions., *Journal of Nuclear Materials* **270**, 322 - 333 (1999).

14. T. Kido, K. Kanasugi, M. Sugano and K. Komatsu, PWR Zircaloy cladding corrosion behavior: quantitative analyses, *Journal of Nuclear Materials* **248**, 281 - 287 (1997).

15. Y. Etoh, S. Shimada and K. Kikuchi, Irradiation Effects on Corrosion Resistance and Microstructure of Zircaloy-4, *Journal of nuclear Science and Technology*, **29**, p.1173-1183, (1992).

Mater. Res. Soc. Symp. Proc. Vol. 1665 © 2014 Materials Research Society
DOI: 10.1557/opl.2014.648

# Evolution of corrosion parameters in a buried pilot nuclear waste container in el Cabril

Carmen Andrade[1], Samuel Briz[1], Javier Sanchez[1], Pablo Zuloaga[2], Mariano Navarro[2] and Manuel Ordoñez[2]
[1]Institute of Construction Science "Eduardo Torroja" (IETcc), CSIC,
Serrano Galvache 4, 28033, Madrid, Spain.
[2]ENRESA (Spanish Agency for Management of Radioactive Wastes)
Emilio Vargas, 7, 28043, Madrid, Spain.

## ABSTRACT

Modern concrete has a record of good performance of around 120 years although there are structures in perfect conservation made with roman concrete (mixture of lime and natural pozzolans). El Cabril repository has a design life of 300-500 years and therefore, it should keep its integrity much longer than the back experience we have on reinforced concrete structures, which makes necessary a closer monitoring with time on the aging of concrete in real conditions. With this purpose, Enresa has designed in collaboration with IETcc and Geocisa the installation of permanent sensors in a pilot nuclear waste container in buried conditions. The sensors were installed in 1995 for monitoring corrosion parameters and have been working until present. The non-destructive tests (NDT) applied are based in electrochemical measurements (corrosion rate, corrosion potential, electrical resistivity, concrete strains, oxygen availability). Relations between the climatic influence, the buried depth and the corrosion parameters are also presented. The results indicate that temperature is a very relevant variable influencing the measurements. All the other parameters evolve according to seasonal changes. Values of activation energies of the resistivity changes are given although it seems more adequate to model the evolution with time by simply plotting the values registered at $20 \pm 2$ °C.

## INTRODUCTION

El Cabril repository has a design life of 300-500 years with three main periods: i) operational, during the construction, ii) surveillance, where a minimum of monitoring is considered, and iii) the post-surveillance, where no more control is expected. The main cement based structures used as engineering barriers in the repository of El Cabril for low and medium radioactive wastes are the cells, the containers and the mortar filling the gaps between the drums introduced in the containers. Cells and containers are made of the same concrete composition while the mortar was specifically designed to be pumpable and with high impermeability (between $10^{-17}$ and $10^{-18}$ m$^2$). The possible aggressions that the cement based materials can suffer during these periods have been identified to be: carbonation (during the operation only), water permeation (leaching) and reinforcement corrosion. More unlikely might be the bio-attack. Chlorides are not in the environment but they are inside the drums as part of analytical wastes. The description of the installations has already been reported before [1].

The design life of 300-500 years is longer than the existing experience on this kind of structures because the oldest structures in reinforced concrete were built about 120-150 years ago. Much older concretes have survived to present, but they are unreinforced with steel and based in mixes of lime and pozzolan which are not clean analogues of modern cements. In addition the concrete in El Cabril will be in contact with the atmosphere until it is finally buried. In order to gain experience on this buried conditions; a pilot container was built in 1994 and was instrumented with corrosion sensors (figure 1). In the present paper the results registered from the beginning are shown and, due the strong relative influence of temperature, an attempt to start to model the evolution with time is made by calculating the "apparent activation energy" of the different parameters measured. Only the case of the resistivity is mentioned in present paper.

**Figure 1**. Left: sensors attached to the reinforcement and the aspect of the buried container. Middle: the container being moved after fabrication and the sensors placed in the three levels of reinforcing. Right: instrumented drum before being placed inside the container.

## DESCRIPTION OF PILOT CONTAINER AND ITS INSTRUMENTATION

The containers were prefabricated in a specially devoted plant. Several views and its instrumentation are shown in figure 1. The reinforcements were placed in the modules and then, concrete was cast by robotic means, to be finally steam cured with temperatures lower than 60°C. Characteristic strengths between 50-60 MPa are typically achieved at 28 days. 6 drums, also shown in figure 1, were introduced in the container and the spaces filled with the "filling mortar" [1].

This pilot container was fabricated in 1994 and instrumented by embedding 27 sets of electrodes (figure 2). The container was buried leaving aside a chamber where the corrosion data logger (Geologger measurement system) and a reference probe were placed beside the container. All the details were described in a previous paper [2].

**Figure 2.** Position of the sensors in the container and in the drums (Grupo= Group of sensors).

## Instrumentation and Sensors

The Geologger potentiostat has 50 available channels. The parameters controlled are: temperature, concrete deformation, corrosion potential, resistivity, oxygen availability and corrosion rate. The impact of temperature on several of the parameters is remarkable, and therefore, care has to be taken when interpreting on-site results. Some of the sensors embedded are shown in figure 3.

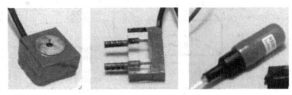

**Figure 3.** Electrochemical sensors embedded in the buried container. Left: sensor for corrosion rate and resistivity. Middle: sensor for resistivity (additional method). Right: sensor for corrosion potential with its cap.

For measuring the corrosion potential and the corrosion rate, direct metallic contact was made to a main rebar of the container or by removing the surface of the drum. The rebar and the drum were made to act as working electrodes. In order to obtain the real value of the corrosion rate, the measurement method used was the Polarization Resistance [3, 4]. However, as the surface of the working electrode is bigger than the counter (small disk) a technique based on the measurement of the slope of the transient pulse after application of a current step, has been used. This technique is not as accurate as the "sensorized guard ring" [5, 6]. However, as the technique is continuously recording, any scatter or wrong measurements can be easily identified. On the other hand, it has the advantage of being very quick, and it is a non-destructive technique. Obviously, a method based in the use of a guard ring around the small counter disk entails great difficulties in order to operate embedded in the concrete. Therefore, an alternative methodology had to be used in spite of the loss accuracy.

Regarding resistivity [7], it is measured by means of the current interruption method from a galvanostatic pulse. The oxygen flow at the rebar level is measured by applying a cathodic constant potential of about −750 mV(SCE) and measuring the current of reduction of oxygen [8].

## RESULTS

From the 27 groups of sensors installed, only less than 10% failed. The rest show a good response even ten years after their installation. As an example, figure 4 shows the temperature values from the end of 1994 for all the sensors embedded in the container. From the figure it is possible to deduce that the temperature evolves seasonally. Figure 5 gives the maximum, minimum and average values of certain groups placed in the top, medium and lower levels of the pilot container and that embedded in the reference specimen placed in the chamber that contains the Geologger beside the container.

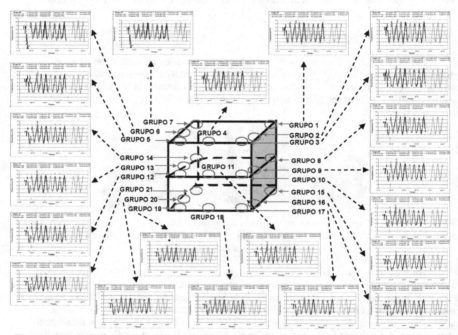

**Figure 4**. Evolution with time of temperature in all the sensors embedded in the pilot container. Each diagram represents an electrode with the temperature in the Y-axis and the time in the X-axis).

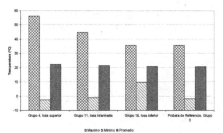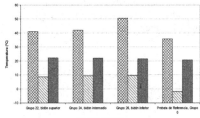

**Figure 5**. Maximum (in green), minimum (red) and average temperature(blue) in the three levels of height (see figure 2) of the container (group 4 and 22 are placed in the upper part, group 11 and 24 are placed in the intermediate height and group 18 and 26 in lower part of the container). The reference specimen (numbered 0) is placed in the chamber besides the container. Left figure gives the groups of the container. Right figure the groups in the drums.

These records indicate that, although the average values are similar in all levels, the maximum temperature is much higher in the top part reaching around 50°C while in the lower level the maximum is around 35°C. This behaviour is not shown however by the drums where the maximum temperatures do not show the same trend, being similar in the three levels.

Figure 6 shows the evolution of micro-strains and of oxygen availability. The micro-strains show initially the sharp shrinkage produced and its slower increase with time. The values will enable the modelling and prediction in time of this phenomenon that seems not to stop completely.

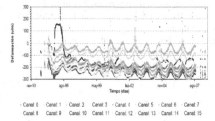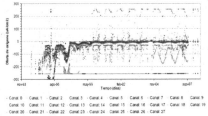

**Figure 6**. Evolution with time. Left: micro-strains (deformacion). Right: oxygen availability (oferta de oxigeno).

Regarding oxygen evolution, it also suffers and abrupt change during the first years of recording as it is relatively high at the beginning while the cathodic current is zero or even positive later. This behaviour is not clear as it would represent that an anaerobic ambient is being produced. However it can be analysed together with the values of corrosion potential and resistivity. The corrosion potential evolves slowly but steadily towards more positive values indicating that a lack of oxygen is not being produced (figure 7). The trend of the oxygen then may be justified with the increasing of the resistivity which can indicate and continuous decrease in the amount of liquid water in the pores, likely due to the progressive hydration. The lack of electrolyte would justify the almost zero or positive values of the oxygen availability. However this hypothesis has to be confirmed in the future when several different humidity conditions will be applied to the container.

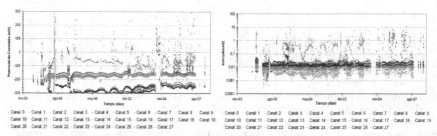

**Figure 7**. Evolution with time. Left: Corrosion Potential (potencial de corrosion). Right: Corrosion rate, $I_{corr}$.

As mentioned before the corrosion potential almost does not change although a slow progression towards more positive values is clear (figure 7). What is more interesting is to verify that the corrosion potential is different (more positive for around 100 to 150 mV) in the reinforcement embedded in the container than that of the steel of the drum embedded in the filling mortar. With respect to the corrosion rate, the values below 0.1 $\mu$A/cm$^2$ indicate a perfect passivity of the steel. It is worth noting however that the temperature seasonal cycles can be well identified in both the corrosion potential and rate.

Finally, regarding the resistivity, it increases with time and shows the effect of the seasonal evolution of temperature in a higher proportion and the time progresses (figure 8 Left). This has to be analysed together with the previous appreciation of the progressive hydration and then it would imply that a drier concrete will show more pronounced resistivity changes due to temperature than a more humid concrete. The increase with time of the resistivity is better appreciated in figure 8-Right, where the evolution of the resistivity values at 20±2 °C is plotted. This graph reveals that at the beginning the evolution is exponential but later a straight line can be plotted.

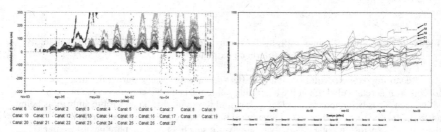

**Figure 8**. Evolution with time of resistivity (resistividad). Left: all values, and right: values between 18 and 22 °C.

## DISCUSSION

The existing models for predicting service life [9-12] are of different nature: empirical, analytical or numerical. None of the existing models for predicting corrosion of reinforcements have been validated during enough long time. In general all of them are based in considering diffusion as the main mechanism of transport of carbon dioxide and chlorides. Very scarce is the number of models considering water absorption in addition to diffusion [12]. For the sake of present results, it is the effect of temperature in the evolution of the different parameters measured what is studied and also its effect on possible dimensional changes (shrinkage). Only the preliminary analysis of the resistivity is made in present paper.

In figure 9 the representation of the resistivity versus the temperature is made. From the plot it can be calculated an "apparent activation energy": $A = A_0 e^{(-Ea/R \cdot T)}$. From the figure it is very interesting to note that at 20 °C there is a significant change in the slopes. All of them are shown in table 1. Below 20 °C the slopes do not change very much, although they show an increase in the resistivity with age. However above 20 °C the slopes change very significantly with time. It might be deduced that as the concrete is drier with time, the change in resistivity is more pronounced with temperature. This trend can be related to the oxygen availability and calls for further testing by producing artificially different humidity states in the container in order to verify the hypothesis.

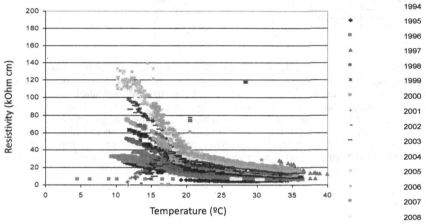

**Figure 9**. Resistivity values of group 3 sensors versus temperature that allows to calculate the Activation Energy.

**Table I.** Apparent Activation Energies, $E_{ap}$, of the different slopes shown in figure 9.

| N° | Eap > 20 °C (J/mol) | Eap < 20 °C (J/mol) | Average R2 Eap > 20 | Average R2 Eap < 20 |
|---|---|---|---|---|
| 1 | 31390.291 | 59968.961 | 0.888 | 0.793 |
| 2 | 78552.776 | 50004.315 | 0.899 | 0.881 |
| 3 | 37777.635 | 87358.217 | 0.949 | 0.904 |
| 4 | | | | |
| 5 | 39477.362 | 57462.029 | 0.929 | 0.941 |
| 6 | | | | |
| 7 | 35355.421 | 77513.328 | 0.674 | 0.814 |
| 8 | 30245.416 | 82646.018 | 0.972 | 0.927 |
| 9 | 34475.459 | 30095.728 | 0.854 | 0.765 |
| 10 | | | | |
| 11 | | | | |
| 12 | 38524.552 | 90782.394 | 0.957 | 0.962 |
| 13 | 40306.981 | 44634.395 | 0.912 | 0.870 |
| 14 | 33450.406 | 56639.846 | 0.925 | 0.894 |
| 15 | 39921.175 | 96788.769 | 0.889 | 0.885 |
| 16 | 76736.424 | 40940.460 | 0.961 | 0.921 |
| 17 | 129828.817 | 70651.644 | 0.946 | 0.906 |
| 18 | 58528.728 | 42445.518 | 0.932 | 0.895 |
| 19 | 133026.563 | 56083.054 | 0.955 | 0.958 |
| 20 | 126592.825 | 37470.356 | 0.965 | 0.879 |
| 21 | 91492.450 | 44405.932 | 0.967 | 0.952 |
| 22 | 39073.778 | 36962.698 | 0.962 | 0.870 |
| 23 | 29862.174 | 33838.515 | 0.928 | 0.857 |
| 24 | 24734.584 | 25095.363 | 0.704 | 0.793 |
| 25 | 30407.063 | 29919.905 | 0.816 | 0.831 |
| 26 | 27238.966 | 38195.922 | 0.932 | 0.929 |
| 27 | 37917.152 | 32925.933 | 0.502 | 0.439 |
| Average | 54126.826 | 53166.491 | 0.888 | 0.864 |
| Average total | 53646.659 | 53646.659 | 0.876 | 0.875 |

If the hypothesis is correct that would mean that the concrete continues to hydrate with time at expenses of the pore solution and then, it will become drier in spite of being buried. This drying will have an impact in the shrinkage, the resistivity and the oxygen available to be reduced at cathodic potentials. These facts should be incorporated into the models of service life.

With respect to the values of the slopes or Apparent Activation Energies, it has to be commented by looking at table 1, that the values are very high as the sensors embedded in the container show average values higher than 53.646 kJ/mol for temperatures higher than 20 °C while it is normal in electrolytes is to show values around 20 kJ/mol and other authors [13, 14] have measured this range of values, which only are recorded here in the drums.

## CONCLUSIONS

The establishment of a suitable model of service life of cementitious materials used in repositories of radioactive wastes has the main difficulty of the lack of proper calibration of existing models and their use for much shorter service lives. In addition, there are still important items to be measured and collected for the characterization of the environment and for the correct identification of key parameters involved in long term durability.

The results of several corrosion sensors embedded in a pilot container from 1994 indicate that:

1. The temperature in the concrete changes according to seasonal variations and the buried depth attenuates the extreme changes.
2. All the parameters evolve according to these seasonal changes which facilitates the analysis through the calculation by means of an "apparent activation energy", $E_{ap}$.
3. The $E_{ap}$ are in general around and much above 20 kJ/mol.
4. The resistivity shows a marked change of regime with temperature, below and above 20 °C, being the changes more pronounced above that temperature.
5. The evolution with time can be modelled by simply plotting the values registered at 20±2 °C.

## ACKNOWLEDGMENTS

Recognition is made to the funding provided by ENRESA to support the investigation. The authors would like to acknowledge the financing of the Ministry of Science and Innovation for the INGENIO 2010-CONSOLIDER Project on "Safety and Durability of Structures: SEDUREC" and BIA2010-18863.

## REFERENCES

1. C. Andrade, I. Martinez, M. Castellote, and P. Zuloaga, "Some principles of service life calculation of reinforcements and in situ corrosion monitoring by sensors in the radioactive waste containers of El Cabril disposal (Spain)," *Journal of Nuclear Materials,* vol. 358, pp. 82-95, Nov 2006.
2. C. Andrade, J. Rodriguez, F. Jimenez, J. Palacio, and P. Zuloaga, "Embedded sensors for concrete structures instrumentation," in *OECD-NEA Workshop on Instrumentation and Monitoring of Concrete Structures Brussels*, ed Belgium, 2000.
3. C. Andrade and C. Alonso, "Test methods for on-site corrosion rate measurement of steel reinforcement in concrete by means of the polarization resistance method," *Materials and Structures,* vol. 37, pp. 623-643, 2004.
4. C. Andrade, C. Alonso, J. Gulikers, R. Polder, R. Cigna, O. Vennesland, *et al.*, "Test methods for on-site corrosion rate measurement of steel reinforcement in concrete by means of the polarization resistance method," *Materials and Structures,* vol. 37, pp. 623-643, Nov 2004.
5. C. Andrade, J. Sanchez, J. Fullea, N. Rebolledo, and F. Tavares, "On-site corrosion rate measurements: 3D simulation and representative values," *Materials and Corrosion,* vol. 63, pp. 1154-1164, 2012.
6. C. Andrade and J. A. González, "Quantitative measurements of corrosion rate of reinforcing steels embedded in concrete using polarization resistance measurements," *Materials and Corrosion,* vol. 29, pp. 515-519, 1978.
7. K. R. Gowers and S. G. Millard, "Measurement of concrete resistivity for assessment of corrosion severity of steel using Wenner technique," *Aci Materials Journal,* vol. 96, pp. 536-541, 1999.

8.     O. E. Gjorv, O. Vennesland, and A. H. S. Elbusaidy, "Diffusion of dissolved-oxygen through concrete," *Materials Performance,* vol. 25, pp. 39-44, Dec 1986.

9.     D. J. Naus, M. W. Johnston, C. Andrade, H. Ashar, Z. Bittnar, H. Breulet, *et al.,* "RILEM TC 160-MLN: Methodology for life prediction of concrete structures in nuclear power plants - Progress report - August 1999," *Materials and Structures,* vol. 33, pp. 98-100, Mar 2000.

10.    J. R. Clifton, "Predicting the service life of concrete," *Aci Materials Journal,* vol. 90, pp. 611-617, Nov-Dec 1993.

11.    J. Rodriguez, L. M. Ortega, J. Aragoncillo, D. Izquierdo, and C. Andrade, *Structural assessment methodology for residual life calculation of concrete structures affected by reinforcement corrosion* vol. 16, 2000.

12.    P. Zuloaga, M. Ordonez, C. Andrade, and M. Castellote, "Ageing management program for the Spanish low and intermediate level waste disposal and spent fuel and high-level waste centralised storage facilities," in *Amp 2010 - International Workshop on Ageing Management of Nuclear Power Plants and Waste Disposal Structures.* vol. 12, V. Lhostis, K. Philipose, R. Gens, and C. Galle, Eds., ed, 2011.

13.    M. Koster, J. Hannawald, and W. Brameshuber, "Simulation of water permeability and water vapor diffusion through hardened cement paste," *Computational mechanics,* vol. 37, pp. 163-172, 2006.

14.    M. Raupach, "Results from laboratory tests and evaluation of literature on the influence of temperature on reinforcement corrosion," in *Corrosion of Reinforcement in Concrete: Monitoring, Prevention and Rehabilitation,* J. Mietz, B. Elsener, and R. Polder, Eds., ed, 1998, pp. 9-20.

Mater. Res. Soc. Symp. Proc. Vol. 1665 © 2014 Materials Research Society
DOI: 10.1557/opl.2014.649

# Towards a more Realistic Experimental Protocol for the Study of Atmospheric Chloride-Induced Stress Corrosion Cracking in Intermediate Level Radioactive Waste Container Materials

A.B. Cook[1], B. Gu[1], S.B. Lyon[1], R.C. Newman[2] and D.L. Engelberg[1, 3]

[1]Corrosion and Protection Centre, School of Materials, The University of Manchester, Oxford Road, Manchester M13 9PL, UK.

[2]Department of Chemical Engineering and Applied Chemistry, University of Toronto, 200 College Street, Toronto, Ontario, M5S 3E5, Canada.

[3]Research Centre for Radwaste and Decommissioning, The University of Manchester, Oxford Road, Manchester M13 9PL, UK.

## ABSTRACT

The occurrence of Atmospheric chloride-Induced Stress Corrosion Cracking (AISCC) under wetted deposits of $MgCl_2$ or sea-salt at 70°C has been investigated at various Relative Humidities (RH). The appearance of AISCC is a function of the environmental RH. At 33% RH (the deliquescence point of $MgCl_2$), AISCC generated under $MgCl_2$ or sea-salt deposits is of a similar appearance with regards to the number of cracks produced and average crack length. At 50% RH sea-salt seems to be more aggressive at least in terms of crack frequency. This observation may highlight the significance of carnallite ($KMgCl_3.6H_2O$) in promoting AISCC in types 304L and 316L stainless steels. The use of accelerated testing methods to validate apparent thresholds in chloride deposition density and other critical factors that influence the initiation and propagation of AISCC is briefly discussed.

## INTRODUCTION

Prior to disposal in a purpose built Geological Disposal Facility (GDF), the UK's Intermediate Level radioactive Waste (ILW) is expected to be housed for many decades in interim surface stores, within containers fabricated from Types 304L and 316L austenitic stainless steel. Many of these storage facilities are located close to marine environments; hence it is likely that the outer surfaces of ILW containers in these locations will be subject to deposition of particulate matter from sea-salt aerosol. Sea-salt deposits contain chloride species that deliquesce at and remain wet above certain values of RH producing a thin-layer chloride-containing electrolyte, in which localized corrosion (pitting and/or crevice attack) may be supported. Should any such corrosion phenomena develop in areas containing sufficiently high tensile stress then a transition to AISCC may develop. Knowledge of the environmental parameter space in which AISCC initiates, and subsequently propagates is, therefore, of particular importance.

The most often cited article regarding the occurrence of room temperature AISCC in austenitic stainless steel is that of Shoji and Ohnaka [1], although it should be noted that an observation of AISCC in type 302 stainless steel under wetted zinc chloride deposits at 40°C (well below the then considered threshold of 60°C) was reported by Truman and Pirt [2] several years earlier. Nevertheless, the former workers were the first to demonstrate the occurrence of AISCC in 304L and 316L at ambient temperature and provided valuable information regarding the RH range over which wetted deposits of artificial sea-salt and its various chloride-containing constituents produce cracks [1]. Since this work there have been numerous studies of AISCC in austenitic stainless steels [3-8]. It is probably fair to say that the experimental conditions employed in most of these studies have not been specifically tailored to the needs of the nuclear industry and vary widely in terms of what may be regarded as the critical factors affecting the initiation and propagation of AISCC, such as the nominal tensile stress, chloride type and deposition density, environmental RH and temperature. For instance, in many laboratory investigations U-bend specimens are employed to provide the nominal tensile stress, $MgCl_2$ is often chosen as a surrogate for sea-salt and the chloride salt deposition density and temperatures are often far higher than those likely to be encountered during storage of ILW. As such it may be realistically argued that conditions used in laboratory tests are far more aggressive than those likely to develop in store.

Assessing container integrity during interim storage of ILW clearly require a means of predicting the tendency of ILW containers towards localized corrosion and AISCC over a wide range of different environmental conditions. Knowledge of the effects of changes in material and environmental variables on mechanistic and kinetic quantities such as the induction time (time to cracking) and crack growth rate will be crucial to the development of testing methods and/or models to predict damage evolution in the storage condition over the long timescales expected to be associated with interim storage. In this paper we discuss environmental and materials issues pertinent to the development of AISCC, consider the relative merits of using accelerated testing methods and present some initial results with regards to the suitability of using data obtained using $MgCl_2$ to predict the Cl⁻ threshold for AISCC in the presence of sea-salt. We also consider the likelihood of the sea-salt evaporate, carnallite ($KMgCl_3.6H_2O$), playing a significant role in promoting AISCC in 304L and 316L stainless steels.

## EXPERIMENTAL DETAILS

### Materials

Solution annealed types 304L and 316L austenitic stainless steels obtained from Goodfellows Metals Cambridge as 0.914 mm thick sheet were used in this work. Their chemical compositions are given in Table 1. U-bend specimens were prepared according to ASTM G30-97 [8] with a bending radius of 9 mm from rectangular specimens with length 120 mm and width 15 mm. Prior to U-bend fabrication the metal strips were abraded to 1200 grit with silicon carbide paper and degreased ultrasonically in acetone and then ethanol to remove surface contamination.

**Table I.** Steel Composition (wt%)

|  | C | Si | Mn | P | S | Ni | Cr | Mo | Al | Cu | Nb | Ti | V | Fe |
|---|---|---|---|---|---|---|---|---|---|---|---|---|---|---|
| 304L | 0.025 | 0.51 | 1.28 | 0.028 | 0.003 | 7.79 | 17.87 | 0.43 | 0.01 | 0.24 | 0.02 | 0.02 | 0.09 | Balance |
| 316L | 0.022 | 0.47 | 1.32 | 0.032 | 0.003 | 9.67 | 16.43 | 2.18 | 0.01 | 0.45 | 0.03 | 0.01 | 0.09 | Balance |

### Environmental Exposure

Two nominally dry deposits of either artificial sea-salt or $MgCl_2$ were placed on the outer apex of each of the U-bends from evaporation at room temperature of droplets applied from 5 µl aliquots of artificial sea water or 10.9 g dm$^{-3}$ MgCl$_2$.6H$_2$O. This value was chosen to provide approximately the same amount of MgCl$_2$ per unit area on drying from both MgCl$_2$.6H$_2$O and artificial sea water. The value was calculated on the assumption that the concentration of Mg$^{2+}$ in sea water is approximately 0.054 mol dm$^{-3}$. The application technique produced surface droplets of ca. 3 mm diameter with good reproducibility and a nominal MgCl$_2$ deposition density of 360 μg cm$^{-2}$ (270 μg cm$^{-2}$ Cl$^-$). Chloride-laden U-bends were placed within air-tight vessels containing a reservoir of a saturated salt solution to maintain the required RH. These containers were subsequently placed in an oven set to control the temperature at 70°C (this temperature was chosen to produce results over a short timescale rather than because it reflects the likely value for the in store condition). Three values of environmental RH were employed in this work: namely; 33% RH (maintained with saturated MgCl$_2$ – slightly higher than the expected RH of 27% [8]), 50% RH (maintained with saturated NaBr [8]), and 75% RH (maintained with saturated NaCl [8]). The environmental RH and temperature were monitored using a Lascar temperature and RH probe. Optical images of the exposed surfaces were obtained at regular intervals over the test period of seven weeks using an Olympus SZH Stereo Zoom Microscope. During imaging specimens were removed from their environment for no more than 30 minutes.

**RESULTS AND DISCUSSION**

The occurrence of AISCC in types 304L and 316L stainless steels is known to be influenced by a wide range of environmental and materials variables, including:

- The type of chloride salt – AISCC at temperatures less than 60°C has been observed under deposits of artificial sea-salt and Mg, Ca, Li and Zn chlorides [1,3,4].
- Chloride deposition density – there is some evidence that there exists a threshold chloride deposition density below which AISCC does not initiate [7].
- Environmental RH and hence the electrolyte concentration – AISCC has been observed over a wide range of RH values; the RH range over which AISCC occurs is, however, temperature dependent with a narrower range at ambient temperature than at elevated temperature [1,3,4,6].
- Environmental temperature – affects the extent, severity and the RH range over which AISCC occurs [1].
- Heat treatment and surface finish [6].

To our knowledge the only work to date that is specifically designed to predict damage accumulation as a result of AISCC in ILW containers during storage is the ACSIS (Atmospheric Corrosion of Stainless Steel In Stores) model of King and co-workers [10]. This model uses information regarding in-store time evolution of the environmental RH and temperature together with information regarding the nature of the contaminant and chloride deposition density to determine whether pitting and/or AISCC will initiate during times of wetness. In instances where initiation is predicted to occur, the extent of damage in terms of pit, crevice or crack depth is calculated. The model assumes AISCC will only occur above a certain threshold value of the chloride deposition density and over a defined range of RH. Threshold values for the chloride deposition density are taken from the work of Albores-Silva et al. [7] who reported that U-bend specimens of 316L taken from an ILW container, loaded with MgCl$_2$, were immune to AISCC below 10 μg cm$^{-2}$ at 50°C and below 25 μg cm$^{-2}$ at 30°C in tests conducted at 30% RH over a 12 week period; higher threshold values were observed at 60% RH. In this work the time to cracking (induction time) at each deposition density was not monitored, hence it is possible that AISCC was not observed below the threshold values because the tests were not conducted over a sufficiently long timescale. For instance Cook et al. [5] have observed an average induction time at 40°C of ca. 13 weeks in 304L [5] and ca. 20 weeks in 316L [11] tensile specimens loaded to approximately yield stress at a chloride loading of 340 $\mu g$ $cm^{-2}$. At such a deposition density a U-bend specimen taken from an ILW drum would, according to the work of Albores-Silva et al. [7] have developed AISCC within a 12 week period; it is suspected that the longer induction time observed by Cook et al. [5] is probably the result of the effect of the lower applied stress on the induction time. Long induction times may also be inferred from the work of Shoji and Ohnaka [1]. For example, cracks produced at room temperature and 30% RH in U-bends of 304L and 316L under wetted $MgCl_2$ deposits at a $Cl^-$ deposition density of ca.1300 $\mu g$ $cm^{-2}$ display minimum induction times of between 4-8 weeks and 8-12 weeks, respectively. Minimum induction times under sea-salt deposits may be inferred as between 4-8 weeks for 304L and between 6-12 months for 316L. As such it may be prudent to consider the development of accelerated testing methods; that is, to work at temperatures above that expected in store to complement and guide work under more realistic storage conditions. In this way temperature dependent trends in induction times and crack growth rates as a function of various critical experimental parameters may be obtained over relatively short timescales. The accuracy of accelerated testing in predicting these parameters under storage conditions may then be examined, by performing a range of more representative experiments at lower temperature for comparison. It should also be considered whether thresholds based on chloride-salt deposition density conducted under controlled conditions in a laboratory situation is actually a suitable means of predicting any tendency towards AISCC in an actual storage situation. The parameter simply represents an average amount of salt deposited per unit area. It does not necessarily provide an accurate measure of the dimensions of any contaminated areas (unless of course salt contamination is uniform) or the size of individual salt-deposits therein and hence the dimensions or continuity of any atmospherically produced electrolyte layer.

The use of a single salt, such as $MgCl_2$, to assess the potential for AISCC in the storage environment rather than sea-salt is another practice that may not be altogether appropriate since the thin layer electrolyte formed on wetting sea-salt at any particular RH and temperature will likely have a far more complex chemistry. For example, during drying prior to exposure testing sea-water becomes more concentrated and solid phases will precipitate as their solubility is exceeded. The least soluble species will precipitate first, usually the calcium containing salts ($CaCO_3$, $CaMgCO_3$, $CaSO_4.2H_2O$ and $CaSO_4$), followed by halite (NaCl) then the potassium and magnesium containing salts such as carnallite ($KMgCl_3.6H_2O$) and finally bischofite ($MgCl_2.6H_2O$). At 25°C halite wets to form a saturated solution at around 75% RH, carnallite at 59% RH and bischofite at 33% RH [12], hence it is likely that sea-salt deposits will wet, to some degree at least, over a large RH range. In previous work [13] we have shown using EDX analysis, that on drying artificial sea-salt forms discrete crystals of NaCl that are surrounded by amorphous looking Mg and Ca salts. On increasing the environmental RH, the areas where $MgCl_2$ is prevalent wet first, at ca. 35% RH, followed by the NaCl crystals at ca. 70% RH.

To investigate the suitability of using $MgCl_2$ to promote AISCC rather than sea-salt, U-bend specimens laden with either artificial sea-salt or $MgCl_2$ salt deposits were exposed to RH values of 33, 50 and 75% at 70°C. Optical micrographs highlighting typical differences in observed AISCC behaviour of 304L under these conditions are shown in Figure 1. At 33% RH both sea-salt and $MgCl_2$ deposits (Figure 1A and B) were found, on average, to produce a similar number of cracks (2 – 3) per deposit. At 50% RH, however, sea-salt appeared to be more aggressive species towards AISCC than $MgCl_2$, at least in terms of the apparent crack frequency (Figure 1D) – typically producing a larger number of shorter cracks than observed in the presence of $MgCl_2$ (Figure 1C). As 50% RH is likely to be very close to the deliquescence point of carnallite at the environmental temperature we have employed, our observations suggests that this particular sea-salt evaporite may be of importance in promoting AISCC under certain conditions of RH and temperature. At 75% RH no AISCC was observed in any specimen; at this RH the dominant cationic solution species is $Na^+$. Tests on 316L produced similar results. As only three specimens were used in each condition a caveat should be inserted here inasmuch as further work is required to ensure that observations like the one depicted in Figure 1D are statistically valid. It should also be noted this work was performed at 70°C and it must not automatically be assumed that the same trends will be observed at 40°C or lower. Work is currently in progress to investigate whether or not the same behaviour is observed at lower temperature.

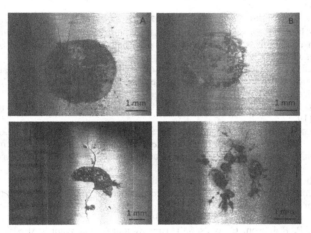

**Figure 1.** Optical micrographs of AISCC that develops in the presence of wetted $MgCl_2$, at a nominal deposition density of 360 µg cm$^{-2}$, (A and C) and artificial sea-salt (C and D) deposits at 70°C after five weeks of exposure testing at: A 33% RH; B 33% RH; C 50% RH; D 50% RH.


## CONCLUSIONS

The appearance of AISCC that develops under wetted deposits of $MgCl_2$ (deposition density 360 µg cm$^{-2}$) or sea-salt at 70°C is a function of the environmental RH. At 33% RH (the deliquescence point of $MgCl_2$), both sea-salt and $MgCl_2$ generate AISCC of a similar appearance with regard to the number of cracks and average crack length, suggesting that a $MgCl_2$ electrolyte controls cracking in each environment. However at 50% RH (close to the deliquescence point of $KMgCl_3$), sea salt seems to be more aggressive at least in terms of the number of cracks produced. This observation may be due in part to the way in which $MgCl_2$ is distributed in the sea-salt deposits, but more likely highlights the significance of $KMgCl_3$ in promoting AISCC in 304L and 316L stainless steels. Most significantly the result indicates that $MgCl_2$ is not necessarily a good substitute for sea-salt deposits at a RH of 50%.

## ACKNOWLEDGMENTS

DLE for financial support by EPSRC grant EP/I036397/1 and NDA contract NPO004411A-EPS02.

## REFERENCES

1. S. Shoji and N. Ohnaka, Boshoku Gijutsu (Corrosion Engineering JP, English Edition), **38** (1989) 111.
2. J. Truman and K. Pirt, Corros. Sci., **17** (1977) 71.
3. A. Iversen and T. Prosek, "Atmospheric stress corrosion cracking of austenitic stainless steels in conditions modeling swimming pool halls", Eurocorr 2007, EFC, Freiburg im Breisgau, Germany, Paper No. 1142.
4. T. Prosek, A. Iversen and C. Taxen, "Low Temperature Stress Corrosion Cracking of Stainless Steels in the Atmosphere in Presence of Chloride Deposits", Corrosion 2008, New Orleans, LA, Paper No. 08484.
5. A. Cook, J. Duff, N. Stevens, S. Lyon, A. Sherry and T.J. Marrow, ECS Transactions, **25** (2010) 119.
6. N.D. Fairweather, N. Platts and D.R. Tice, "Stress corrosion crack initiation in type 304 stainless steel in atmospheric environments containing chloride: Influence of surface condition, relative humidity, temperature and thermal sensitization", Corrosion 2008, New Orleans, LA, Paper No. 08485.
7. O.E. Albores-Silva, E.A. Charles and C. Padovani, CEST, **46** (2011) 124.
8. ASTM International G 30 – 97, "Standard practice for making and using U-Bend stress-corrosion test specimens" (2009).
9. L. Greenspan, J. Res. NBS., **81A** (1977) 89.
10. F. King, P. Robinson, C. Watson, J. Burrow and C. Padovani, "ACSIS – A model to Assess the Potential for Atmospheric Corrosion of Stainless Steel ILW Containers During Storage and the Operational Phase of a UK Geological Disposal Facility", Corrosion 2013, Orlando, FL, Paper No. 02717.
11. A. Cook, unpublished results.
12. R.S. Lillard, D.G. Kolman, M.A. Hill, M.B. Prime, D.K. Veirs, L.A. Worl and P. Zapp (and references contained therein), Corrosion **65**, (2009) 175.
13. A. B. Cook, N. Stevens, J. Duff, A. Mschelia, T. S. Leung, S. Lyon, J. Marrow, W. Ganther and I. S. Cole, "Atmospheric-Induced Stress Corrosion Cracking of Austenitic Stainless Steels under Limited Chloride Supply", Paper No. 427, 18th International Corrosion Congress, 2011, Perth, Australia.

# High level waste

Mater. Res. Soc. Symp. Proc. Vol. 1665 © 2014 Materials Research Society
DOI: 10.1557/opl.2014.650

## Fast/Instant Radionuclide Release: Effects inherent to the experiment

B. Kienzler, A. Loida, E. González-Robles, N. Müller, V. Metz
Karlsruhe Institute of Technology (KIT), Institute for Nuclear Waste Disposal (INE),
Hermann-von-Helmholtz Platz 1, D-76344 Eggenstein-Leopoldshafen, Germany

## ABSTRACT

The release of radionuclides measured during washing cycles of spent nuclear fuel samples in a series of experiments using different solutions are analyzed with respect to the fission products Cs, Sr, and Tc and the actinides U, Pu, and Am. Based on the concentrations of the dissolved radionuclides, their release rates are evaluated in terms of fraction of inventory in the aquatic phase per day. The application of this information on the fast/instant radionuclide release fraction (IRF) is discussed and following issues are addressed: Duration of the wash steps, solution chemistry, and radionuclide sorption onto surface of the experimental vessels. Data for the IRF are given and the correlation between the mobilization of the various elements is analyzed.

## INTRODUCTION

A recent publication by Johnson et al. [1] summarized results on the release of $^{137}$Cs and $^{129}$I from high burn-up $UO_x$ and MOX fuels in the range between 58 and 75 MWd/kgU over leaching periods of ~100 days. According to this publication, the ratio of fission gas release (FGR) to the fractional release of $^{137}$Cs and $^{129}$I was related to the power history of the spent nuclear fuels (SNF) and resulting heterogeneities in the SNF. Johnson et al. concluded, increased power ratings resulted in higher releases of fission gases, $^{137}$Cs and $^{129}$I. Still, the contribution of the rim region (fission gas , $^{137}$Cs and $^{129}$I accumulation) and the contribution of grain boundary release remains unclear. Besides these conclusions related to fuel inherent properties, the authors addressed effects caused by the experimental set-up:
- The preparation of the SNF samples significantly influenced the determination of IRF / FGR, e.g. experiments with SNF fragments resulted in relatively high IRF values.
- Presence of cladding lead to an underestimation of the IRF.

In previous studies a relatively small fraction of $^{90}$Sr was found to be fast released in comparison to the IRF of $^{137}$Cs and $^{129}$I, e.g., [2-5]. For evaluation of SNF leaching experiments, in many cases the $^{90}$Sr release is used as an indicator for the dissolution of the $UO_x$ matrix of the fuel. Since the IRF of $^{90}$Sr depends on the power history of the studied fuel, it has to be discussed, whether $^{90}$Sr is an adequate proxy for $UO_x$ matrix dissolution of the specific SNF under investigation.

The EURATOM FP7 Collaborative Project "Fast / Instant Release of Safety Relevant Radionuclides from Spent Nuclear Fuel (CP FIRST-Nuclides)" addresses the influence of experimental conditions on IRF determination and the relation between uranium release from the SNF matrix and $^{90}$Sr release, beyond other issues. During discussions of the project work program with waste management organizations the question of the effects of solution chemistry, especially of higher salinity, on the fast / instant release processes, were brought forward. A comparison of the long-term release of radionuclides in different solutions have been published

[6]. As the dissolution based fast release investigations in CP FIRST-Nuclides are performed with various SNF samples in solutions of low ionic strength, the existing data from previous investigations with one SNF type are re-evaluated in the present contribution.

## MATERIALS

Considering the published literature, dissolution based radionuclide release studies have been performed mainly in different types of artificial groundwater and using a wide variety of SNF samples with respect to enrichment, burn-up, linear heat generation rate and power history characteristics. In contrast, the experiments to be discussed in this contribution, were conducted with SNF pellets from a single fuel rod segment denoted as SBS 1108 N0203. The additive-free $UO_2$ fuel was fabricated by Kraftwerk Union AG using the short-term fast sintering "NIKUSI" process ([7] [8]). The fuel rod segment was irradiated together with the adjacent segment N0204 in the pressure water reactor of Gösgen, Switzerland and the characteristic data are listed in table I (Metz et al., 2013 [9],).

**Table I.** Characteristic of SNF segments of KKG fuel rod SBS 1108.

| | |
|---|---|
| average burn-up (MWd/kg$_{HM}$) | 50.4 |
| average linear heat generation rate (W/cm) | 260 |
| maximum linear heat generation rate (W/cm) | 340 |
| number of cycles | 4 |
| initial rod diameter (mm) | $10.75 \pm 0.05$ |
| initial radial gap (mm) | 0.17 |
| cladding wall thickness (mm) | 0.725 |
| initial enrichment % $^{235}U$ | 3.8 |
| discharge date | 27/05/1989 |
| duration of irradiation (days) | 1226 |

Pellets were cut from this fuel rod. The incisions were located at the gaps between the pellets. Complete pellets (in some cases half pellets) together with the zircaloy cladding were used for leaching experiments. The pellets were stored for several years under inert gas conditions in metal casks before starting the experiments denoted as K1, K2, K5, K6, K8, K9, K11a, K11b, K14, and K18/19. In the case of K9, the experiment was directly started after preparation of the pellet. The leaching experiments covered the effects of solid, aqueous and gaseous system components on SNF dissolution, such as the effects of magnetite, chloride and hydrogen. Prior to the start of a static phase of the experiments, the pellets were pre-treated in so-called washing cycles. Characteristic data of the wash cycles are listed in Tab. II. In the present work, results obtained during the wash cycles are discussed.

The washing solutions were prepared in presence of air and then bubbled with Ar for several hours. Solutions for K11a/b were prepared under Ar atmosphere. The wash cycles were performed under Ar atmosphere with the same leachant as in the static experiments (i.e. deionized water, DIW, NaCl brine and Evolved Cement Water, ECW). After each washing cycle, the solution was completely exchanged and analyzed.

**Table II.** Evaluated wash cycles under inert gas conditions (Ar atmosphere).

| Pellet | Solution | Specifics | Number of wash cycles | Duration (days) | pH$_m$[e] | Eh (mV) |
|--------|----------|-----------|----------------------|-----------------|-----------|---------|
| K1 | DIW | | 2 | 50 | 6.41 | +264 |
| K2 | DIW | | 2 | 50 | 6.50 | +261 |
| K5 | 5.6 m NaCl | magnetite | 2 | 4 | 10.5 | -386 |
| K6 | 5.6 m NaCl | HA[#] | 5 | 85 | 7.2 | +420 |
| K8 | 5.6 m NaCl | | 7 | 293 | 7.1 | +385 |
| K9 | 5.6 m NaCl | | 2 | 42 | 6.6 | +295 |
| K11a | ECW[*] | ½ pellet | 3 | 76 | 12.5 | |
| K11b | ECW[*] | ½ pellet | 3 | 76 | 12.5 | |
| K14 | 5.6 m NaCl | magnetite | 4 | 106 | 8.0 | +500 |
| K18/19 | 5.0 m NaCl | 2 pellets | 3 | 9 | 7.5 | |

* Evolved Cement Water: Na$^+$ 16 mmol/l, K$^+$ 0.2 mmol/l, Ca$^{2+}$ 15 mmol/l [10].
# Hydroxylapatite.
[e] Molal H$^+$ concentrations (pH$_m$) are derived from measured pH using the method of Altmaier et al. [11]

## RESULTS

In this chapter, several notations are used: fission gas release (%), concentrations (mol dm$^{-3}$), FIAP (fraction of inventory in the aquatic phase) and release rates (FIAP per day). The FIAP is obtained by adding the measured concentrations of each wash step dividing by total inventory of the respective element/isotope. The release rate is determined from the incremental FIAPs divided by the duration of each wash step. For detailed explanations see [12].

### Fission gas release (FGR)

The FGR from the adjacent fuel rod segment SBS 1108 N0204 was determined. It was calculated as ratio (in %) from the experimentally determined amount of Xe and Kr extracted from the segment and the total inventory of fission gases generated in the fuel over the irradiation time (calculated using the KORIGEN code). The result of the fission gas release was found to be 7.0 % of total Kr, and 8.5 % of total Xe being the total fission gas release (Xe+Kr) 8.4 % [13].

FGR measurements for other PWR fuels with an average burnup of ~50 MWd/kg$_{HM}$ (e.g. Ringhals 2 and 4 [14] and Vandellos (Roth and Puranen, 2013 [9])) show mean values of about 2–8% ranging up to 15 %. It was found that the FGR mainly depends on the linear power of the fuel rod. In the case of Ringhals 2, Ringhals 4 and Vandellos fuels, the average linear power rate was 220.5, 178.3 and 136 W/cm, respectively, whereas the Gösgen SNF reached a higher value of about 260 W/cm. The measured FGR of the Gösgen SNF is in the expected range for PWRs.

### Fission products (FP)

In figures 1 - 3 the measured release rates (FIAP per day) of $^{137}$Cs, $^{90}$Sr, $^{99}$Tc are plotted as a function of time. The release rate of $^{99}$Tc in DIW (experiments K1 and K2) could not be determined due to the fact that the concentration in solution was below the detection limit. Figs. 1 - 3 show that most curves have similar time dependences, starting from fast release rates that decrease considerably within about 50 days. After the first seven weeks, release rates decrease much less. The release rates in high ionic strength solutions for $^{137}$Cs and $^{90}$Sr decrease to $1 \times 10^{-5}$ FIAP/day. In the case of $^{99}$Tc, the asymptotic value is about one order of magnitude lower. In regards to other leachants, the release rates reveal different behaviours. The $^{137}$Cs release rates in ECW and in concentrated NaCl solutions show the same trends. On the contrary, the $^{90}$Sr release rates in ECW dropped about 1.5 orders of magnitude below the NaCl data.

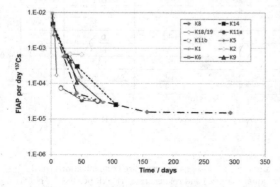

**Figure 1.** $^{137}$Cs release rate as function of time

In the experiments K5, K6 and K14 adsorbing substrates (magnetite or hydroxylapatite) were already present during the washing steps which did not influenced significantly the release rates. The fact that K9 was started directly after sample preparation did not change the release rates of $^{137}$Cs and $^{99}$Tc (only one data point available).

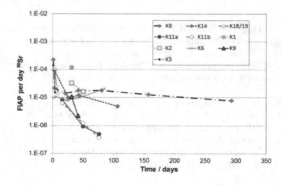

**Figure 2.** $^{90}$Sr release rate as function of time.

## Actinides

Figures 4 - 6 show the measured concentrations (mol dm$^{-3}$) of U, Pu, and Am in the wash cycles. The reason for showing concentrations instead of FIAP or FIAP per day is attributed to the potential sorption on to vessel walls or precipitation of solids phases of the actinides. In few cases the concentrations in solution of U, Pu and Am were found below the detection limits; these measurements are not included in the figures.

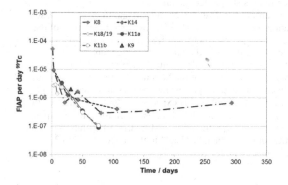

**Figure 3.** $^{99}$Tc release rate as function of time.

The U concentration decrease in within the first 50 days in the experiments K6, K8, K14 and K18/19 in concentrated NaCl solution (figure 4). In presence of magnetite (K6) and hydroxyapatite (K14), similar U concentrations were measured as in the experiments in NaCl brine without these solids (K8 and K18/19). The concentrations suggest that U is in the hexavalent state. Solubility of U(VI) in the NaCl systems is controlled by formation of meta-schoepite below $pH_m = 7$ and by formation of $Na_2U_2O_7{:}H2O$ for pH values above this value, resulting in U(VI) concentrations lower than $1 \times 10^{-6}$ molal [15]. In the K8 experiment the initial pH was 6.4 (measured), it increased to 7.5 after 21 days and decreased afterwards to 6.3 until 293 days. It is suggested that a meta-schoepite / $Na_2U_2O_7{:}H2O(s)$ phase controlled the decrease in the aqueous U concentration. For experiment K9 only one data point is available showing a significantly lower concentration in comparison to the other NaCl experiments. Finally, the concentrations of U in solution are close to the meta-schoepite saturation [16] in deionized water at pH 6.5 (K1 and K2). The lowest U concentrations were observed in the experiments K11a and K11b conducted in ECW at $pH_m = 12.5$ (figure 4).

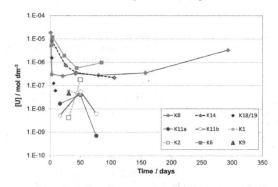

**Figure 4.** Measured U concentrations in the aquatic phase as function of time. After 293 days: FIAP(K8): $2 \times 10^{-4}$ FIAP(K8) per day: $2 \times 10^{-7}$

The measured plutonium concentrations are plotted as function of time in figure 5. In experiments K11a and K11b, Pu concentration are below detection limit and thus not discussed in the following. Though Pu concentrations scatter by several orders of magnitude, most of the results show a decrease of the concentrations, except experiment K8 with a "long-term" wash cycle, in which the Pu concentration increased. In the studied $pH_m$ range (see table II), concentrations of dissolved Pu(IV) species are expected to be controlled by the solubility of Pu(IV) intrinsic colloids below $10^{-8}$ M, largely independent on the ionic strength and $pH_m$. Due to the high dose rate in the SNF experiments, oxidizing species may be present in the solution forming Pu(V). In the $pH_m$ range of the experiments, $PuO_2^+$ is expected to dominate the aquatic Pu speciation whereas $PuO_{2+x}(s,hyd)$ is the solubility limiting solid phase. In the range $6 \leq pH_m \leq 9$, the concentration of Pu(V) is between $10^{-5.5}$ ($pH_m = 6$) and $10^{-8}$ M ($pH_m = 9$). Solubility experiments by Rai et al. [17] using $PuO_2$(am, hyd) in 4.4 m NaCl in the range $pH_m$ 5 to 8 showed only a slight dependence on the ionic strength.

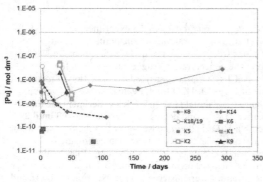

**Figure 5.** Measured Pu concentration in the aquatic phase as function of time ($^{238}Pu+^{239/240}Pu+^{241}Pu$). After 293 days: FIAP(K8): $6\times10^{-5}$ FIAP(K8) per day: $1\times10^{-7}$

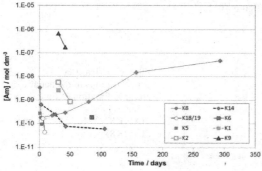

**Figure 6.** Measured Am concentration in the aquatic phase as function of time. After 293 days: FIAP(K8): $6\times10^{-4}$ FIAP(K8) per day: $3\times10^{-6}$

Measured Am concentrations scatter by several orders of magnitude (figure 6). All experiments - with the exception of the "long-term" wash cycle (K8) - showed a decrease in the Am concentration with time. The interpretation of the observed differences in concentrations between the experiments in terms of solubility equilibria is complicated, because the solubility of $Am(OH)_3(am)$ varies by orders of magnitude in the relevant pH range ($6 \leq \log m_{H+} \leq 9$). Furthermore, the solubility of $Am(OH)_3(am)$ depends strongly on the ionic strength. Thermodynamic data of Am(III) in the NaCl system was developed by Neck et al. [18]. In the considered $pH_m$ range, the $Am(OH)_3(am)$ solubility is above $10^{-1}$ M at $pH_m = 6$ and around $10^5$ M for $pH_m = 9$. Since the observed Am concentrations are significantly below the $Am(OH)_3(am)$ solubilities at the respective pH values, we hypothesise that sorption of Am(III) onto surfaces of the vessel wall may control the decrease in aqueous Am concentrations.

## DISCUSSION

The goal of the wash-cycles of the studied experiment was rather the removal of oxidized actinide / technetium surface layers and the removal of fast released radionuclides from the SNF samples than the determination of radionuclide release rates or IRF. The extensive data base of radionuclide measurements in the wash cycles of experiments K1, K2, K5, K6, K8, K9, K11a, K11b, K14, and K18/19 are used to deal with following questions:

- Does the solution chemistry, in particular the ionic strength, affect the instant / fast release of radionuclides as well as the release of actinides?
- Can the data of the wash cycles be used for comparison with forthcoming results of experiments in a low ionic strength solution, conducted in the framework of CP FIRST-Nuclides?

### Timespan of the wash steps
The washing steps for K14 and K5 experiments had a duration of less than 10 days, most of the others up to 73 days, and K8 up to 300 days. The measured concentrations and derived FIAPs indicate that for the fission products an approximately constant release rate established after 50 to 100 days.

### Presence of sorbing solids
Apparently, the release of the fission products $^{137}Cs$, $^{90}Sr$ and $^{99}Tc$ is not influenced by magnetite and hydroxoapatite under the studied conditions. However, relatively low Pu and Am concentrations were measured in the wash cycles of K5, K6 and K14 where magnetite and hydroxoapatite were added to the experiments. The concentrations of the Pu and Am were found below the detection limits in experiments K11a and K11b in alkaline ECW. It is suggested that rather sorption than solubility phenomena limit the aqueous Pu and Am concentrations in those experiments.

### pH effects
Despite differences in the pH of the wash cycles, the release of $^{137}Cs$ was about the same in the experiments. Similarly, $^{90}Sr$ and $^{99}Tc$ release show no univocal correlation to the pH. The extraordinary decrease of the $^{90}Sr$ release rate (figure 2) is related to interactions of $^{90}Sr$ with dissolved Ca, $SO_4^{2-}$ and traces of stable Sr of the ECW. Under high pH conditions, Pu and Am were found below the detection limits in some experiments. The concentrations of actinides are expected to react most sensitive on pH changes, such as in the case of Pu and Am in the K8 wash steps. U shows a much less pronounced effect.

**Ionic strength effects**

The washing steps were performed in DIW, ECW and NaCl brine (table II) covering a wide range of ionic strength values. Fission products and actinides show release rates in the same order of magnitude in the case of DIW and in concentrated NaCl systems.

**Atmospheric conditions**

All of the wash steps were performed under anaerobic conditions in Ar atmosphere. Fast release experiments under reducing conditions (Ar, $H_2$) are planned within the CP FIRST-Nuclides, and the results are not yet available.

**Instant release fractions**

For performance assessment of a SNF repository, IRF values of [137]Cs, [90]Sr and [99]Tc are derived from the washing cycles data (see figure 7). These data are in good agreement with published results which were compiled and analyzed within the project FIRST-Nuclides [19].

**Interactions with the vessel walls**

In the course of the limited samplings during the washing cycles, rather U, Pu, and Am precipitation/sorption phenomena were observed than mobilization/dissolution phenomena. SNF in contact with air during the sample preparation undergoes oxidation forming oxidized layers which dissolve within hours or days of the first washing step. Retention processes may have different kinetics depending on the concentrations of dissolved elements. Such processes do not only happen in the systems where magnetite and hydroxoapatite were added, but also in the experiments where the SNF samples were immersed in pure solutions. In these cases, sorption could take place onto the vessel walls. Determination of the amount of adsorbed actinides onto the vessel walls was possible only after the termination of the static phase of each experiment and not during the washing steps. Details on the sorption onto the vessel wall have been reported previously [20].

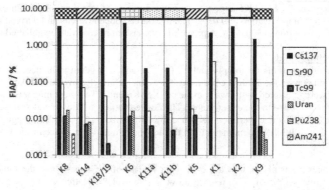

**Figure 7.** Percentage of [137]Cs, [90]Sr, [99]Tc, U, [238]Pu and [241]Am released in the aquatic phase for the leaching period of t ≤ 50 days in the different experiments. The bar above the diagram indicates the experimental systems:

saturated NaCl solution      Evolved Cement Water
sat. NaCl solution with magnetite      deionized water
sat. NaCl solution with hydroxoapatite

**Surface of the SNF samples**

In the case of the fission products, the measured concentrations increased continuously with time. The concentrations and FIAPs of $^{137}$Cs, $^{90}$Sr and $^{99}$Tc were found more or less in the same range for the respective radionuclides in all experiments (Figure 7). Slight variations (by a factor of ±2) can be explained. In experiments K11a and K11b, parts of pellets cut in two parts were used. These half pellets exposed only one inter-pellet gap side to the leachant and relatively low FIAPs are observed.

**Relations between the release rates of $^{90}$Sr and actinides**

For evaluation of SNF leaching experiments, in many cases the $^{90}$Sr release is used as indicator for fuel matrix dissolution. However, the question is still open, whether $^{90}$Sr is an adequate proxy for the fuel matrix dissolution process. For this reason, correlations between the release rates (FIAP per day) have been analysed for the different wash cycles with the same SNF material. It is clear, that correlations make sense only, if the concentrations are not effected by precipitation or sorption effects. As previously discussed, the U concentration is controlled by solubility phenomena already during the wash steps. However, in the case of the experiments K8 and K14 in NaCl solutions a positive correlation exists. Plutonium tends to form colloids or undergoes sorption and cannot be considered in this context.

## CONCLUSIONS

The conclusions are related to a single kind of PWR SNF. Fuel sample preparation is not discussed, as pellets were used in the experiments. Following conclusions can be drawn from reevaluating the measured radionuclide concentrations in the washing cycles of the studied experiments:

- With respect to burn-up and burn-up history, as well as its fission gas release, the SNF used for these investigations correspond to modern PWR SNF.
- The instant release fractions derived from the washing steps are in good agreement with published results IRF data.

## ACKNOWLEDGMENTS

The data were obtained partly by various European Community Programmes. The presented study has received funding from the European Atomic Energy Community's Seventh Framework Programme (FP7/2007-2011) under grant agreement no. 295722, the FIRST-Nuclides project. We also acknowledge gratefully the financial support by ONDRAF/NIRAS, Brussels (B) for the K11a,b experiments.

## REFERENCES

1.    Johnson, L., et al., *Rapid aqueous release of fission products from high burn-up LWR fuel: Experimental results and correlations with fission gas release.* Journal of Nuclear Materials, 2012. **420**: p. 54-62.

2.	González-Robles Corrales, E., *Study of radionuclide release in commercial $UO_2$ Spent Nuclear Fuel - Effect of burn-up and high burn-up structure*, 2011, Universitat Politècnica de Catalunya: Barcelona. p. 109.

3.	Johnson, L., et al., *Estimates of the instant release fraction for UO2 and MOX fuel at t=0*, in *A Report of the Spent Fuel Stability (SFS) Project of the 5th Euratom Framework Program*2004, NAGRA: Wettingen, CH.

4.	Poinssot, C., et al., *Spent fuel radionuclide source term model for assessing spent fuel performance in geological disposal. Part II: Matrix alteration model and global performance*. Journal of Nuclear Materials, 2005. **346**: p. 66-77.

5.	Roudil, D., et al., *Gap and grain boundaries inventories from pressurized water reactor spent fuels*. Journal of Nuclear Materials, 2007. **362**(2–3): p. 411-415.

6.	Loida, A., B. Kienzler, and H. Geckeis, *Corrosion behavior of pre-oxidized high burnup spent fuel in salt brine*, 2003: Karlsruhe. p. 6.

7.	Kutty, T.R.G., et al., *Characterization and densification studies on ThO2–UO2 pellets derived from ThO2 and U3O8 powders*. Journal of Nuclear Materials, 2004. **335**(3): p. 462-470.

8.	Stratton, R.W., et al. *A comparative irradiation test of $UO_2$ sphere-pac and pellet fuel in the Goesgen PWR*. in *Int. Topical Meeting on LWR Fuel Performance* 1991. Avignon, France, 21-24 April 1991.

9.	Kienzler, B., et al., *$1^{st}$ Annual Workshop Proceedings of the Collaborative Project 'FIRST-Nuclides'*, 2013, Karlsruhe Institute of Technology (KIT): Karlsruhe.

10.	Loida, A., et al., *Corrosion behavior of spent nuclear fuel in high pH solutions – Effect of hydrogen*. MRS Online Proceedings Library, 2012. **1475**: p. null-null.

11.	Altmaier, M., V. Neck, and T. Fanghanel, *Solubility of Zr(IV), Th(IV) and Pu(IV) hydrous oxides in CaCl2 solutions and the formation of ternary Ca-M(IV)-OH complexes*. Radiochimica Acta, 2008. **96**(9-11): p. 541-550.

12.	Grambow, B., et al., *Long-term safety of radioactive waste disposal: Chemical reaction of fabricated and high burnup spent $UO_2$ fuel with saline brines. Final Report*, 1996, Forschungszentrum Karlsruhe.

13.	González-Robles, E., et al. *Physical characterisation of spent nuclear fuel: First steps to further Instant Release Fractions investigations*. in *EURADWASTE '13*. 2013. Vilnius, Lithuania, 14-17 October 2013: European Commission

14.	Nordström, E., *Fission gas release data for Ringhals PWRs*, in *SKB TR-09-26*2009.

15.	Altmaier, M., V. Neck, and T. Fanghänel. *Solubility of uranium(VI) in dilute to concentrated NaCl, $MgCl_2$ and $CaCl_2$ solutions*. in *$12^{th}$ Internat.Symp.on Solubility Phenomena and Related Equilibrium Processes (ISSP 2006)*. 2006. TU Bergakademie Freiberg, July 23-28, 2006.

16.	Grenthe, I., et al., *Chemical Thermodynamics of Uranium*. Chemical Thermodynamics Series, ed. OECD/NEA1992: North Holland Publisher.

17.	Rai, D., et al., *Thermodynamics of the $PuO_2^+-Na^+-OH^--Cl^--ClO_4^--H_2O$ system: use of $NpO_2^+$ Pitzer parameters for $PuO_2^+$*. Radiochim. Acta, 2001. **89**: p. 491-498.

18.	Neck, V., et al., *Thermodynamics of trivalent actinides and neodymium in NaCl, $MgCl_2$, and $CaCl_2$ solutions: Solubility, hydrolysis, and ternary Ca-M(III)-OH complexes*. Pure and Applied Chemistry, 2009. **81**(9): p. 1555-1568.

19.	Kienzler, B. and E. González-Robles. *State-of-the-art on instant release of fission products from spent nuclear fuel*. in *$15^{th}$ International Conference on Environmental*

*Remediation and Radioactive Waste Management (ICEM2013)*. 2013. Brussels, September 8-12, 2013: ASME.

20.     Loida, A., B. Kienzler, and H. Geckeis. *Corrosion behavior of pre-oxidized high burnup spent fuel in salt brine*. in *Mat. Res. Soc. Symp. Proc. Vol. 807*. 2004. Materials Research Society.

Mater. Res. Soc. Symp. Proc. Vol. 1665 © 2014 Materials Research Society
DOI: 10.1557/opl.2014.651

## Preparation and characterization of UO$_2$-based AGR SIMFuel

Zoltan Hiezl[1], David Hambley[2] and William E. Lee[1]
[1]Imperial College London, Centre for Nuclear Engineering,
London, SW7 2AZ, UK
[2]Spent Fuel Management and Disposal, UK National Nuclear Laboratory (NNL),
Springfields, PR4 0XJ, UK

### ABSTRACT

Preparation and characterization of a Simulated Spent Nuclear Fuel (SIMFuel), which replicates the chemical state and microstructure of Spent Nuclear Fuel (SNF) discharged from UK Advanced Gas-cooled Reactor (AGR) after a cooling time of 100 years is described. Thirteen stable elements were added to depleted UO$_2$ and sintered to simulate the composition of fuel pellets after burn-ups of 25 and 43 GWd/tU and, as a reference, pure UO$_2$ pellets were also investigated. The fission product distribution was calculated using the Fispin code provided by NNL. SIMFuel pellets exhibit a microstructure up to 92% TD. During the sintering process in H$_2$ atmosphere Mo-Ru-Rh-Pd metallic precipitates and grey-phase ((Ba, Sr)(Zr, RE)O$_3$ oxide precipitates) formed within the UO$_2$ matrix. These secondary phases are present in real PWR and AGR SNF, although they are smaller in size than those examined in this study. The grain size of the produced SIMFuel is in good agreement with literature references.

### INTRODUCTION

To engineer a safe geological disposal facility (GDF) for high level nuclear waste (HLW), the behavior of the materials that will be placed in it have to be well-understood. As many different types of nuclear power plant (NPP) exist, a range of wastes arise from their operation and the effects of their different behavior and possible interactions of their degradation products with the environment need characterizing.

SNF is one of the types of material that could be consigned to a GDF. As actual SNF is highly radioactive, simulated SNF (SIMFuel) has been developed [1] to provide a convenient method to study high burn-up fuel in the laboratory, without the complication of the intense radiation field. This material is produced by doping a natural or depleted uranium dioxide (UO$_2$) matrix with a series of non-radioactive elements in appropriate proportions to replicate the chemical composition and phases in irradiated fuels. These compositions can be varied to reflect the effects of different burn-ups and cooling times. Nonradioactive elements in SIMFuel can represent most fission products (FP) present in an irradiated SNF. A number of these FPs will segregate from UO$_2$ matrix once their solubility limit is exceeded, forming various metallic and oxide precipitates or being released from the fuel entirely. However, other FPs with high solution limit remain dissolved in the UO$_2$ matrix.

While there are a few, slightly different, e.g. [2, 3], classifications for FP typically they are:
1. Metallic precipitates: Ru, Pd, Rh, Tc, Mo, Ag, Cd
2. Oxide precipitates: Ba, Zr, Mo, (Rb, Cs, Te, I)
3. Oxides dissolved in the UO$_2$ matrix: Sr, Zr, Y, La, Ce, Sm, Nd, Pu, Np
4. Inert gases and volatile elements: Xe, Kr, He; I, Br, (Rb, Cs, Te)

Some elements, such as Rb, Cs, Te, have dual behavior and can be either volatile or form ionic solids with oxygen or with other anions ($CsI$, $Cs_2O$, $Cs_2Te$, $Cs_2I_2$) and accumulate close to the pellet-clad interface. [4] Using standard ceramic processing techniques, the chemical state and microstructure of the first three classes can be replicated. However, inert gases and volatile elements must be implanted into the simulant SNF ceramics by irradiation.

To date, research has focused on LWR (Light Water Reactor) SNF since LWR is the most common type of reactor from which SNF will be directly disposed. However, due to differences between UK AGR and LWR design, such as use of graphite moderator and steel cladding in the former rather than water moderator and zircalloy cladding and that during operation the fuel sees some differences in heat generation and temperatures, research of direct relevance to UK AGR is needed. This work is being carried out by a consortium of researchers from Imperial College London, Cambridge University and Lancaster University [5].

The properties of $UO_2$ ceramics, the metallic precipitates and the oxide precipitate are well-known from experiments as well as from modeling studies, e.g. [6]. Irradiated fuel and SIMFuel dissolution studies, and ion irradiation tests have been conducted and thermal properties evaluated for LWR fuels, e.g. [7] However, for the UK indigenous AGR spent nuclear fuel these studies are only now being done.

The stoichiometry of the $UO_2$ has a great effect on the FP behavior [8]. If the SNF is hyper-stoichiometric, the solution energy of certain elements into $UO_2$ (such as Y, Nd) will be positive and so these elements will most likely accumulate in the grey-phase rather than remain in $UO_2$ as would be the case for stoichiometric $UO_2$.

This paper focuses on the preparation and main characteristics of UK AGR SIMFuel and compares LWR and AGR SIMFuel.


**EXPERIMENT**

To determine the composition of a spent AGR nuclear fuel after certain cooling periods, FISPIN 10.0.1 program was used by NNL and calculations performed for a typical case (burn-up of 25 GWd/t U) and a peak case (burn-up 43 GWd/t U) with inventories calculated for a range of cooling times up to 100,000 years. The output from the calculations contains information on isotopic composition, radioactive emission rates for individual FPs and gamma-spectra for one tonne of SNF. These datasets have been analyzed focusing on the exact composition of spent AGR fuel after various times to be able to produce a simulated SNF that will mimic as far as possible actual SNF. FPs are present in a wide range of concentrations in spent AGR fuel. In this study, isotopes higher than $10^{-2}$ mol within 1 tonne of SNF are taken into consideration. As the FP distribution does not differ significantly after various cooling periods, only the 100 year cooling time dataset was selected for further study for both the mean and peak case.

For the experimental work, 13 metal oxides and carbonates were selected as FP surrogates. Table I lists these oxides and their concentration in the SIMFuel. Depleted $UO_2$ was provided by NNL. It contains 0.3 g $^{235}U$/100 g $UO_2$ and it has the particle size of 3.80±0.25 µm. All experimental work was conducted at the UKs National Nuclear Laboratory (NNL), Preston Laboratory. 60 g blend of each composition (peak and typical) in addition to a pure uranium reference blend were prepared. The FP surrogates were weighed and ball milled for 12 h using $ZrO_2$ milling media to achieve a submicron particle size and good mixing. The dopant blends were then homogenized with depleted uranium-dioxide (DUD). The next step was precompaction of the powder (at 75 MPa) to produce granules, after subsequent sieving through a 1.14 mm sieve, to achieve a good flow of the material. 0.2 wt% of zinc stearate was added as a granulate lubricant, and slowly mixed with the batch for 5 minutes using a rotation mixer. Granulates were uniaxially pressed (length: ~10.27 mm, diameter: ~11.45 mm) at 400 MPa.

Pellets were sintered at the same time at a heating rate of 5 °C/min to 300 °C and then 15°C/min to 1730 °C. Sintering time was 300 min in an atmosphere of 99.5 Vol% $H_2$ and 0.5 Vol% $CO_2$ and cooling rate was 15 °C/min to room temperature. From the 60 g blend of each composition, 10 reference pellets, 9 low-doped and 9 high-doped pellets were prepared. The dimensions of one of each set of pellets were measured before sintering and the geometric green density was calculated ($\rho_{green\ pellet}$ = $m_{pellet\ before\ sintering}$ / $V_{pellet\ before\ sintering}$). After sintering the length and diameters of all pellets were measured along with the bulk density using Archimedes method. One pellet from each composition (pure $UO_2$ pellet, low-doped and high-doped pellets) was mounted in resin, polished and prepared for SEM-EDX analysis. Porosities were determined using ImageJ image analyzer software based on image grey level on reflected light optical microscopy images. The average grain size was calculated using grain boundary interception method.

**Table I.** SIMFuel composition after 100 years cooling time for both 25 and 43 GWd/t U burn-up of UK AGR fuel as well as previously studied SIMFuel LWR compositions calculated by ORIGEN[2], for comparison.

| SIMFuel composition (wt%) | 100 years cooling time | | ORIGEN calculations (For LWR) | |
|---|---|---|---|---|
| Oxides: | 25 GWd/t U (2.60 wt% burn-up) | 43 GWd/t U (4.44 wt% burn-up) | 3 at% burn-up | 6 at% burn-up |
| $UO_2$ | 97.587 | 95.870 | 97.68 | 95.31 |
| $Nd_2O_3$ | 0.483 | 0.827 | 0.460 | 0.912 |
| $ZrO_2$ | 0.369 | 0.602 | 0.339 | 0.601 |
| $MoO_3$ | 0.334 | 0.566 | 0.359 | 0.73 |
| $RuO_2$ | 0.257 | 0.454 | 0.364 | 0.764 |
| $BaCO_3$ | 0.244 | 0.435 | 0.147 | 0.311 |
| $CeO_2$ | 0.193 | 0.329 | 0.285 | 0.526 |
| PdO | 0.090 | 0.199 | 0.149 | 0.440 |
| $Rh_2O_3$ | 0.038 | 0.056 | 0.028 | 0.034 |
| $La_2O_3$ | 0.096 | 0.160 | 0.106 | 0.194 |
| SrO | 0.032 | 0.050 | 0.072 | 0.110 |
| $Y_2O_3$ | 0.041 | 0.064 | 0.041 | 0.061 |
| $Cs_2CO_3$ | 0.191 | 0.309 | - | - |
| $TeO_2$ | 0.044 | 0.079 | - | - |

## RESULTS AND DISCUSSION

Densities (Table II) suggest that the sintering method gave uniform, sufficiently dense samples, but when dopants were added to the system, the density of the samples decreased significantly. This density change arises from secondary phase formation and the increase of porosity is either due to carbon dioxide from carbonates or water vapor generation during oxide reduction. Dopant mixes were not calcined prior to sintering. These SIMFuel pellets are simplified in terms of temperature, burnup, radiation damage and geometry.

Porosity data (Table III) clearly show that as the dopant level increase, the porosity of the sample also increases. This is consistent with density measurements between reference and low-doped pellets, but shows a small discrepancy in trend when the pellets are highly doped.

**Table II.** Densities of green and fired reference and doped samples

| | Theoretical Fired Density [TD] (g/cm$^3$) | Geometric Green Density (g/cm$^3$) | Relative Green Density (geometric) (%) | Mean Fired Bulk Density (Archimedes) (g/cm$^3$) | Relative Fired Density (Archimedes) (%) |
|---|---|---|---|---|---|
| Reference pellet | 10.94 | 6.01 | 54.94 | 10.69±0.004 | 97.7±0.04 |
| Low-doped pellet | 10.76 | 5.96 | 55.39 | 9.82±0.017 | 91.2±0.16 |
| High-doped pellet | 10.62 | 5.91 | 55.65 | 9.74±0.021 | 91.7±0.20 |

**Table III.** Pore diameter and percent porosity variations.

| Porosity of the samples | | |
|---|---|---|
| Description | Average Pore Size | % Area |
| Reference | 21.2 µm | 2.50% |
| Low-doped | 18.9 µm | 11.30% |
| High-doped | 42.7 µm | 15.70% |

The effect of the dopants on the grain structure is clearly visible in SEM images (not shown). The difference in grain size between doped and undoped samples is the most significant (Table IV). In actual AGR SNF the grain size varies between 3-30 µm, but it is more usually measured at between 10-20 µm [9].

**Table IV.** Average grain size for the undoped and the low doped samples

| Sample: | Average size (µm) | Confidence/ Error |
|---|---|---|
| Undoped pellet | 9.664 | ±2.215 |
| Low-doped pellet | 2.646 | ±0.556 |

These observations suggest different sintering mechanisms for the reference sample and the two doped samples. For the reference pellets, it is solid state sintering as the melting point of the $UO_2$ is above 2800 °C. For the doped samples, the sintering mechanism is more complicated as a number of reactions occur between numerous compounds. It is assumed, that noble metal oxides are reduced to metals in $H_2$ atmosphere. Binary, ternary or quaternary systems may form eutectics and so melting can occur, which leads to a liquid phase sintering. This difference in sintering mechanism is responsible for the lower density and smaller grain size of the doped samples.

The behavior of those FPs, which form oxide precipitates has been analyzed and the most representative examples are now described. FPs, such as Ba, Zr, Sr are less soluble in $UO_2$, so they partially accumulate and form precipitates embedded into the matrix material. In SIMFuel, these elements are represented as oxides, with significant particle size (>5 μm) even after milling, so it is expected that those dopant grains will initiate secondary phase formation. Examples of such typical oxide precipitates are given and briefly described below.

Oxide precipitates predominantly form together with large (10-20 μm diameter) pores or forms spherical bodies inside the DUD matrix with low level of surface connection between the precipitate and the matrix material, indicating that they do not prefer to be in close contact with each other. This is due to different crystal structures and solubility between these two phases.

Figure 1A reveals one of the secondary phase regions. EDX analysis of this globular body shows the dark region is rich in zirconium and barium. This is consistent with previous studies [2, 3]; this particle is likely an oxide precipitate (perovskite phase = `grey phase`; $BaZrO_3$). The point analysis from the same particle but from the brighter inner region shows higher U content indicating that U may be in solid solution in the grey phase. Within this brighter region lower Ba content is detected. The U-rich region also contains Nd which is expected as Nd is most likely to go into solution in the U matrix due to the higher oxygen potential of the mixed oxide [3].

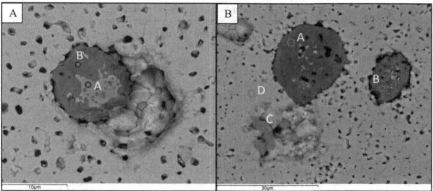

**Figure 1**. A: SEM-BS image of a typical oxide precipitate present in the low-doped sample; B: Different type of secondary phases in the highly doped SIMFuel sample.

In Figure 1B 2 secondary phases are visible, also there are 2 variations of the oxide precipitates. Submicron pores, which are distributed uniformly throughout the $UO_2$ matrix, also exist in the secondary phases. The light grey region is the $UO_2$ matrix. At the bottom of the image, point "C" is a metallic precipitate which contains Mo, Ru and Pd. The dark region "B" is a $BaZrO_3$ containing oxide precipitate. The larger, darker region ("A") is another variation of oxide precipitate, which contains significant amount of $SrZrO_3$. According to EDX, the $UO_2$ matrix ("D") contains some Nd. Semi-quantitative EDX analysis results are available for eleven grey phase particles, which show that the Sr and U content is rather small and the dominant elements are Ba, Zr and O with a ratio of 1:1:3, respectively. These measurements clearly show that BaO, derived from $BaCO_3$, and $ZrO_2$ prefer to accumulate together in a stoichiometric form.

As previously mentioned, brighter inner regions exist within the grey phase. These regions are depleted in Ba, but rich in U. Based on the ternary phase diagram, it is likely to be t-$(U,Zr)O_2$ or c-$(U,Zr)O_2$ phase, depending on U content, with low Ce and Nd as contamination. Due to overlapping energy peaks for Ba, Ce and Nd, quantitative EDX analysis, supported by independent measurements, is yet to be done.

Atomistic modelling using empirical pair potentials support these observations and predict that under stoichiometric conditions both $BaZrO_3$ and $SrZrO_3$ readily precipitate from solid solution in $UO_2$. This is not the case for $UO_{2+x}$ under equilibrium conditions. Partition energy calculations for FPs show that $Ce^{4+}$ and $Y^{3+}$ will remain in the $UO_2$ phase, whereas, $Ru^{4+}$, $Mo^{4+}$, $Nd^{3+}$ and $La^{3+}$ will preferentially segregate to the $BaZrO_3$ phase. [8, 10, 11]. This trend is slightly different for $SrZrO_3$, where partition energies predict than $Nd^{3+}$ and $La^{3+}$ prefer to be dissolved into $UO_2$. The concentration of Y and La were below the detection limit of EDX, so although the predicted precipitation of BaZrO3 and SrZrO3 has been shown, further work must be done to examine the segregation of FPs to the grey phase.

Metallic precipitates form spherical bodies on the surface or fill pores within the DUD matrix. 28 metallic precipitates were characterized using EDX. Although the composition varies between precipitates the molybdenum content tends to be 35-45 at%; the ruthenium content is generally below 20 at% and the palladium makes up the 100 at%. Phase diagrams at different temperatures are available for this system. According to previous studies e.g. [12], the system is characterized by the 3 main phases. These are the phases arising from the large solubility of molybdenum and palladium in the hcp ruthenium and the two intermediate phases of the boundary systems, the σ-phase $Mo_5Ru_5$ and the ε-phase $Mo_9Pd_{11}$, which exist only in limited temperature regions, 1143-1915 °C and 1370-1720 °C, respectively. Both isothermal sections are important because of the slow cooling rate, 15 °C/min, which after sintering could potentially allow phase change. The theoretical compositions (43.81 at% Mo, 38.06 at% Ru and 18.14 at% Pd for the 43 GWd/t U) of these metallic precipitates are calculated based on the metal oxides weighed in for both burn-ups, however precipitates from the high burnup SIMFuel samples were only analysed. Comparing these calculations with the EDX results, it can be seen that the Mo percentage is close to theoretical in the majority of the cases in contrast to the Ru and Pd content. The reason for this is currently unknown. Several causes can be ruled out: it cannot be evaporation or other type of material loss during production, as ICP-MS measurements conducted by NNL has proved that the appropriate amounts of additives are in the pellets. One possible explanation is the small number (28) of particles analyzed here. The Rh content of these precipitates is not considered, as EDX rarely detected Rh whose peaks are heavily overlapped by Ru and Pd peaks. Rh detection was therefore problematic.

## CONCLUSIONS

From the experimental observations detailed in the above sections, the following conclusions about the AGR SIMFuel can be extracted.

SEM-EDX analysis revealed metallic and oxide precipitate (grey phase) formation. The main components of the metallic precipitates are Mo, Rh, Ru and Pd, whereas in the grey-phase Ba, Zr and Sr are found. Several fission product surrogates are dissolved in the $UO_2$ matrix, such as Ce and Nd. The formation of these secondary phases is consistent with previous experimental and modeling studies.

## ACKNOWLEDGMENTS

The author is grateful to the UK EPSRC (grant EP/I036400/1), the Radioactive Waste Management Directorate of the UK Nuclear Decommissioning Authority (contact NPO004411A-EPS02) and the Armourers & Brasiers' Gauntlet Trust for financial support.

## REFERENCES

1.  S. Strausberg and E.W. Murbach. (1963). Multicycle Reprocessing and Refabrication Experiments on Sintered $UO_2$-fissia Pellets. *I&EC Process Design and Development, 2*(3), 228-231.
2.  P.G. Lucuta, R.A. Verral, H. Matzke, and B.J. Palmer. (1991). Microstructural Features of SIMFUEL-Simulated High-burnup $UO_2$-based Nuclear Fuel. *Journal of Nuclear Materials, 178*, 48-60.
3.  M.A. Mignanelli and T.L. Shaw, Information on Phase Chemistry in UK Spent AGR Fuels, NNL Report, (2009).Tech. No: 9734.
4.  A.N. Shirsat, M. Ali, S. Kolay, A. Datta, and D. Das. (2009). Transport Properties of I, Te and Xe in Thoria–Urania SIMFUEL. *Journal of Nuclear Materials, 392*(1), 16-21.
5.  N. Rauff-Nisthar, C. Boxall, I. Farnan, Z. Hiezl, W. Lee, C. Perkins, and R. Wilbraham. (2013). Corrosion Behavior of AGR Simulated Fuels – Evolution of the Fuel Surface. *Corrosion in Nuclear Energy Systems: From Cradle to Grave "ECS Transactions", 53*.
6.  R.W. Grimes and C.R.A. Catlow. (1991). The Stability of Fission Products in Uranium Dioxide. *Philosophical Transactions of the Royal Society A: Mathematical, Physical and Engineering Sciences, 335*(1639), 609-634.
7.  P.G. Lucuta, H. Matzke, and R.A. Verral. (1995). Thermal Conductivity of Hyperstoichiometric SIMFUEL. *Journal of Nuclear Materials, 223*, 51-60.
8.  M.W.D. Cooper, S.C. Middleburgh, and R.W. Grimes. (2013). Partition of Soluble Fission Products Between the Grey Phase, $ZrO_2$ and Uranium Dioxide. *Journal of Nuclear Materials, 438*(1-3), 238-245.
9.  S. Morgan, Grain Size and Porosity Variations in Unfailed Spent AGR Fuel, NNL Memorandum, (2011).
10. S.C. Middleburgh, D.C. Parfitt, R.W. Grimes, B. Dorado, M. Bertolus, P.R. Blair, L. Hallstadius, and K. Backman. (2012). Solution of Trivalent Cations into Uranium Dioxide. *Journal of Nuclear Materials, 420*(1-3), 258-261.
11. M.W.D. Cooper, S.C. Middleburgh, and R.W. Grimes. (2013). Swelling due to the partition of soluble fission products between the grey phase and uranium dioxide. *Progress in Nuclear Energy, In Press.*
12. H. Kleykamp. (1989). Constitution and Thermodynamics of the Mo-Ru, Mo-Pd, Ru-Pd and Mo-Ru-Pd Systems. *Journal of Nuclear Materials, 167*, 49-63.

Mater. Res. Soc. Symp. Proc. Vol. 1665 © 2014 Materials Research Society
DOI: 10.1557/opl.2014.652

# Increased Molybdenum Loading for Vitrified High Level Waste

Nick R. Gribble[1], Rick J Short[1], Barbara F. Dunnett[1] and Carl J. Steele[2]
[1]National Nuclear Laboratory, Sellafield, Seascale, Cumbria, UK, CA20 1PG
[2] Sellafield Ltd, Sellafield, Seascale, Cumbria, UK, CA20 1PG

## ABSTRACT

The solubility of molybdenum in borosilicate glasses is low. The UK National Nuclear Laboratory has developed a new glass formulation containing calcium and zinc for the vitrification of high molybdenum containing waste arising from the Post Operational Clean Out of the Highly Active Storage Tanks at Sellafield that will decrease the number of product containers required, reducing both production and disposal costs. The new formulation increases the quantity of molybdenum that can be vitrified through the formation of a durable $CaMoO_4$ phase once the solubility limit of molybdenum in the glass has been exceeded. Extensive laboratory trials confirmed the potential to increase the Mo loading significantly. Recently full scale testing has been performed on the Vitrification Test Rig using highly active liquor simulants to determine the maximum $MoO_3$ loading that can be achieved. This paper explores the full scale testing and product quality of the glass manufactured during this study.

## INTRODUCTION

The Highly Active Liquor (HAL) stored in the Highly Active Storage Tanks (HASTs) at Sellafield contains a quantity of solids. The major solid phases that precipitate or crystallise within HAL are barium-strontium nitrate, zirconium molybdate (ZM) and caesium phosphomolybdate (CPM), but it is expected that under the current HASTs conditions most of the CPM will have converted to ZM. Zirconium phosphates are also present as a solid phase, but only a small proportion will have settled to the base of the tanks due to its flocculent nature.

Prior to decommissioning, the Post Operational Clean Out (POCO) of the HASTs will result in waste streams with relatively high molybdenum contents. It is likely that these POCO wastes will be blended with typical THORP and Magnox reprocessing wastes prior to vitrification. The standard alkali borosilicate base glass crizzle (MW – "Mixture Windscale") used for the immobilisation of High Level Waste (HLW) streams within the Sellafield Waste Vitrification Plant (WVP) can tolerate some Mo, but at concentrations >5.25 wt% $MoO_3$ phase separation becomes readily apparent. The yellow phase formed is a mixture of alkali, alkaline earth and rare earth molybdates and chromates, is undesirable in the vitrified product as it is partially water soluble and is corrosive to the Nicrofer melting crucible.

The number of product containers required for vitrification of POCO waste will be dependent on the incorporation of $MoO_3$ into the glass. Hence, a new glass formulation shown in Table I containing calcium and zinc, known as 'Ca/Zn' has been developed for the vitrification of high molybdenum waste. The Ca/Zn formulation increases the quantity of molybdenum that can be vitrified through the formation of a durable $CaMoO_4$ phase once the solubility limit of molybdenum in the glass has been exceeded.

**Table I.** Compositions of MW and Ca/Zn glasses

|                                | MW wt% | Ca/Zn wt% |
|--------------------------------|--------|-----------|
| $B_2O_3$                       | 22.4   | 23.41     |
| $Li_2O$                        | 5.4    | 4.20      |
| $Na_2O$                        | 11.0   | 8.56      |
| $SiO_2$                        | 61.2   | 47.63     |
| $Al_2O_3$                      | -      | 4.20      |
| CaO                            | -      | 6.00      |
| ZnO                            | -      | 6.00      |

Extensive laboratory trials examined compositional variation in the waste, incorporation rates and formation conditions. Durability (Soxhlet and Product Consistency Test (PCT)) and viscosity studies together with macro and micro analysis confirmed the suitability of the base glass to accommodate significantly increased levels of molybdenum without detrimental phase separation. This paper reports the results of recent full scale trials on the Vitrification Test Rig (VTR) where POCO simulant was blended with HAL simulant representing waste from Magnox fuel and a 50:50 blend of Oxide and Magnox high level waste (50o:50m) and compares the results obtained with the Ca/Zn glass with those of MW base glass.

## EXPERIMENT

The VTR is a full scale replica of the WVP feed, calcination, vitrification and primary off-gas systems. It is used to perform representative experiments to gather operability and product quality data prior to implementing changes on the active plant. The HAL simulant used in this work comprised Magnox or 50o:50m which was blended with varying amounts of the POCO simulant such that there was between 8 and 12 wt% $MoO_3$ in the glass products at 38 wt% HAL waste oxides (HWO) from the combined HAL + POCO feed (see Table II). There was a significant amount of Li in the POCO feed due to the use of $LiMoO_4$ as a source of Mo. As a result the composition of the glass frit was modified to remove the $Li_2O$ so that the product glass contained the correct amount of alkali and maintained the target Li:Na ratio of 1:1.

It should be noted that whilst the chemical composition of the feeds to the VTR was tailored to provide the desired glass composition, the physiochemical form of the elements within the feed were unrepresentative of the expected active POCO waste stream. In particular the zirconium and molybdenum are likely to be present as zirconium molybdate (ZM) and therefore operability data for both the calciner and the off-gas system were not considered to be representative of active feeds. Future studies using ZM of the correct morphology are planned.

For each experiment the simulants were blended appropriately and two pours of 190 kg were produced at the target $MoO_3$ concentration. The target melt temperature was usually 1050°C, the normal WVP operating temperature. However, more challenging operating conditions were also applied at the highest $MoO_3$ loading for each waste type to assess the effect on product quality, i.e. melt temperature was reduced to below 1020°C and mixing by air sparge was reduced from 3x150 l/h to 3x50 l/h, so as to provide an operational envelope for the active plant.

Table II. Composition of simulants used during vitrification trials

| Element | Concentration in solution (g/l) | | | Element | Concentration in solution (g/l) | | |
|---|---|---|---|---|---|---|---|
| | Oxide | Magnox | POCO | | Oxide | Magnox | POCO |
| Li | - | - | 8.51 | Mo | 12.88 | 7.53 | 66.63 |
| Na | - | 0.52 | - | Ru | 3.96 | 4.59 | - |
| Mg | - | 22.00 | - | Te | 1.85 | 1.04 | - |
| Al | 0.77 | 18.21 | - | Cs | 14.25 | 7.84 | 2.52 |
| Cr | 2.96 | 3.18 | - | Ba | 8.15 | 3.36 | 18.46 |
| Fe | 12.21 | 16.00 | - | REE | 77.81 | 3.83 | - |
| Ni | 2.15 | 2.26 | - | $NH_4NO_3$ | 77.81 | 5.3 | - |
| Sr | 3.56 | 1.86 | 1.62 | Other | 2.20 | 0.82 | - |
| Zr | 18.45 | 8.18 | 27.56 | | | | |

Product quality was judged by visual examination of the pour to enable a decision to be made on the target $MoO_3$ concentration for the subsequent experiment. Unacceptable product quality was defined as excessive yellow phase in the product, compared to those previously made on the VTR, and thus relied on the experience of the glass technologists to make that assessment. All of the products were considered to be of acceptable quality.

Scanning Electron Microscopy/Energy Dispersive Spectroscopy (SEM/EDS) was performed with a JEOL JSM-5600 SEM fitted with a Princeton Gamma-Tech Prism Digital Spectrometer. Digital image processing in backscattered mode was used to determine the percentage of crystalline material in the relevant samples. X-ray Diffraction (XRD) analysis of laboratory samples was carried out using an INEL Equinox 1000 X-ray diffractometer with a CuKα source and a curved positive sensitive detector ($2\theta = 5 - 110°$). The diffractometer was operated at 35 kV and 40 mA. The XRD patterns obtained for each material were compared with standard patterns obtained from the Joint Committee for Powder Diffraction Studies (JCPDS) X-pert database using Match software for phase identification. Glass monoliths were leached in Soxhlet apparatus with deionised water at ~97°C for 28 days. Leach solutions were analysed by Inductively Coupled Plasma Optical Emission Spectrometer (ICP-OES). Heat treatment of samples was performed at 750°C for 2 weeks (the temperature that produced the highest amount of crystallisation in the glass). Glass density was measured by Archimedes method. Viscosity measurements were performed using a Theta Rheotronic II high temperature rotating viscometer. Measurements were made every 10 seconds as the melt was cooled from 1200°C to 900°C at a rate of 2°C/min. Measurements of Tg were made using a Perkin Elmer Pyris Diamond thermogravimetric / differential thermal analyser.

## DISCUSSION

Initial studies with a representative POCO simulant were performed in 2010 to assess the potential of Ca/Zn glass to incorporate pure POCO waste or to co-process POCO and Magnox waste using MW glass. In the former case 8.7 wt% $MoO_3$ was incorporated prior to significant yellow phase formation and in the latter 3.75 wt% was the target within a product containing 25 wt% HWO which was also achieved without yellow phase formation. Subsequent co-processing vitrification trials showed that phase separation became unacceptable above about 5.3 wt% $MoO_3$, Figure 1, although in this case the HWO incorporation rate had been increased to 38 wt%.

**Figure 1.** Phase separation at 5.6 wt% $MoO_3$ in MW glass

Further studies with MW glass were performed in advance of the current work to assess the amount of phase separated material that occurred at substantially increased (up to 12 wt%) $MoO_3$ at 38 wt% Magnox+POCO HWO in the product glass (Figure 2). Work is underway to quantify the amount of phase separated material.

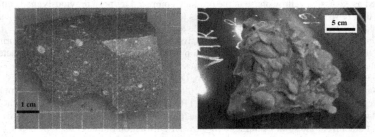

**Figure 2.** Bulk glass (left) and top surface of the pour (right) at 12 wt% $MoO_3$ target in MW glass showing excessive phase separation

The laboratory studies with Ca/Zn base glass indicated that as the HWO derived from HAL increased the maximum amount of $MoO_3$ that could be incorporated before phase separation occurred reduced, see Figure 3. The results suggested that the VTR experiments, which tend to produce better quality products due to scale, elaboration time and mixing efficiency, would be expected to produce good products at between 8 and 10 wt% $MoO_3$ in the product glass.

Table III provides the target and measured $MoO_3$ concentration in the VTR product glasses and the Soxhlet bulk leach rate data for both as cast and heat treated glasses. The analysed $MoO_3$ levels were generally slightly lower than target, which was suspected to be due to the rapid settling of the solids within the POCO simulant and the highly viscous nature of the sediment that formed, making resuspension difficult. However, good product quality was achieved with $MoO_3$ concentrations at or above the expected maximum and the Soxhlet bulk leach rate test results were all within the range of 2 to 6 $gm^{-2}d^{-1}$ which is typical of 38 wt%

product glasses manufactured from non-POCO containing HAL with MW base glass. Elemental leach rates and PCT results were not available at time of press, but the results for glasses made in the laboratory were within the normally expected ranges for products not containing POCO waste.

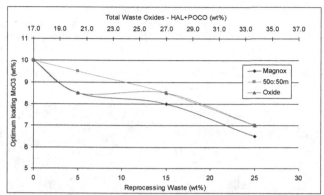

**Figure 3.** Variation of optimum $MoO_3$ loading with waste incorporation rate and waste type

**Table III.** Soxhlet bulk leach rates (BLR) data

|  | Measured (target) $MoO_3$ concentration in product (wt%) | Soxhlet BLR $(gm^{-2}d^{-1})$ as cast | Soxhlet BLR $(gm^{-2}d^{-1})$ heat treated |
|---|---|---|---|
| **Magnox + POCO** | 7.5 (8) | 3.01 | 2.91 |
|  | 9 (9) | 6.06 | 2.69 |
|  | 8.6 (10) | 4.43 | 3.64 |
|  | 10.5 (12) | 4.81 | 4.99 |
|  | 10.6 (12)* | 4.90 | 4.61 |
| **50o:50m + POCO** | 9.4 (10) | 3.25 | 2.28 |
|  | 11.3 (12) | 2.81 | 2.39 |
|  | 11.6 (12)* | 3.20 | 2.89 |

*Intentionally lower target temperature and poorer mixing conditions

Laboratory made glasses at $MoO_3$ concentrations >7 wt% displayed relatively high viscosity below 1015°C which was attributed to the presence of the separate $CaMoO_4$ phase and it was postulated that at 10 wt% $MoO_3$ pouring problems from the WVP melter crucible could become evident. However, the pouring rates from the VTR even at the highest $MoO_3$ concentrations and the lowest temperature were significantly higher than those currently achieved on WVP with MW glass and lower HWO incorporation rates.

The visual appearance of the VTR HAL+POCO glasses differed from products without POCO. They exhibited a more rock-like appearance generated by many thousands of small flat faces at fracture surfaces that gave it a sparkly appearance, compared to the shiny smooth appearance of standard products. When breaking up the pours, they were reported to be much "harder" and tended to form pieces with less sharp edges. This was probably due to the high concentration of crystalline material in the product inhibiting direct crack propagation through the matrix. The colour of the Ca/Zn product also varied depending on its position within the glass container. Pieces from near the edges (up to ~1cm deep) tended to be grey, changing to uniform dark green then mottled light and dark green as the centre of the container was approached (see Figure 4). This was attributed to the rate of cooling of the glass, i.e. the mottling occurred within the relatively slow cooled glass towards the centre of the container allowing a secondary phase to precipitate and grow.

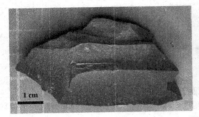 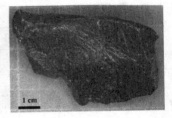

**Figure 4.** Ca/Zn with 50o:50m+POCO product glass showing (left) grey and (right) mottled light and dark green glass

Micro scale analysis of the products identified zirconium/cerium and $RuO_2$ phases which are regularly found in simulated product glasses made with MW base glass and were expected in this product. Spinel phases rich in Fe and Cr are also found in HLW products and were seen in the Ca/Zn HAL+POCO products, although this phase appeared to be predominantly Cr based. The presence of a calcium molybdate phase (most likely $CaMoO_4$) was the major difference between the Ca/Zn glasses and similar products made with MW base glass. This phase is desirable in the product as it "mops up" the insoluble Mo ions and prevents the formation of yellow phase. The scheelite crystalline family, to which $CaMoO_4$ belongs, is capable of accommodating a wide range of ions on the A site of the $A(MoO_4)$ unit cell, hence other elements such as Sr and Nd are sometimes associated with the $CaMoO_4$. As long as any associated ions are not alkali metals, the resulting crystalline phase will have a very low solubility in water and can thus be tolerated in the HLW product.

The amount of crystalline material covering the surface of the Ca/Zn as cast product was higher than is usually seen with HLW products made with MW base glass, probably as a result of the $CaMoO_4$ phase being present. This quantity increased upon heat treatment. An additional Mg-rich phase also developed upon heat treatment of the Ca/Zn Magnox+POCO glasses, that could not be quantified using the image analysis technique applied to the other crystalline phases for surface coverage analysis, as it was too dark to give sufficient contrast to the glass matrix[1].

---

[1] The crystalline quantification technique traditionally applied to simulated HLW glasses relies on the contrast exhibited by heavier elements to the lighter glass matrix under backscattered imaging mode

The densities of the glasses were 2.89 & 3.07 +/-0.06 g/cm$^3$ for Magnox+POCO & 50o:50m +POCO respectively (cf 2.79 & 2.91 g/cm$^3$ for 38wt% Magnox & 50o:50m MW products) and the glass transition temperatures ranged from 513 to 530°C (cf ~535°C for 38wt% non-POCO MW products) which were well above the expected centreline temperature during storage.

Whilst the Ca/Zn products exhibited some different properties to MW based HLW products, these differences did not appear to negatively influence the most important measure of product quality, i.e. the chemical durability. As indicated in Table III, the soxhlet durability of the Ca/Zn product was similar to 38 wt% Magnox and 50o:50m blend products made with MW base glass.

There were no operability issues encountered during the processing of the simulants. The losses from the calciner to the off-gas system were less than 1%, which is normal and there was no apparent increase in Mo loss relative to the other elements present in the feed to the plant. However, it must be restated that the feed was not fully representative of real POCO material both in terms of the lack of ZM and the increased level of Li compared to standard feeds. Li is known to act as a binder and reduces losses to the off-gas system. Additional extended operability studies will be performed at a later date using fully representative simulants.

## CONCLUSIONS

This work has shown that the recently developed Ca/Zn base glass can be used to vitrify blended HAL+POCO waste streams at MoO$_3$ concentrations of up to at least 10.6 wt% in the product, without significant alkali molybdate phase formation. Whilst visual appearance and the type and amount of crystalline content differed from previous co-vitrified POCO and non-POCO products made with MW base glass, there was no detriment to product quality.

## ACKNOWLEDGMENTS

The authors thank Sellafield Limited, a Nuclear Management Partners company, operated under contract to the UK Nuclear Decommissioning Authority, for funding the work.

---

on the SEM. Mg has a similar atomic number to the glass matrix average, thus does not exhibit the required contrast.

Mater. Res. Soc. Symp. Proc. Vol. 1665 © 2014 Materials Research Society
DOI: 10.1557/opl.2014.653

# Effects of matrix composition and sample preparation on instant release fractions from high burnup nuclear fuel

O. Roth[1], J. Low[1], K. Spahiu[2]
[1]Studsvik Nuclear AB, Hot Cell Laboratory, SE-611 82 Nyköping, Sweden
[2]SKB, Box 250, SE-101 24, Stockholm, Sweden.

## ABSTRACT

The rapid release of fission products segregated either to the gap between the fuel and the cladding or to the $UO_2$ grain boundaries from spent nuclear fuel in contact with water (often referred to as the instant release fraction - IRF) is of interest for the safety assessment of geological repositories for spent fuel due to the potential dose contribution. In September 2012 a study was initiated with the aim of comparing the instant release behavior of fuels with and without additives/dopants. Preliminary results from this (ongoing) study indicate that the release of uranium during the first contact periods was higher than during the tests with fuel segments, even though the fuel was cut open recently [1]. This could be due to the sample preparation method which included axial cutting of the cladding in order to remove the fuel fragments used in the study. In the present work, leaching data from both studies are presented and the releases are discussed comparing the two sample preparation methods and considering the effect of matrix composition. The leaching studies have been performed in air using 10 mM NaCl + 2 mM $NaHCO_3$ as leaching solution.

## INTRODUCTION

The performance assessment of geological repositories for direct disposal of spent fuel requires understanding of the processes involved in the release of radionuclides from the spent fuel upon contact with ground water. The majority of the radionuclides are distributed within the $UO_2$ matrix of the fuel and their release will be governed by the matrix dissolution.

However, a fraction of certain volatile and segregated fission products will be leached from the gap between the fuel and the cladding, from the $UO_2$ grain boundaries or from separate phases within the fuel. The leaching rate varies depending on the origin of the nuclide (matrix/gap/grain boundary etc.) and the leaching behavior of a specific fission product will be strongly dependent on its chemical state. Elements leached from the gap and grain boundaries (i.e. I and Cs) are generally released rapidly upon contact with water and therefore referred to as the "Instant Release Fraction" (IRF). Previous studies have shown that the IRF is often comparable to the released fission gases (FGR) as measured in gas release testing of fuel rods [2-5]. This relationship can be explained by the fact that under reactor operation temperatures volatile and segregated fission products are usually in gaseous state which facilitates their migration to the grain boundaries and the fuel/cladding gap.

The new fuel types used today (i.e. additive fuels) generally have lower FGR than traditional standard fuels when operated under the same conditions and it is reasonable to believe that also the IRF should differ between the two fuel types. In order to investigate this behavior a study, with the aim of comparing the instant release behavior of fuels with and without additives/dopants was initiated in September 2012 using an ADOPT (Advanced DOped Pellet

Technology) fuel [6], produced by Westinghouse, containing additions of $Cr_2O_3$ and $Al_2O_3$ and a standard $UO_2$ fuel with similar burnup and power history.

Preliminary results from this (ongoing) study showed that for the non-doped fuel sample, the release of $^{129}I$ is lower than that of $^{137}Cs$ [7]. This contradicts findings in earlier studies where $^{137}Cs$ is generally released at a lower rate than $^{129}I$, a relationship which is sometimes attributed to the lower diffusion coefficient of cesium [8]. Furthermore, the preliminary results of the study started in 2012 indicated that the release of uranium during the first contact periods was higher than during the tests with fuel segments, even though the fuel was cut open a short time before the test start [7]. This could be due to the sample preparation method which included axial cutting of the cladding in order to remove the fuel fragments used in the study. The axial cutting using a saw may cause local heating of the sample resulting in pre-oxidation of the matrix and/or loss of iodine. This could also be an explanation for the lower iodine releases observed.

In order to investigate the role of sample preparation a parallel experimental series has been initiated where the fuel fragments have been removed by crushing the fuel instead of cutting the samples axially. The fuel samples used are from pellets in close proximity to the samples in the first study.

## EXPERIMENTAL

Sample preparation:
Sample preparation (cutting, cladding detachment, fragment collection) and leaching tests were carried out under aerated conditions in hot cells; for the leaching tests, a hot cell dedicated to leaching experiments was used. More details on the experimental setup can be found in [8] where similar experiments are described. In Table I, the fuel rod data are listed for the samples used in the leaching experiments.

**Table I.** Fuel rod data.

| Sample designation (Studsvik) | 5A2 | C1 |
|---|---|---|
| Reactor | Oskarshamn 3 | |
| Fuel type | Standard $UO_2$ | Al/Cr doped $UO_2$ |
| Initial enrichment [wt% $U^{235}$] | 3.5 | 4.1 |
| Rod average burnup [MWd/kgU] | 57.1 | 59.1 |
| Fission gas release [%] | ~2.5 | ~1.5 |

Leaching experiments:
The samples were leached as fuel fragments + separated cladding. The samples were prepared by cutting ~20 mm long pieces of each rod. The cut samples were prepared by cutting the cladding longitudinally by sawing on opposite sides of the segment. Force was applied to the halves until the fuel broke away from the cladding. The crushed samples were prepared by gently using a hammer on the outside of the cladding until the fuel was fragmentized and removed from the cladding. The cladding, together with detached fuel fragments were weighed, collected in a glass vessel with glass filter bottom (100-160 μm pores) and immersed into 200 ml leaching solution (10 mM NaCl + 2 mM NaHCO$_3$). The total weight of each sample (fuel + cladding) was approximately 12 g. The results of the leaching of the samples for the 4 initial contact periods with a cumulative contact time of ~ 90 days are reported here.

### Radionuclide analyses

After each contact period, samples were collected for ICP-MS (Inductively Coupled Plasma-Mass Spectrometry) isotopic and γ-spectrometric analyses, as well as for pH and carbonate determination (pH and carbonate results were fairly constant during all the tests). [129]I and [127]I were analyzed by ICP-MS equipped with a Dynamic Reaction Cell (DRC) [9]. Oxygen was used as reaction gas to remove the interfering isotope [129]Xe on mass 129. Details on analyses and corrections (e.g. isobaric interferences) that are applied to the raw ICP-MS data can be found in Zwicky et al. [1]. Release fractions are calculated by dividing the total amount of a nuclide of concern in the analyzed solution by the total amount in the corroding fuel sample. The total amount of each nuclide present in the fuel sample was determined by general model calculations (ORIGEN code) for fuel with burnup ~59 MWd/kgU. Cumulative release fractions are the sum of release fractions up to a certain cumulative contact time.

## RESULTS AND DISCUSSION

Here results from four experiments are reported: 1. ADOPT fuel – cut, 2. Standard fuel – cut, 3. ADOPT fuel – crushed, 4. Standard fuel – crushed. Preliminary [137]Cs and [129]I releases from experiment 1 and 2 have been reported previously [7]. In Figure 1 and 2 data for [137]Cs and [129]I are shown. The uncertainty in the experimental data is estimated to be around 5-10 %, 10% deviation is shown by error bars in the figures. For [137]Cs it is quite clear that the cumulative release (Figure 1a) from the doped fuel is lower than that from the standard fuel. This is consistent with the difference in FGR between the two fuel types. It can also be noted that the cumulative release from the crushed fuel samples is higher than for the corresponding cut samples. This can probably be attributed to the difference in fragment size between the two methods which will be discussed further below.

In figure 1b [129]I cumulative release fractions are presented. As can be seen in the figure, for the crushed fuel samples, the [129]I release from the doped fuel is lower than that from the standard fuel (similar behavior as shown for [137]Cs above), whereas the relationship is reversed for the cut fuel samples. For both fuel types, the release from the crushed samples is higher than from the cut samples.

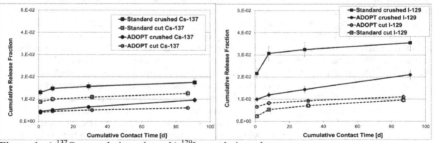

**Figure 1.** a) [137]Cs cumulative release b) [129]I cumulative release

As mentioned above, previous studies have shown that the release of [137]Cs is generally lower than the release of [129]I [8]. Comparing the [137]Cs and [129]I release fractions in figure 1a and 1b it can be seen that all experiments follow this trend except for the cut standard fuel samples (as presented previously [7]). It can also be noted that the [129]I release from the cut standard fuel sample is much lower than would be expected based on the FGR of the rod and previous data [8]. One explanation for the very low [129]I release from this particular sample is that iodine may have been lost during the sample preparation by sawing.

In Figure 2 the fractional release rates of [137]Cs and [129]I are shown. For [137]Cs it can be noted that the difference between the doped samples and the standard samples are more pronounced in the early stage of the experiment, at longer time scales no difference between the fuel types can be seen. The behavior of [129]I is very similar to that of [137]Cs, except for the first measuring point for the cut standard fuel. This supports the argument that loss of iodine during sample preparations gives rise to the low cumulative release rate. When performing the axial cutting, the blade of the saw could have penetrated through the cladding into the fuel matrix. This may have caused local heating and in turn loss of iodine from the exposed surface. The effect of this would be a very low initial release rate, followed by an increase in release rate as water reached the non-heated surfaces. Given the much higher volatility of iodine than e.g. cesium [10], selective loss of iodine is expected.

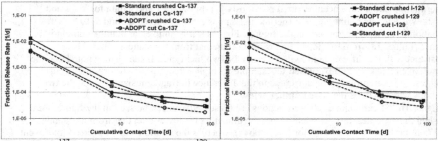

**Figure 2.** a) [137]Cs fractional release rate b) [129]I fractional release rate

As mentioned earlier, one of the motives for investigating the effect of the sample preparation method was that it was noticed that the releases of uranium for the cut fuel samples (experiment 1 and 2) are relatively high compared to previous data obtained for fuel segments and it was suspected that the axial cutting of the fuel could induce further pre-oxidation of the matrix [7]. In Figure 3a, the fractional release of [238]U is shown for the four fuel samples. As can be seen in the figure, the cumulative uranium release from the crushed fuel is slightly higher than for the samples prepared by axial cutting. The release from the doped samples seems to be slightly lower than from the standard fuel, although the difference may be within the experimental uncertainty estimated to 5-10% Also for other elements, which are generally considered as being leached by matrix dissolution (e.g. Eu and Nd) the release from the doped fuel seems to be lower compared to the standard fuel (shown for [144]Nd in Figure 3b). However, comparing the release of these elements from the cut and crushed fuel samples, the cumulative release from the cut fuel is higher than that from the crushed fuel.

Investigating the rate of radionuclide release as a function of leaching time gives further information on the release behavior. In Figure 4 the [238]U and [144]Nd release rates as a function of cumulative contact time is shown. In this figure it can be seen that the difference in uranium

264

release rate between the two sample types is very low initially and becomes larger at longer contact times.

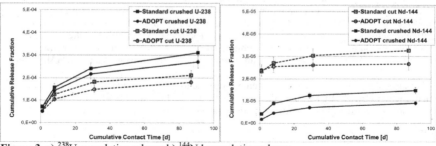

**Figure 3.** a) $^{238}$U cumulative release b) $^{144}$Nd cumulative release

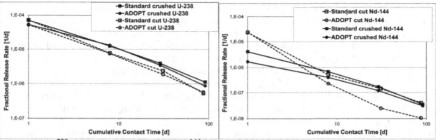

**Figure 4.** a) $^{238}$U fractional release rate b) $^{144}$Nd fractional release rate

The larger release from the crushed fuel samples at longer time scales can most likely be explained by a difference in fragment size. A smaller fragment size gives a larger surface area exposed to the leaching solution which in turn should lead to a faster fuel dissolution and radionuclide release. At shorter time scales, it is possible that the axial cutting produces some pre-oxidezed fine fuel grains/powder which gives rise to the initially high release from the cut fuel. During the cutting process it is possible that temperatures higher than 280 °C are achieved at the place where the saw contacts fuel surface. This would make possible the oxidation of a few fuel grains all the way to $U_3O_8$ [11] enhancing the leaching. This effect is however very minor and is only observed at very short time scales (~1 day leaching time). Many radionuclides analyzed in the leaching solution (e.g. Tc, Mo, Sr, Ba, Np) exhibit similar behavior as shown for uranium. Ocular examination of the fuel fragments from the different experiments confirms that the fragment size is somewhat smaller for the crushed samples; however no measurement of the fragment size has been performed.

The higher release rate of some elements (e.g. Nd, Eu or Pu) from the crushed fuel samples can probably also be explained by the fragment size and sample preparation (formation of pre-oxidized fines). These elements all show very high initial release rates from the cut fuel samples (as shown for $^{144}$Nd in Figure 4b) whereas the release at longer time scales are similar between the cut and crushed samples. This observation agrees with the discussion concerning uranium above, regarding possible effect of the production of small grains/powders during axial cutting. However, for Nd, Eu and Pu the difference in initial release rate is much larger between the two sample types than observed for uranium and for this reason the cumulative release fraction fuel samples are higher for the cut samples than for the crushed for Nd, Eu and Pu. The larger effect of cutting on initial release of the Nd, Eu, Pu elements can be partly explained by the enrichment of these elements in the fuel rim [12] (due to higher burn-up). Any effect of the axial cutting should lead to preferential leaching of the rim particles (as it is the rim zone that may be affected by the saw blade).

It should also be noted that these elements are very prone to adsorption and the majority of the leached fraction is generally found on the glass vessel walls. This explains why the release fractions are ~ one order of magnitude lower than the release fraction for uranium.

## SUMMARY

The results of this study have shown that:

- The relative release rates vary with contact time hence, investigating cumulative release fractions only could be misleading. In this study the release rates from the cut samples are generally higher than from the crushed samples at the early stages of the experiment. This can be attributed to the production of small oxidized fuel particles during the axial cutting of the fuel. At longer time scale the release from the crushed samples tends to be higher than from the cut samples, which is likely due to a smaller fragment size.
- ADOPT fuel has lower IRF of iodine and cesium than the standard fuel for both sample preparation methods. The only exception is in the case of iodine for the cut fuel samples, which can probably be explained by iodine loss during fuel cutting due to its higher volatility.
- Also for other elements the release from the ADOPT fuel is generally slightly lower that from the standard fuel.

The results represent a work in progress and should be interpreted with care.

## REFERENCES

1. H-U. Zwicky, J. Low, E. Ekeroth, SKB TR 11-03, (2011).
2. L. H. Johnson and D. F. McGinnes, Nagra Technical Report NTB 02-07, (2002).
3. L. Johnson, C. Poinssot, C. Ferry, P. Lovera, Nagra Tech.Rep. NTB 04-08, (2004).
4. L. H. Johnson and J. C. Tait, SKB TR 97-18, (1997).
5. C. Ferry, J-P. Piron, A. Poulesquen, C. Poinssot, MRS Symp. Proc. 1107, 447–454 (2008).
6. K. Backman et. al., in: IAEA-TECDOC-1654, pp. 117-126 (2010).
7. O. Roth, J. Low, M. Granfors, K. Spahiu, MRS Symp. Proc. 1518, 145-150 (2013).
8. L. Johnson et. al., J. Nucl. Mater., 420, 54-62 (2012).
9. A. Izmer, S. F. Boulyga and J. S. Becker, J. Anal. At. Spectrom. 18 1339-1345 (2003).
10. W. Haynes, CRC handbook of chemistry and physics: a ready-reference book of chemical and physical data. Boca Raton, FL: CRC Press, 2011.
11. L. Thomas, R. Einziger, H. Buchanan, J. Nucl. Mater. 201, 310-319 (1993).
12. H. Matzke, J. Spino, J. Nucl. Mater. 248 170-179 (1997).

Mater. Res. Soc. Symp. Proc. Vol. 1665 © 2014 Materials Research Society
DOI: 10.1557/opl.2014.654

## Effect of cement water on UO$_2$ solubility

C.Cachoir[1], Th. Mennecart[1], and K. Lemmens[1]
[1]SCK•CEN, Boeretang 200, B-2400 Mol, Belgium

## ABSTRACT

To assess the stability of spent fuel in the highly alkaline chemical environment of the Belgian *Supercontainer design*, static leach experiments were performed with depleted UO$_2$ and $^{238}$Pu-doped UO$_2$ at different SA/V ratios for 1.5 years in cement waters (11.7< pH < 13.5) at ambient temperature and under argon atmosphere. The influence of the calcium concentration on the uranium release was also investigated. While the ultrafiltered U(IV) concentration was 10$^{-9}$-10$^{-8}$ mol.L$^{-1}$ and independent of the pH, the U(VI) release from the UO$_2$ surface was enhanced by the OH$^-$ concentration, leading to soluble U concentrations up to 10$^{-5}$ mol.L$^{-1}$ at high SA/V and under the influence of radiolysis. Together with the high Ca concentration, this can lead to the formation of Ca-U(VI) colloids as precursor of Ca-U(VI) secondary phases, decreasing the soluble U concentration. The precipitation of Ca-U secondary phases was however not clearly evidenced by surface analyses.

## INTRODUCTION

The *Supercontainer design* is the current reference design for the geological disposal of spent nuclear fuel in Belgium [1]. In the case of spent UOX fuel, four assemblies are placed inside a Supercontainer with a 30 mm thick carbon steel overpack and a 540 mm thick concrete buffer. The Supercontainer will thus provide a highly alkaline chemical environment. After perforation of the overpack, the high pH of the infiltrating water may have an impact on the radionuclide release from the spent fuel. As the majority of published data related to radionuclide release is reported at neutral pH [2-3], a research program was started at the Belgian Nuclear Research Centre (SCK•CEN), supported by ONDRAF/NIRAS, to evaluate the stability of UO$_2$, as analogue of real spent fuel, in such alkaline environment. Our experimental program was defined to determine the UO$_2$ dissolution rate, the UO$_2$ solubility and the influence of α-radiation on UO$_2$ behavior. Complementary experiments with fresh spent fuel (SF) were done by KIT-INE [4, 5]. The current paper summarizes the results of static experiments with depleted UO$_2$ and Pu-doped UO$_2$ performed at SCK•CEN to study the effect of high pH on the UO$_2$ solubility.

## EXPERIMENTAL

All experiments were performed at 25 – 30°C under Ar atmosphere with pO$_2$ below 1 ppm and pCO$_2$ below 0.1 ppm. Static experiments were performed with depleted and $^{238}$Pu-doped UO$_2$ powders (table I), previously annealed (Argon/5% H$_2$ gas at 1000 °C), at four different ratios of fuel surface area to leachant volume (SA/V of 6, 17, 130 and 257 m$^{-1}$) and for more than one year with and without metallic iron as redox buffer. Three different cement waters were used for the experiments (table II), i.e. Young Cement Water with Calcium (YCWCa-pH 13.5-[Ca] 6.5×10$^{-4}$ mol.L$^{-1}$), Evolved Cement Water (ECW-pH 12.5-[Ca] 1.3×10$^{-2}$ mol.L$^{-1}$) and Old Cement Water (OCW-pH 11.7-[Ca] 7.3×10$^{-4}$ mol.L$^{-1}$). As Ca may strongly interact with uranium [4-7], a variant composition at pH 13.5 (Young Cement Water with less calcium –YCW- pH 13.5-[Ca] 3.8×10$^{-5}$

mol.L$^{-1}$) was also tested. All details concerning the Pu-doped fuels, the cement water composition/ preparation, and the setup parameters were previously described in [8].

The released $^{238}$U for depleted UO$_2$ and $^{233}$U for Pu-doped UO$_2$ was measured by ICP-MS and alpha-spectrometry, respectively to follow the uranium release as function of time. To better understand the UO$_2$ dissolution process, surface analysis (XPS) was performed, the uranium speciation of the ultrafiltered solutions was determined and the colloids were characterized by different filtrations, to differentiate the total uranium concentration (Not Filtered/NF), the large colloidal U fraction (> 200 nm), the small colloidal U fraction (2.6 to 200 nm), and the soluble U fraction (< 2.6 nm/ultrafiltered/UF). The total uranium fraction may vary depending on the handling/sampling of the leach solution, which was shaken before the sampling to stir up precipitated colloids (Not Filtered - Shaken/NF-S or Not Filtered - Not Shaken/NF-NS). The speciation of the uranium (U(IV)/U(VI)) was determined for the soluble/ultrafiltered fractions by anion-exchange chromatography in HCl medium [9-10]. All speciations and all filtrations were carried out in an Ar glove box to limit oxidation of the samples.

**Table I.** Description of tested UO$_2$ batches

| Fuel type | $^{233}$U Weight fraction[a] | $^{238}$Pu Weight fraction[b] | α-activity [MBq/g UO$_2$] | grain size / average grain size (μm) | Surface area x 10$^{-2}$ m$^2$.g$^{-1}$ |
|---|---|---|---|---|---|
| F1 | 3.92.10$^{-3}$ | 3.98.10$^{-4}$ | 2400 | 100-200 / 150 | 1.4 |
| F2 | 4.22·10$^{-3}$ | 5.52·10$^{-5}$ | 36 | < 100 / 75 | 3.2 |
| F4 | 4.22·10$^{-3}$ | 2.49·10$^{-5}$ | 17 | <100 / 75 | 3.2 |
| Depleted UO$_2$ | | | 0.01-0.04 | 100-200 / 150 | 1.4 |
| | | | | 50-100 / 75 | 3.2 |

a: mass of $^{233}$U/mass of doped UO$_2$, b: mass of $^{238}$Pu/mass of doped UO$_2$, c: based on literature survey [11] and calculated with the average grain size of the powder fraction.

**Table II.** Composition of the cement waters (YCWCa, ECW and OCW) in mol.L$^{-1}$.

| | Na | K | Ca | Al | Si | SO$_4^{2-}$ | CO$_3^{2-}$ | pH |
|---|---|---|---|---|---|---|---|---|
| YCWCa | 1.4×10$^{-1}$ | 3.3×10$^{-1}$ | 6.5×10$^{-4}$ | < 7×10$^{-6}$ | 2.0×10$^{-4}$ | 2×10$^{-3}$ | 9.3×10$^{-4}$ | 13.5 |
| YCW | 1.4×10$^{-1}$ | 3.3×10$^{-1}$ | 3.8×10$^{-5}$ | < 7×10$^{-6}$ | 2.0×10$^{-4}$ | 2×10$^{-3}$ | 9.3×10$^{-4}$ | 13.5 |
| ECW | 1.6×10$^{-2}$ | 2.2×10$^{-4}$ | 1.3×10$^{-2}$ | 1.1×10$^{-5}$ | 1.8×10$^{-4}$ | 3×10$^{-6}$ | 4.8×10$^{-4}$ | 12.5 |
| OCW | 3.7×10$^{-3}$ | | 7.3×10$^{-4}$ | 3.7×10$^{-6}$ | 1.5×10$^{-4}$ | 4×10$^{-5}$ | 1.4×10$^{-4}$ | 11.7 |

## RESULTS AND DISCUSSION

The α-activity and the SA/V ratio did not influence the pH, and iron was ineffective as reductant at such alkaline pH [8]. Hence, we did not further distinguish the uranium concentrations with and without redox buffer. Moreover, in this paper, the variation of the uranium concentrations with time is not discussed explicitly. Instead we focus on the range within which the concentrations vary. For the tests with depleted UO$_2$, the $^{238}$U concentrations are shown. For the tests with α-doped UO$_2$, the normalized $^{233}$U concentrations are indicated, i.e. the measured molar $^{233}$U concentration, divided by the $^{233}$U mole fraction in the UO$_2$.

### Experiments in absence of alpha-radiation

The total uranium concentrations with depleted UO$_2$ varied between 10$^{-9}$ up to 3 ×10$^{-5}$ mol.L$^{-1}$ at all pH-values, and are quite equivalent for all media, at constant SA/V (figure 1A). Higher uranium concentrations generally correspond to higher SA/V ratios and can be related to the shaking of the solution before the sampling. The concentrations of $10^{-7}$ up to $10^{-5}$ mol.L$^{-1}$ in the stirred solutions are caused by the presence of large colloids (> 200 nm) as revealed by the different filtrations. The ratio of colloidal/total uranium was smaller in YCW with very low Ca concentration (series 'no Ca' of figures 1 and 2), as well as in systems with little (pre-)oxidized U(VI) (using UO$_2$ pellets) compared to systems with high SA/V ratios and a large (pre-)oxidized surface (using UO$_2$ powder). For test series that were performed simultaneously, the soluble uranium concentration was lower in ECW (with much Ca) than in YCWCa or OCW (with less Ca). This is illustrated in figure 4 for the tests with Pu-doped UO$_2$ F1 at 6 m$^{-1}$ (see next section). These are indications that the colloids formed in the cement water are Ca-U(VI) associations.

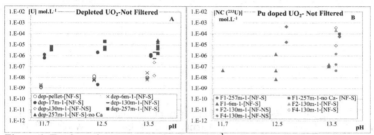

**Figure 1.** Total uranium concentration in mol.L$^{-1}$ as function of pH. **A**- without α-radiation, **B**- with α-radiation.
[NF-S]: Not Filtered/Shaken concentration. [NF-NS]: Not filtered/Not Shaken concentration. SA/V ratio increases from 6 to 257 m$^{-1}$ from the left to the right for each pH value.

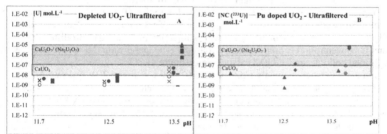

**Figure 2.** Soluble uranium concentration (UF) in mol.L$^{-1}$ as function of pH and solubility range of U(VI) phases. **A**- without α-radiation, **B**- with α-radiation.
SA/V ratio increases from 6 to 257 m$^{-1}$ from the left to the right for each pH value.

The soluble uranium concentration was in the range between $10^{-9}$ to $10^{-7}$ mol.L$^{-1}$ at low SA/V, but up to $10^{-5}$ mol.L$^{-1}$ at high SA/V (figure 2A). The major factors influencing the concentration were SA/V, Ca concentration (see previous paragraph) and the pH. The Pu doping is another important factor (see next section). Concerning the pH effect, the results of the uranium speciation indicate that U(VI) dominates the soluble uranium at all pH values (and all SA/V ratios) (figure 3). The measured U(IV) concentration was close to the U(IV) solubility of $10^{-9}$-$10^{-8}$ mol.L$^{-1}$ and independent of the pH , as suggested by Neck [12]. While the U(IV) concentration was constant, the U(VI) concentration slightly increases when pH increases, especially from pH 12.5 to pH 13.5. These tests were done with depleted $UO_2$, where there should be only little radiolytical oxidation, so the U(VI) in the system should be attributed to (pre-)oxidation by traces of atmospheric oxygen in the glove box. This should be independent of pH. Therefore, we assume that the higher dissolved U(VI) concentrations at higher pH are caused by the increased complexation of U(VI) with OH⁻, as proposed by Tits [13]. The pH effect becomes very pronounced at high SA/V (figures 3C and 2A). Ollila [14] also reports an increasing uranium concentration from $4 \times 10^{-10}$ (pH 11) to $2 \times 10^{-9}$ mol.$L^{-1}$ (pH 13).

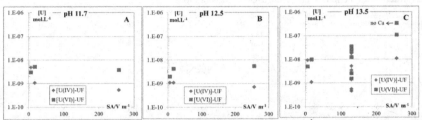

**Figure 3.** [U(IV)] and [U(VI)] soluble concentrations (UF) in mol.$L^{-1}$ as function of the SA/V ratio, for the static tests with depleted $UO_2$ powder in the different cement waters. **A**: OCW-pH 11.7, **B**: ECW-pH 12.5 and **C**: YCWCa-pH 13.5.

## Experiments with alpha-radiation

The total uranium concentrations were enhanced by a factor of 10-100 ($10^{-8}$-$10^{-3}$ mol.$L^{-1}$) in presence of alpha-radiation (Pu-doped $UO_2$) due to oxidative dissolution (figure 1B). In figure 1, it can be seen that apart from radiolysis, also an increase of the SA/V ratio raises the U concentration, due to the larger $UO_2$ surface. Comparison of figure 1B (total [U]) with figure 2B (ultrafiltered [U]) shows that the total U concentration is predominantly colloidal. The larger ratio (NF-S)/UF in ECW suggests that the formed colloids are Ca-U(VI) associations (figure 4).

The soluble concentrations range mostly from $10^{-8}$ to $10^{-5}$ mol.$L^{-1}$ (figure 2B). At low SA/V, they are lower (figure 4). Although the uranium speciation was not determined in the experiments with alpha-radiation, we assume that the higher uranium concentration is due to radiolysis, and therefore is mainly U(VI). Indeed, the soluble uranium concentration was in most cases higher than the U(IV) solubility of about $3 \times 10^{-9}$ mol.$L^{-1}$ at pH 7-14 [12].

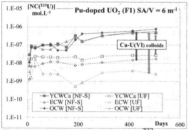

**Figure 4.** Normalized [U] (NC[233U]) as a function of time for the tests with 238Pu-doped UO$_2$ F1 at 6 m$^{-1}$.
Not Filtered-Shaken concentration [NF-S] and Ultrafiltered concentration [UF].

### Comparison with the literature data

In most static experiments with depleted UO$_2$, the ultrafiltered concentrations stopped increasing with time. This suggests that uranium in solution reached equilibrium with the depleted UO$_2$ surface or – for higher U concentrations - with U(VI) secondary phases, which may be formed with Ca, Na and K from the cement waters. Based on thermodynamic considerations, calcium uranate (CaUO$_4$) is the only stable U(VI) phase able to precipitate in a system with Na and Ca in the pH range 12-14, with a uranium equilibrium concentration around $10^{-14}$ mol.L$^{-1}$ [15]. According to literature, the experimentally measured solubility of most U(VI) phases in high pH/cement conditions is, however, a few orders of magnitude higher than $10^{-14}$ mol.L$^{-1}$, and seems pH dependent (table III). Hence, a possible range for solubility of U(VI) phases would be $10^{-8}$-$10^{-5}$ mol.L$^{-1}$ in the cementitious pH range such as in our systems.

**Table III.** Solubility of U(VI) phases at pH > 10

| Phases | Solubility mol.L$^{-1}$ | pH range | Remark | References |
|---|---|---|---|---|
| CaUO$_4$ | $10^{-7}$- $10^{-8}$ | 9-13.5 | | [16] |
| CaUO$_{4(am)}$ | $4 \times 10^{-6}/10^{-5}$ | 12.2 / 13.5 | U(VI) complexation increase with OH$^-$ | [13] |
| CaU$_2$O$_7$.3H$_2$O$_{(cr)}$ | $10^{-5}$ | 9-13.5 | No CaUO$_4$ with 0.1 < [Ca] M < 5 | [17] |
| α-Na$_2$U$_2$O$_7$ | $5 \times 10^{-7}$ | 13.4 | [Ca]= $10^{-3}$ M, Eh= -300 mV$_{/SHE}$ | [18] |
| Na$_2$U$_2$O$_7$ | $10^{-7}/10^{-6}/10^{-5}$ | 11/ 12/ 13.5 | anoxic conditions | [19] |
| K$_2$U$_2$O$_7$ | $2 \times 10^{-7}$ | pH > 10 | brines (thermodynamic calculations in KOH system) | [20] ([21]) |
| U(VI)-Ca-Si | $10^{-9}$ - $10^{-7}$ | pH > 11 | [Si] > $10^{-3}$ M | [16] [22] |

Using the solubility data from the literature (table III), we can evaluate the possible solubility controlling phases in our tests with depleted UO$_2$. The soluble uranium concentrations with depleted UO$_2$, range from $10^{-9}$ to < $10^{-8}$ mol.L$^{-1}$ at pH 11.7 and at pH 12.5, and from $10^{-9}$ to $10^{-5}$ (at high SA/V) mol.L$^{-1}$ at pH 13.5 (figure 2A). According to table III, solutions at pH 11.7 are probably undersaturated with respect to all U(VI) phases. In ECW at pH 12.5 and low SA/V ratio, the solution still remains undersaturated with respect to the U(VI)-phases, but the uranium

271

concentration may approach saturation with $CaUO_4$ at high SA/V ratio (figure 2A). At pH 13.5, the solutions are probably (over)saturated w.r.t. the mentioned phases.

The influence of Ca, decreasing the $UO_2$ solubility, was also observed in $UO_2$ leach tests performed in cement water of pH 12.7 ($NaOH$, $Ca(OH)_2$ and $KOH$), compared to $UO_2$ leach tests using KOH water [6], but no phases were proposed to explain the uranium concentration evolution. The oversaturation with respect to a Ca-U(VI)-phase may explain the colloid formation, assuming that these colloids can be considered as suspended precipitates. The colloid formation in YCWCa (pH 13.5) may be explained by formation of a Ca-,Na-, or K-uranate. Ca and Na uranates are more likely to be formed than K-uranate, as the K-uranate was not obviously proven in cement systems. The correlation between the Ca concentration and the colloidal fraction supports the formation of Ca-uranate rather than Na-uranate in this system. Some of the Ca-U phase may have precipitated on the $UO_2$ surface, as suggested by surface analysis (XPS) of our depleted $UO_2$ samples showing the presence of Ca on the $UO_2$ surface, but Ca-uranate was not unambiguously identified on the $UO_2$ surface. In absence of Ca (YCW) at pH 13.5, uranium concentrations could be controlled by $Na_2U_2O_7$ (figure 2A) [18, 19]. The potential presence of Ca-U(VI) and Na-uranate phases was also suggested by thermodynamic calculations with Geochemist's Workbench [23] and our in-house database (Moldata- [24]).

Based on the reasoning made for tests without radiation and the measured uranium concentrations of $10^{-8}$ mol.$L^{-1}$ at pH 11.7, $10^{-9}$ -$10^{-8}$ mol.$L^{-1}$ at pH 12.5, and $10^{-8}$-$10^{-5}$ mol.$L^{-1}$ at pH 13.5, we assume that $CaUO_{4(am)}$ represents the most probable solubility limiting phase for tests with Pu-doped $UO_2$ at low SA/V and all pH. For test at pH 13.5, and at high SA/V, rather Na- or Ca-uranate seem to represent the solubility controlling phase, depending on the Ca concentration (figure 2B). At the present time, we cannot confirm these hypotheses because the interpretation of the XPS analyses of the Pu-doped $UO_2$ are not yet performed, so far.

Precipitation of Ca-containing solids has been observed in tests with fresh spent fuel in ECW, giving a precipitated layer that may lead to a partial protection against further fuel oxidation-dissolution [4-5]. This was also suggested by tests with $UO_2$ in Ca-containing solutions at neutral pH [25]. Calcium may also absorb on the $UO_2$ surface and may so inhibit the oxidation/dissolution of the fuel [7]. In our tests, these effects were probably small, because the presumed secondary phases were formed in the solution, rather than on the $UO_2$ surface.

## CONCLUSION

The behavior of the $UO_2$ matrix in alkaline cement waters was studied in static experiments. The high pH did not increase the soluble U(IV) concentrations, but it increased the soluble U(VI) concentrations due to formation of hydroxocomplexes. It increased less in the presence of Ca. The total [U] concentrations increased with SA/V and alpha dose, but were not much influenced by the pH or Ca concentration. In tests with $UO_2$ powders, almost all U is colloidal, except when Ca is absent. The colloids are considered as a precursor of precipitation of Ca-U(VI) phases such as $CaUO_4$ or Ca-uranate, but macroscopic evidence for this is still lacking. The colloid formation is enhanced by SA/V, and is much less observed with realistic fuel geometries (pellets). Further investigations are needed to corroborate these preliminary results.

## ACKNOWLEDGEMENTS

This work was performed as part of the program of the Belgian Agency for Radioactive Waste and Enriched Fissile Materials (NIRAS/ONDRAF) on the geological disposal of high-level/long-lived radioactive waste. The authors gratefully acknowledge the technical support from Ben Gielen and Regina Vercauter.

## REFERENCES

1. J.J.P. Bel *et al.*, Mater. Res. Soc. Symp. Proc. Vol 932, pp.23-32, (2006).
2. B. Grambow *et al.*, Report JRC-ITU-SCA-2005/01, (2005).
3. B. Grambow *et al.*, Final Project Report: MICADO final report, (2010).
4. A. Loida *et al.*, Mater. Res. Symp. Proc. Vol 1193, pp 597-604, (2009).
5. A. Loida *et al.*, Mater. Res. Symp. Proc. Vol. 1475, pp 119-124, (2012).
6. D. Cui *et al.*, Mater. Res. Soc. Symp. Proc. Vol. 757, pp. 359-364, (2003).
7. B.G. Santos *et al.*, Journal of Nuclear Materials 350, pp 320-331, (2006).
8. T. Mennecart *et al.*, Mater. Res. Symp. Proc. Vol. 1475, 293-298, (2012).
9. M. Hussonnois *et al.*, Mater. Res. Soc. Symp. Proc. Vol. 127, pp 979-984, (1989).
10. K. Ollila, Posiva report, POSIVA-96-01, (1996).
11. V. M. Oversby, SKB Technical report TR-99-22 (1999).
12. V. Neck *et al.*, Radiochimica Acta 89, pp 1-16, (2001).
13. J. Tits *et al.*, Mater. Res. Soc. Symp. Proc. 1107, pp 467-474, (2008).
14. K. Ollila, Posiva report 2008-75, 2008.
15. L. Wang *et al.*, NIROND-TR report 2008-23 (2009).
16. L.P. Moroni *et al.*, Waste Management Vol. 15, N° 3, pp 243-253, (1995).
17. M. Altmaier *et al.*, Migration 2005, Avignon- France Septembre 18-23, (2005).
18. U. Berner *et al.*, PSI report PSI Bericht -99-10, (1999).
19. S. Meca *et al.*, Radiochimica Acta 96, pp 535-539 (2008).
20. M. Sutton, PhD. Thesis, Loughborough University (1999).
21. H. Yamazaki. *et al.*, Radioactive Waste Management vol 2., pp1607- 1611, (1992).
22. D.M. Wellman *et al.*, Cement and Concrete Research 37, pp 151-160, (2007).
23. C.M. Bethke, S. Yeakel, The Geochemist workbench release 9.0. December 2011.
24. S. Salah *et al.*, First Full Draft. SCK•CEN report, ER-198, (2012).
25. J. M. Cerrato *et al.*, Environ. Sci. Technol. 46, pp 2731-2737 (2012).

Mater. Res. Soc. Symp. Proc. Vol. 1665 © 2014 Materials Research Society
DOI: 10.1557/opl.2014.655

# Interpretation of Knudsen Cell Experiments to determine the Instant Release Fraction in Spent Fuel Corrosion Scenarios by using a Mechanistic Approach: the Caesium Case

Daniel Serrano-Purroy[1], Laura Aldave de las Heras[1], Jean-Paul Glatz[1], Ondrej Benes[1], Jean Yves Colle[1], Rosa Sureda[2], Ernesto González-Robles[2,*], Joan de Pablo[2], Ignasi Casas[2], Marc Barrachin[3], Roland Dubourg[3], Aurora Martínez-Esparza[4]

[1]European Commission, Joint Research Centre, Institute for Transuranium Elements, P.O. Box 2340, D-76125 Karlsruhe, Germany
[2]CTM Centre Tecnològic, Avda. Bases de Manresa 1, 08240 Barcelona, Spain
[3]Institut de Radioprotection et de Sûreté Nucléaire (IRSN), B.P. 3, F-13115 St Paul lez-Durance, France
[4]ENRESA, C/Emilio Vargas 7, 28043 Madrid, Spain
* Present address: Institute for Nuclear Waste Disposal (INE), Karlsruhe Institute of Technology, P.O. Box 3640, D-76021 Karlsruhe, Germany.

## ABSTRACT

The Knudsen Effusion Mass Spectrometer (KEMS) and the mechanistic MFPR (Module for Fission Product Release) code are tools which seem particularly interesting to support studies of the Instant Release Fraction (IRF) of Cs from spent nuclear fuel in a final repository. With KEMS, the thermal release of $^{137}$Cs and $^{136}$Xe were analysed by annealing up to total vaporization (2500K) of high burn-up (60 GWd/tU) Spent Nuclear Fuel (SNF) samples. Powder samples from the centre of the fuel, without high burn-up structure, were used. To determine the IRF, samples were analysed before and after being submitted to corrosion experiments in bicarbonated aqueous media.

MFPR was applied to determine the localization of Cs and fission gases in the SNF at the end of irradiation; the results are compared and supported by dedicated thermodynamics calculations performed for equilibrium conditions at various temperatures and fuel oxygen potentials by the non-ideal thermodynamic MEPHISTA (*Multiphase Equilibria in Fuels via Standard Thermodynamic Analysis*) database. A possible mechanism for Cs release during thermal annealing is proposed, taking into account inter-granular release and Cs oxide vaporization, atomic diffusion, ternary oxide phase formation and bubble release.

Differences in KEMS release profiles before and after submitting the samples to aqueous corrosion are attributed to the IRF and to changes in the vaporisation mechanism because of differences in the oxygen potential ($pO_2$). The IRF of Cs estimated from the KEMS spectra, consisting on the part located at the grain boundaries and in inter-granular bubbles, is not significantly different from that corresponding to the experimental results found using classical static leaching experiments.

New experimental campaigns are being designed to confirm our interpretation proposed after this first run.

## INTRODUCTION

The Instant Release Fraction (IRF) is an important parameter in Performance Assessment exercises of nuclear waste repositories. Among others, Cs is an important element to be taken into account in both, short-term ($^{137}$Cs) and, to a minor extend, long-term ($^{135}$Cs) release scenarios.

In the last decade several IRF studies have been carried out at ITU using commercial high burn-up $UO_2$ and MOX spent nuclear fuels (SNF) under oxidative conditions. Experiments have been carried out in different leaching solutions, i.e. bicarbonate water, bentonitic-granitic water, sea-water [1-7] and using samples with different morphologies: cladded segments, powder from the centre and from the periphery (enriched in high burn-up structure) of the fuel and fragments. Cs IRF was calculated as the difference between the Fraction of Inventory in the Aqueous Phase (FIAP) for Cs minus the FIAP for the matrix U. In all cases, results presented two different contributions:

An "instant" release, of up to a few days, attributed to the fraction of Cs segregated from the matrix and present at the gap and/or at the grain boundaries immediately in contact with the leaching solution at the beginning of the experiments.

A "fast" release (months, years) attributed to the Cs fraction segregated from the matrix present at the grain boundaries not-directly accessible to water at the beginning of the experiment, the so-called internal grain boundaries. This contribution slows down with time due to the less accessibility to the remaining grain boundaries.

Whereas a clear number can be given for the first contribution it is difficult to determine the second contribution and therefore give a total IRF contribution.

A tool, which seems particularly interesting to study the IRF of Cs, is the Knudsen Effusion Mass Spectrometer (KEMS). Previous experiments carried out at ITU with non-leached samples proposed a release mechanism of gases during the annealing process [8-11]. However nothing is reported about effusion mechanism of Cs. In addition no experiment has been carried out using pre-leached samples.

In the present study, release experiment of Cs and gas by thermal annealing up to total vaporization (2500K) were performed on powder samples from the centre of high burn-up (60 GWd/t U) SNF samples. In order to determine the IRF, samples were analysed before and after being submitted to corrosion experiments in bicarbonated aqueous media.

In order to interpret the results, the mechanistic MFPR (*Module for Fission Product Release*) code [12-13] which models the behaviour of the fission products (gaseous and chemically-active) in close connection with the evolution of the microstructure in irradiation and in annealing regime was applied to reproduce the KEMS vaporisation experiments of Cs and gases in the SNF samples.

From the fuel power history during normal reactor operations, MFPR calculations were carried out to determine the localization and speciation of Cs and fission gases before the simulation of the heat up during KEMS experiments in the SNF at the end of irradiation. This localization is a key point to interpret the KEMS vaporization profiles of Cs and Xe release. The results from MFPR are compared and supported by dedicated thermodynamics calculations performed for equilibrium conditions at various temperatures and fuel oxygen potentials by the non-ideal thermodynamic MEPHISTA (*Multiphase Equilibria in Fuels via Standard Thermodynamic Analysis)* database [14]. The Cs behaviour is quite close to that of Xe at high temperatures were both are gaseous and for comparison Xe profiles were also calculated. A possible mechanism for Cs release during thermal annealing is proposed, taking into account inter-granular release and Cs oxide vaporization, atomic diffusion, ternary oxide phase formation and bubble release.

In addition KEMS vaporization experiments were also carried out with the same sample after aqueous corrosion. Differences in the release profiles were attributed to the IRF released in the aqueous media and changes in the vaporization mechanism because of differences in the oxygen potential ($pO_2$).

## EXPERIMENTAL
### Spent Nuclear Fuel sample
The selected SNF was irradiated in a commercial PWR nuclear power plant under normal operating conditions, discharged beginning of 2001, with a mean burnup of 60 GWd/tU. The Fission Gas Release (FGR) related to the release of Kr and Xe measured in a puncturing test, was of (15±1) % (14.6% of Xe and 14.8% of Kr isotopes in relation to the calculated inventories). In order to simplify the study the sample was selected from the centre

of the pellet without the high burn-up structure. The preparation of the sample was described previously [1, 4].

**Leached sample**

Part of the powder sample was leached for more than one year in bicarbonate water under oxidative conditions [1, 4]. After more than one year and taking into account the total fraction of Cs released at the end of the experiment including the washing pre-treatments, the experimental Cs IRF was $5.7 \pm 0.9$ %.

**KEMS**

The experimental set-up combining a Knudsen cell with a quadrupole mass spectrometer was described earlier [9]. Two annealing experiments were carried out using approximately 6 mg of non-leached and leached powder sample. All the annealing experiments were performed in vacuum in the temperature range 300-2500K. A heating rate of 10 K/min was used in the experiments. All fission products, fission products compounds as well as helium and actinides oxides were measured. In addition Gas fission products were measured by Q-GAMES [15].

**MODELLING**

In this study, two mechanistic approaches (MFPR, *Module for Fission-Product Release*, and MEPHISTA, *Multiphase Equilibria in Fuels via Standard Thermodynamic Analysis*), originally developed for PWR studies by the French 'Institut de Radioprotection et Sûreté Nucléaire' in close collaboration with the Russian Nuclear Safety Institute (IBRAE) and the THERMODATA/CNRS/INPG Laboratory, are used to analyse KEMS experiments. MFPR is a computer code mechanistically modelling Fission Product behaviour in irradiated UO2 fuel in normal and accident situations. It describes the evolution of the various defects of the fuel microstructure and their interaction with fission-gas atoms and bubbles. The model also includes the chemistry of FPs in the temperature range 500–3000 K by considering the chemical equilibrium at the grain boundary of the multicomponent and multiphase U–O–FP system. MEPHISTA is a self-consistent non-ideal thermodynamic database designed for thermochemical equilibria calculations. The CALPHAD *(Calculation of Phase Diagram)* approach [16] is used to obtain the assessed parameters of the Gibbs energies of all the phases in the chemical system Ba–Cs–La–Mo–O–Pu–Ru–Sr–U–Zr + Ar–H from the crystallographic structure of the phases and all the experimental thermodynamic and phase boundary data available.

**RESULTS AND DISCUSSION**

As can be seen in Figure 1, where the KEMS release curves before and after leaching vs. temperature for $Xe^{136}$ are shown, there is a significant effect of the leaching pre-treatment on the effusion mechanism, especially at temperatures between 1100 and 1800K.

In order to explain this significant effect in the KEMS spectra, the fuel behaviour during normal irradiation conditions was calculated by MFPR. In addition, the annealing conditions during the KEMS experiments were also simulated changing some parameters in the calculation such as the migration activation energy and the gas atom capture trying to take into account changes in fuel $pO_2$ during the leaching experiments.

In the case of the non-leached sample, three gas migration steps were accounted for: (I) diffusion and venting of gas located on grain boundaries and dislocation networks (800 to 1250K); (II) volume diffusion from the body to the grain boundaries (1250 to 1900K), in this case gas capture is prevented if we consider that bubbles are not totally relaxed and over-pressurized, and (III) release of gas previously trapped in almost immobile intragranular bubbles due to the progressive sublimation of the sample (1900-2500K). These results are in good agreement with previous assumptions found in the literature [8].

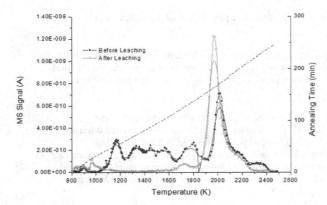

**Figure 1.** Mass spectrometry signal normalized by sample weight: release curve for $Xe^{136}$ before and after leaching vs. temperature. Right axis represents the temperature ramp.

However, major changes for the first two migration steps were identified after submitting the sample to a leaching pre-treatment: (I) no significant release of Xe located at the grain boundaries was found. This fraction, of approximately 6 % of the initially trapped Xe, is expected to be released during the leaching experiment. This fractional release would correspond to the total IRF, including the internal grain boundary contribution, and can be determined after submitting the sample to KEMS for some hours, thus giving a reliable and fast method for gas IRF determination, (II) no diffusion from the body to the grain boundaries was detected. Taking into account the oxidative nature of the leaching experiment, this behaviour was reproduced in MFPR by increasing the $pO_2$ in the sample. This creates defects in the uranium lattice which increase bubble relaxation and gas atom capture and will hinder diffusion to the grain boundaries. As seen in Figure 2, where the effusion profile for uranium and uranium oxides in both samples are shown, the $UO_3$ content after leaching is significantly higher indicating an increase of the oxygen potential in the sample, most probably from an inhomogeneous oxidation state of the sample (surface oxidation) However, more experiments are needed to clarify this point. And finally (III), as for the non-leached sample the peak at 2100K corresponds to the release of the gas trapped in intragranular bubbles.

In Figure 3 the KEMS effusion curves before and after leaching vs. temperature for $Cs^{137}$ are shown. Similar assumptions as for the Xe were taken into account when reproducing the experimental results by MFPR but this time including the possible formation of Cs compounds like $Cs_2MoO_4$ and CsI. In addition the MEPHISTA database was used to calculate the Cs speciation in function of the fuel $pO_2$ and temperature.

In this case five different migration steps were identified before leaching corresponding to: (I) and (II) release from Cs trapped in intergranular positions either as

(I) oxides form ($Cs_2O$, CsO, $CsO_2$...) (800-1150K) or
(II) in gas intergranular bubbles (1150-1250K),
(III) Cs release by atomic diffusion in the bulk (1250-1750K).
(IV) Release from CsI and/or $Cs_2MoO_4$ vaporization (1750-2000K) and
(V) release from the gas trapped in intragranular bubbles (2000-2250K).

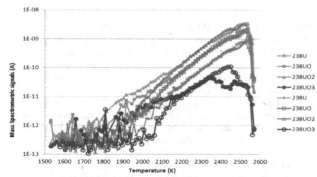

**Figure 2.** Mass spectrometry signal normalized by sample weight: release curve for U²³⁸ and related oxides, before and after leaching vs. temperature. Closed symbols are for sample before leaching and open symbols for sample after the leaching.

Major changes in the Cs release behaviour observed after the leaching treatment are linked to the first peaks: (I) and (II) these peaks, corresponding to the IRF, disappear. This fraction corresponds to approximately 6% of the total released Cs in the untreated sample. The same IRF result was obtained at the end of the leaching experiment, thus confirming the suitability of the KEMS for Cs IRF determinations. (III) Similar to Xe, changes in pO₂ will significantly change the migration mechanism of Cs through the bulk, either by speeding-up the diffusion mechanism and shifting the release from 1400 to 1100K (this effect was already reported in [10,11]) or by increasing the trapping in intragranular bubbles and therefore shifting the release to the end of the experiment. (IV) This peak is not different from the one found in the non-treated sample and could correspond to CsI and/or Cs₂MoO₄ release. MEPHISTA does not predict any formation of other Cs ternary oxides in our conditions. It is also possible that part of the Cs that diffuses through the matrix will react with oxidized Mo to form Cs₂MoO₄. And, finally, (V) the remaining Cs, trapped in intragranular bubbles is released together with the matrix sublimation.

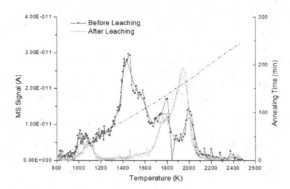

**Figure 3.** Mass spectrometry signal normalized by sample weight: release curve for Cs¹³⁷ before and after leaching vs. temperature. Right axis represents the temperature ramp.

## CONCLUSIONS

IRF was identified in KEMS spectra by means of comprehensive MFPR and MEPHISTA calculations. A similar Cs IRF was found either using a classical static leaching experiment or KEMS, thus proving. KEMS as a good tool for short time determination of total segregated Cs (IRF).

Using MFPR, three different mechanisms were identified for Xe release: intergranular release, atomic diffusion and intragranular bubbles, confirming literature findings

In the case of Cs very similar behaviour was found, although with some particularities due to the lower volatility of Cs and the chemical interaction with other species, such as I or Mo (when oxidised) during irradiation. MFPR needs some improvement in order to predict the complete release profile of Cs; considering its mixed behaviour (gas and chemically active).

The leaching experiment has an effect on the effusion mechanism, mainly in that of the atomic diffusion. Based on some evidence of oxidation of the fuel during the leaching treatment, the effect of increasing $pO_2$ was studied. An hypothesis is proposed where most of the bulk segregated Xe and Cs diffuses to intragranular bubbles and is retained until matrix vaporization.

## REFERENCES

1. D. Serrano-Purroy, F. Clarens, E. González-Robles, J.P. Glatz, D.H. Wegen, J. de Pablo, I. Casas, J. Giménez, A. Martínez-Esparza . J. Nucl. Mat. 427 249–258 (2012).
2. D. Serrano-Purroy, I. Casas, E. González-Robles, J. P. Glatz, D.H.Wegen, F. Clarens, J. Giménez, J. de Pablo, A. Martínez-Esparza.J. Nucl. Mat. 427 249–258 (2012).
3. D. Serrano-Purroy, V.V. Rondinella, D.H. Wegen. Characterisation of Irradiated MOX: Static leaching Report, NF-PRO Deliverable 1.4.10. ITU Technical Report JRC-ITU-TPW-2007/37.
4. E. González-Robles Study of Radionuclide Release in commercials UO2 Spent Nuclear Fuels, Doctoral Thesis, Universitat Politécnica de Catalunya, 2011.
5. V.V., Rondinella D. Serrano-Purroy. J.-P. Hiernaut, D. Wegen, D. Papaioannou, M. Barker, "Grain boundary inventory and instant release fractions for SBR MOX", Proc. International High-Level Radioactive Waste Management Conference, Sept 7-9, 2008, Las Vegas, USA, paper 195780, ANS CD-ROM.
6. J.de Pablo, D. Serrano-Purroy, E. Gonzalez-Robles, F. Clarens, A. Martínez-Esparza, D.H. Wegen, I. Casas, B. Christiansen, J.-P. Glatz, J. Giménez in Scientific Basis for Nuclear Waste Management XXXVII, edited by B.E. Burakov and A.S. Aloy, (Mater. Res. Soc. Symp. Proc. 1193, Warrendale, PA, 2009)pp. 613-620..
7. D. Serrano-Purroy, I. Casas, E. González-Robles, J.P. Glatz, D.H. Wegen, F. Clarens, J. Giménez, J. de Pablo, A. Martínez-Esparza. J. Nucl. Mat. 434,451-460 (2013).
8. J.P. Hiernaut, C.Ronchi. J. Nuc. Mat. 294, 39-44 (2001).
9. J.P. Hiernaut, J.Y. Colle, R.Pflieger-Cuvelier,J. Jonnet, J. Somers, C. Ronchi. J. Nucl. Mat. 344, 246-253 (2005).
10. J.Y. Colle, J.P. Hiernaut, D. Papaioannou, C. Ronchi, A. Sasahara. J. Nucl. Mat. 348, 229-242(2006).
11. J.P. Hiernaut, T. Wiss, D. Papaioannou, R.J.M. Konings, V.V. Rondinella.J. Nucl. Mat.372, 215-225 (2008).
12. M.S.Veshchunov, V.D. Ozrin, V.E. Shestak, V.I. Tarasov, R. Dubourg, G. Nicaise, Nucl. Eng. and Des. 236, 179-200 (2006).
13. Veshchunov M.S., Dubourg R., Ozrin V.D., Shestak V.E., Tarasov V.I., J. Nucl. Mat. 362, 327-335 (2007).
14. B. Cheynet, E. Fischer, http://hal.archives-ouvertes.fr/hal-00222025.

15. J.-Y. Colle, E. A. Maugeri, C. Thiriet, Z. Talip, F. Capone, J.-P. Hiernaut, R. J. M. Konings, T. Wiss, Jean-Yves Colle , Emilio A. Maugeri , Catherine Thiriet , Zeynep Talip, Franco Capone, Jean-Pol Hiernaut, Rudy J. M. Konings, and Thierry Wiss, Journal of nuclear science deans technology Accepted for publication

16. N. Saunders, A.P. Miodownik, CALPHAD calculation of phase diagrams, Elsevier Science Ltd, Oxford (1998)

Mater. Res. Soc. Symp. Proc. Vol. 1665 © 2014 Materials Research Society
DOI: 10.1557/opl.2014.656

# Physico-chemical characterization of a spent $UO_2$ fuel with respect to its stability under final disposal conditions

Ernesto González-Robles[1], Detlef H. Wegen[2], Elke Bohnert[1], Dimitrios Papaioannou[2], Nikolaus Müller[1], Ramil Nasyrow[2], Bernhard Kienzler[1], and Volker Metz[1]

[1] Karlsruhe Institute of Technology (KIT), Institute for Nuclear Waste Disposal (INE), Hermann-von-Helmholtz Platz 1, D-76344 Eggenstein-Leopoldshafen, Germany
[2] European Commission, Joint Research Centre, Institute for Transuranium Elements, P.O. Box 2340, D-76125 Karlsruhe, Germany

## ABSTRACT

Two adjacent fuel rod segments were irradiated in a pressurized water reactor achieving an average burn-up of 50.4 GWd/$t_{HM}$. A physico-chemical characterisation of the high burn-up fuel rod segments was performed, to determine properties relevant to the stability of the spent nuclear fuel under final disposal conditions. No damage of the cladding was observed by means of visual examination and γ-scanning. The maximal oxide layer thickness was 45 μm. The relative fission gas release was determined to be (8.35 ± 0.66) %. Finally, a rim thickness of 83.7 μm and a rim porosity of about 20% were derived from characterisation of the cladded pellets.

## INTRODUCTION

During irradiation in the reactor and also during the cooling time after discharge, the nuclear fuel, as well as the cladding materials, undergoes transformation due to the temperature, external irradiation and fission reactions. As a consequence of the high temperature the cladding can creep and the generated fission products cause a swelling of the fuel [1]. On the other hand, at burn-up (BU) higher than 40 GWd/$t_{HM}$ the formation of porous and fine grained micro-structure is observed , commonly referred as rim structure or high burn-up structure (HBS) [2].

The first step in analysis of irradiated fuel rods is the non-destructive testing (NDT). NDT is an essential set of analyses that allows to consistently acquire reliable data needed for validation of the safety and efficient performance of the fuel rod in pile and to provide a valuable basis of information to plan and implement successful destructive post irradiation examination [3]. Moreover, results of the NDT are of importance for forthcoming studies on the stability of the spent nuclear fuel (SNF) under conditions of deep geological disposal.

This paper focuses on NDT as visual examination, γ-scanning, and oxide layer thickness as well as on destructive analysis which allows the quantification of the fission gases release into the plenum and the characterisation of the HBS.

## EXPERIMENTAL

### Characteristic data of the spent nuclear fuel

Segments N0203 and N0204 of the fuel rod SBS 1108 were irradiated in the pressurised water reactor of Gösgen, Switzerland, during four cycles of 1226 days in total with an average linear power of 260 W/cm achieving an average BU of 50.4 GWd/$t_{HM}$. The fuel rod segments were discharged on the 27th May 1989. Segment N0203 was characterized already after three years cooling time in the "Heiße Zellen" division of KIT (formerly Kernforschungs-zentrum Karlsruhe, KfK); segment N0204 was characterized after 24 years at JRC-ITU.

The description of the SEM and ceramography used for the characterisation of N0203 segment is reported elsewhere [4].

## Visual examination, gamma scanning and oxide layer thickness

Segment N0204 was put horizontally on a metrology bench and pushed at a defined depth into the fixing mandrel centring. With a known speed (2.494 mm/s) the segment was translated in front of a CCD digital video camera at a focal length of about 20 cm.

Afterwards, the segment was transferred in front of a collimator (with aperture of 1.2 mm) and a Ge-detector to perform γ-scanning analysis. Before or after each measurement, keeping exactly the same conditions, the background spectrum was obtained and the total γ-ray intensity was counted. The system was calibrated using $^{137}$Cs and $^{152}$Eu sources at the beginning of each measurement series or when a measurement parameter was changed. The efficiency calibration was carried out with a known, certified source $^{137}$Cs with homogeneous surface activity distribution, with same form, geometry and similar chemical composition as the segment to guarantee identical absorption effects. The quantitative determination of the isotope was performed after subtracting the background signal. The impulses generated by the detector's Ge crystal were simultaneously processed by a multi-channel spectrometer for nuclide identification and by a rate meter to determine the axial distribution of γ-emitters.

In the case of oxide layer thickness, the fuel rod was simultaneously translated and rotated during the measurement, which was carried out by means of eddy current using a punctual coil unit touching the outer surface of the cladding. The spiral rotation was 5 mm advance per rotation, the translation speed was 20 mm per minute and the scanning rate of the data logging was one value per second. The measuring system was calibrated using standards consisting of oxidized rods of the same material under examination. The thickness of the standard's oxide layer is certified by the manufacturer (AREVA-NP). Calibration control is carried out before each fuel rod examination. The precision of the translation is ± 0.05 mm/metre and ± 5° for the rotation. The precision of the oxide layer thickness measurement is ± 2 μm.

## Fission gas release

The plenum of segment N0204 was punctured and gas samples were collected in stainless steel sampling cylinders (SS-4CS-TW-50, Swagelok, Solon, USA). The gas samples were analysed by means of a quadrupole gas mass spectrometer (GAM400, In Process Instruments, Bremen, Germany). Ten scans of each gas sample were measured and the mean value was taken. The calibration was performed in the same pressure range as the respective range for analyses of the samples.

## Preparation of samples

The samples were prepared after NDT in a hot cell under $N_2$ atmosphere ($O_2$ contents below 1%). A cutting machine equipped with a diamond wafering blade (Buehler Isomet ® series 15HC, Buehler ITW Tests and Measurements GmbH, Düsseldorf, Germany) was used for the sectioning of the segment. The dry cutting was performed slowly without any cooling liquid. The complete description of the process is explained in [5].

## RESULTS

### Visual examination, gamma scanning and oxide layer thickness

The optical examination of segment N0204 at three positions (0°, 120°, 240°) shows no large defects, but helical tracks around the rod segment, see figure 1.

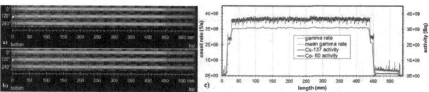

**Figure 1.** Montage of 46 single pictures extracted from three digital video films showing segment N0204 at three positions (a): as recorded inside the hot cell; (b): with colour correction. c) γ-scan showing the total γ-rate and the activities of $^{137}$Cs and $^{60}$Co along the segment.

These tracks were formed before the actual analysis, probably from analyses performed at the reactor site. The longitudinal scratches were most likely formed during the removal of the fuel rod segment from the bundle. In the centre of the segment, at the position corresponding to the location of the fuel pellets, the cladding looks brighter than at the bottom and top end. The boundary between these areas is relatively sharp. The rod segment thus seems to be less oxidised at the top and bottom end, which is in accordance with eddy current measurements of the segment. The slight distortion of the spiral track and the slight change of the colour intensity is an artefact resulting from the montage of 46 single pictures (perspective displacement and slightly inhomogeneous illumination).

The measured total γ-scan rate is displayed in figure 1. The γ-dose rate is relatively low in the region of the bottom end (plug and insulation pellet). At the position of a natural $UO_2$-pellet the γ-dose rate is increased. The next sharp increase marks the start of the enriched $UO_2$-pellet stack. Along this stack the $^{137}$Cs activity remains remarkably constant indicating a homogeneous BU along the stack. Small sharp drops in this area are seen at the pellet/pellet interfaces. The pellets have dished ends (dishing) where the specific activity and therefore also the γ-rate are lower. Near the top end the rate drops at the position of a second natural $UO_2$-pellet. The following drop marks the position of the second insulation pellet. Then the total γ-rate increases because of an increase of the $^{60}$Co-activity resulting from the neutron activation of cobalt as component of the steel spring. Towards the top end plug the γ-rate levels out. The comparison with the optical picture of the segment shows that the brighter area exactly matches the fuel stack.

The oxide profile matches with the optical appearance of the rod segment. Thicker oxide layers (35 to 45 μm) are found in the light grey zone of the pin while at the darker grey zones at both ends the thickness is below 17 μm. The maximal thickness is found on the pin between 250 and 350 mm from the bottom end. The maximum thickness observed during the measurements is below the maximum that could be expected according to [6] between 52.32 and 62.15 µm for fuels in a BU range of 48 to 55 GWd/t$_{HM}$. The oxide thickness is considered as highly relevant to the release behaviour of $^{14}$C, because an enrichment of $^{14}$C in oxide layers of irradiated claddings was observed in previous studies [6 and references therein].

The evolution of the oxide layer thickness together with the gamma rate as a function of the length is shown in figure 2. Besides exposure time, water chemistry, neutron flux and others, the temperature is a main parameter controlling the oxide growth [7]. Looking at the measured oxide profile it can be assumed that despite a relatively constant BU the temperature at the cladding surface was lower towards both ends of the fuel segment.

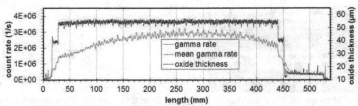

**Figure 2.** Oxide layer thickness and gamma rate along segment.

### Fission gas release

The average values of gases sampled during the puncturing test of segment N0204 are presented in table I. $O_2$ and $N_2$ contents in the gas composition (0.31 and 1.32 vol%, respectively) are related to air contamination during the sampling of the cylinders. The gas released from the plenum of the punctured fuel rod segment contains a considerable content of $CO_2$ ($^{12}C^{16}O_2$ measured at 44 amu). This effect may be related to the presence of impurities during the sintering process of the $UO_2$ fuel. This contribution of $CO_2$ needs to be taken into account for potential carbonate complexation of actinides, in safety assessment studies for SNF disposal in nominal carbonate free host rocks, such as rock salt. In addition to $^{12}C^{16}O_2$, less than 0.001 vol% of $^{14}C^{16}O_2$ is detected at 46 amu, indicating that a small fraction of the SNF $^{14}$C inventory is released as $CO_2$ to the plenum.

**Table I.** Concentration of gases and absolute gas volumes sampled during the puncturing test of fuel rod segment.

| Gas element | Ar | CO$_2$ | N$_2$ | O$_2$ | Kr$_{total}$ | Xe$_{total}$ |
|---|---|---|---|---|---|---|
| Concentration (%)$^*$ | 0.03 | 0.01 | 1.32 | 0.31 | 1.50 | 18.4 |
| Gas volume (cm$^3$)$^*$ | 0.04 | 0.02 | 1.87 | 0.44 | 2.12 | 26.1 |

*Uncertainty in (%): Ar (±4); CO$_2$ (±6); N$_2$ (±1); O$_2$ (±7); Kr (±2); Xe (±1).

The fission gas release (FGR) fraction was calculated from the experimentally determined amount of Xe and Kr extracted from the segment and the calculated total inventory of fission gases generated in the fuel over the irradiation time. The total fission gas inventory was calculated using the KORIGEN code [8]. The result of the fission gas analyses is given in table II.

**Table II.** Determination of the fission gas release (Xe and Kr) released into the plenum.

| Fission gas release (%) | Kr<br>7.03 ± 0.58 | | Xe<br>8.48 ± 0.67 | | Kr+Xe<br>8.35 ± 0.66 | |
|---|---|---|---|---|---|---|
| Isotopic composition of Kr (%)* | $^{80}$Kr<br>9.37x10$^{-3}$ | $^{82}$Kr<br>0.44 | $^{83}$Kr<br>7.53 | $^{84}$Kr<br>36.9 | $^{85}$Kr<br>1.47 | $^{86}$Kr<br>53.6 |
| Isotopic composition of Xe (%)** | $^{128}$Xe<br>4.18x10$^{-3}$ | $^{129}$Xe<br>0.01 | $^{130}$Xe<br>0.31 | $^{131}$Xe<br>3.82 | $^{132}$Xe<br>25.4 | $^{134}$Xe<br>28.2 | $^{136}$Xe<br>42.3 |
| Released volume (at 0°C, 1bar) (cm³) | Kr<br>2.12 ± 0.05 | | Xe<br>26.1 ± 0.35 | | | |

*Uncertainty: $^{80}$Kr (±5.60x10$^{-3}$); $^{82}$Kr (±0.06); $^{83}$Kr (±0.06); $^{84}$Kr (±0.07); $^{85}$Kr (±0.02); $^{86}$K(±0.17).
**Uncertainty: $^{128}$Xe (±0.31x10$^{-3}$); $^{129}$Xe (±0.13x10$^{-2}$); $^{130}$Xe (±0.38x10$^{-2}$); $^{131}$Xe (±0.01); $^{132}$Xe (±0.05); $^{134}$Xe (±0.11); $^{136}$Xe (±0.11).

These results and published FGR data of pressurized water reactor fuel rods [9;10] are plotted as a function of the linear power rate and as function of the BU in figure 3. It is obvious that the FGR is rather depending on the power rate of the high burn-up SNF samples than on the BU of the respective SNF samples. The FGR of the studied fuel rod segment N0204 is in the range of other high burn-up SNF samples.

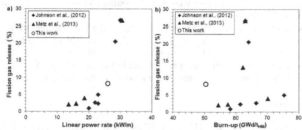

**Figure 3.** Fission gas release as a function of: a) linear power rate; b) burn-up.

### Thickness and porosity of the rim zone

The view of a pellet section and the microscopic observation at the radial r/r$_0$=1 are shown in figure 4. The HBS thickness was calculated based on figure 3b, giving as a result (100 ± 10) µm.

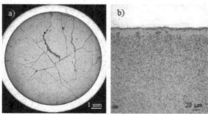

**Figure 4.** a) Macro-overview of a pellet; b) Microscopic observation at the radial position r/r$_0$=1.

This result was compared with a theoretical value that can be deduced using the formula proposed by [11], see equation 1.

$$R_t = 5.44\ BU_R - 281 \tag{1}$$

Where $R_t$ represents the rim thickness in ($\mu$m) and the $BU_R$ is the average BU at the rim zone. According to review made by [12] the $BU_R$ can be estimated as 1.33 times the average pellet BU.

The $BU_R$ for the present fuel is 67.0 GWd/$t_{HM}$ and the Rt obtained after application of equation 1 is 83.7$\mu$m. The differences between experimental and theoretical values of HBS thickness are correlated with the estimation of the $BU_R$. Finally the porosity of the HBS was calculated determining the ratio between the total area of the image and the sum of the areas corresponding to the surface cross section of pores on figure 3b, resulting in a porosity of 20%.

## CONCLUSION

Visual inspection of the fuel segment has shown no larger defects. γ-scanning has shown a homogeneous BU along the pin with a very small decrease towards the bottom end. The measured oxide thickness profile matches well with the visual nature of the outer cladding and gave a maximum value of 45$\mu$m.

It was observed the presence of $CO_2$ in the plenum of the fuel rod that can be a consequence of impurities during the sintering process of the fuel. The relative fission gas release was determined to be (8.35 ± 0.66) % of the fission gas inventory. There is a correlation between the fission gas release and the linear power rate but this is not the case for BU.

Finally, the thickness and the porosity of the HBS were 83.7 $\mu$m and 20%, respectively.

## ACKNOWLEDGEMENTS

We acknowledge F. Weiser (KfK-HZ, KIT) for performing the ceramography of N0203 pellets. The research leading to these results has received funding from the European Atomic Energy Community's Seventh Framework Programme (FP7/2007-2011) under grant agreement no. 295722, the FIRST-Nuclides project.

## REFERENCES

1. D.R. Olander. Technical Report TID-26711-P1, California University (1976).
2. Hj. Matzke, H. Blank, M. Coquerelle, K. Lassmann, I. L. F. Ray, C. Ronchi, C. T. Walker. J. Nucl. Mat. **166**, 165-178 (1989).
3. D. Papaioannou, R. Nasyrow, W. De Weerd, D. Bottomley, V. V. Rondinella. Non-destructive examinations of irradiated fuel rods at the ITU hot cells, 2012 Hotlab conference, 24th-27th September 2012, Marcoule, France (2012).
4. H. Kleykamp. Technical Report KfK-3394 (1983).
5. D.H. Wegen, D. Papaioannou, R. Gretter, R. Nasyrow, V. V. Rondinella, J.-P. Glatz. KIT scientific report 7639, p.193-199 (2013).
6. T. Sakugari, H. Tanabe, E. Hirose, A. Skashita, T. Nishimura. Proc. 15th Intl. Conf. Envir. Remed. Rad. Waste Manag., September 2013, Brussels, Belgium (2013).

7. F. Garzarolli, M. Garzarolli, ed. P. Rudling. High Burnup Fuel Design Isssue and Cosequences (Advanced Nuclear Technology International, Mölnlycke, Sweden, 2012) p. 41.

8. KORIGEN code. http://www.nucleonica.net/Application/Korigen.aspx

9. V. Metz, et al.,. Deliverable No: 1.2, pp. 85 (2013).

10. L. Johnson, I. Günther-Leopold, J. Kobler Waldis, H.P. Linder, J. Low, D. Cui, E. Ekeroth, K. Spahiu, L.Z. Evins. *J. Nucl. Mat.* **420** 54–62 (2012).

11. Johnson, L., Ferry, C., Poinssot, C., Lovera, P. *J. Nucl. Mat.*, **346**, 56-65. (2005)

12. Y.-H. Koo, B.-H. Lee, J.-S. Cheon, D.-S. Sohn. *J. Nucl. Mat.* **295**, 213-220 (2001).

Mater. Res. Soc. Symp. Proc. Vol. 1665 © 2014 Materials Research Society
DOI: 10.1557/opl.2014.657

## Durability studies of simulated UK high level waste glass

Nor E. Ahmad[1,2], Julian R. Jones[3] and William E. Lee[1,3].

[1] Centre for Nuclear Engineering, Department of Materials, Imperial College London, South Kensington Campus, London SW7 2AZ, UK.
[2] Physics Department, Faculty of Science, Universiti Teknologi Malaysia, UTM Skudai, 81300 Johor, Malaysia.
[3] Department of Materials, Imperial College London, South Kensington Campus, London SW7 2AZ, UK

### ABSTRACT

A simulated Magnox glass which is Mg- and Al- rich was subjected to aqueous corrosion in static mode with deionised water at 90 °C for 7-28 days and assessed using X-Ray Diffraction (XRD), Scanning Electron Microscopy (SEM) with Energy X-Ray Dispersive Spectroscopy (EDS) and Inductively Coupled Plasma – Optical Emission Spectroscopy (ICP-OES). XRD revealed both amorphous phase and crystals in the glass structure. The crystals were Ni and Cr rich spinels and ruthenium oxide. After two weeks of incubation in deionised water, the glass surface was covered by a ~11 µm thick Si-rich layer whilst mobile elements and transition metals like Na, B, and Fe were strongly depleted. The likely corrosion mechanism and in particular the role of Mg and Al in the glass structure are discussed.
Keywords: high level waste glass, durability, corrosion mechanism

### INTRODUCTION

In the UK, spent nuclear fuel (SNF) is reprocessed to recover uranium and plutonium. The remnant waste is immobilized by vitrification in an alkali borosilicate glass prior to packaging in steel containers, storage and eventual geological disposal 200-1000 m underground within a suitable geological formation.

There are two sources of HLW glass compositions in the UK: (a) Magnox which arises from reprocessing of Magnox fuel (Mg- and Al- rich); and (b) Blend that arises from other reprocessing activities within the Thermal Oxide Reprocessing Plant (THORP) at Sellafield [1-2]. Glass is used largely because of its random network structure which can accommodate most of the waste components [3]. Glass durability is a key parameter. In the underground repository, the wasteforms will eventually be in water environment. Contact with underground water may lead to corrosion which disrupts the glasses ability to retain the radionuclide inventory.

When glass is in contact with water, rapid ion exchange process occurs between the cations in the glass e.g. $Na^+$, $Ca^{2+}$ and $H^+$ in the water. These ions leach out leaving behind silanol bonds (Si-OH) on the glass surface. The water pH increases and a silica-rich region forms near the glass surface. However, when the water pH goes above 9, it attacks Si-O-Si bonds, total dissolution occurs so dissolving the glass. In some cases at saturation, precipitation of a surface layer occurs. The leaching rate depends on the composition of the glass, pH of the solution and the temperature of the environment [4]. The details of glass corrosion mechanisms have also been discussed extensively [4-6].

Many studies have been done to understand the release of radionuclides from glass wasteforms over time. For HLW glasses, the release is usually assessed through studies of kinetic dissolution behaviour. A large body of data exists from international waste programmes such as those based on COGEMA's R7T7 vitrified product and simulant SON68 [7]. In the UK, most of the data that exists are from Soxhlet tests on simulant glasses, in deionised water at $\geq$ 90°C. However, there are significant differences between the compositions of UK Magnox and blend glasses and those in other programmes. Thus, there are significant uncertainties associated with extrapolating data from international programmes to represent the behaviour of UK glass products [8]. The aim of this project is to study the durability of simulant UK HLW glass and understand the corrosion processes.

## EXPERIMENTAL

Samples of simulated Magnox waste glass with incorporation of 25wt% waste and 75wt% baseline glass were provided by National Nuclear Laboratory (courtesy of C. R. Scales). The glass was pour 36 from VTR (Vitrification Test Rig) campaign 10 produced in February 2010. Oxides were mixed and melted at 1050 °C and poured into a storage canister [9]. The glass composition is given in Table 1.

Leaching tests were performed using the MCC-1 method at a temperature of 90 °C and the sample surface area to water volume ratio (SA/V) is approximately $10m^{-1}$ [10]. The samples were sectioned to about 1 x 1 x 0.2 cm, ground and polished to a 1μm surface finish and cleaned prior to leaching in deionised water (dH$_2$O).The leaching process period varied from 7 to 28 days.

**Table 1.** List of glass compositions

| Oxide | Wt. % | Oxide | Wt. % | Oxide | Wt. % | Oxide | Wt. % |
|-------|-------|-------|-------|-------|-------|-------|-------|
| $SiO_2$ | 46.83 | $BaO$ | 0.48 | $La_2O_3$ | 0.58 | $Pr_2O_3$ | 0.56 |
| $B_2O_3$ | 17.04 | $CeO_2$ | 1.13 | $MgO$ | 4.68 | $RuO_2$ | 0.77 |
| $Na_2O$ | 8.46 | $Cr_2O_3$ | 0.6 | $MoO_3$ | 1.44 | $Sm_2O_3$ | 0.38 |
| $Li_2O$ | 4.12 | $Cs_2O$ | 1.07 | $Nd_2O_3$ | 1.86 | $SrO$ | 0.28 |
| | | $Fe_2O_3$ | 2.93 | $NiO$ | 0.37 | $TeO_2$ | 0.17 |
| | | $Al_2O_3$ | 4.41 | $Y_2O_3$ | 0.18 | $ZrO_2$ | 1.41 |

Aliquots of leachates were taken after cooling and filtered through 0.22 μm filter and diluted by a factor of 10 to enable chemical analysis by inductively coupled plasma atomic emission spectrometry (ICP-OES). The solutions were analysed for all elements but only Li, B, Na, Mg, and Si are shown in this paper. pH measurements were carried out in the undiluted solutions at room temperature to see any pH changes over time.

A PANanalytical X-Pert Pro MPD X-Ray Diffractometer was used to analyse the crystalline phases of the samples. The diffractometer employs Cu Kα radiation with a scan range of 10° – 80° in 1°/min 2 step sizes. The X-ray tube was operated at 40 kV and 40 mA. Initially, the glass samples were crushed and ground in an agate mortar to <100μm before being placed in aluminium sample holders for loading into the diffractometer. XRD traces were peak-matched to crystal phases detailed in the International Centre for Diffraction Data (ICDD) database using Xpert High Score Plus software.

Surface morphology of glass samples was examined using a JEOL JSM 6400 SEM operating at an accelerating voltage of 20 kV and equipped with a Link Analytical Be window EDX unit. Some of the morphologies were also characterized for samples in a wet condition using a Philips Environmental - Scanning Electron Microscope (SEM) coupled with FEI Micro-analysis system.

## RESULTS AND DISCUSSION

XRD of the reference sample (before corrosion, Fig. 1) shows that it was highly amorphous but with several peaks of crystalline phases identifies ruthenium oxide ($RuO_2$), magnesium chromate ($MgCr_2O_4$), and nickel iron oxide ($NiFe_2O_4$) [11]. In leached samples, the crystal peak heights increase with increasing leaching time. This indicates higher levels of crystals in the surface corrosion layers.

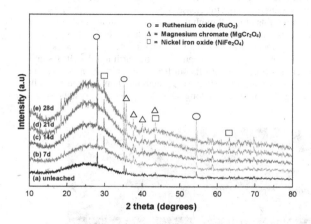

**Figure 1.** XRD traces of the Magnox glass (a) unleached and leached for (b) 7, (c) 14, (d) 21 and (e) 28 days.

SEI and EDX spectra of the reference glass (Fig. 2) show it is homogeneous and also reveal the crystal morphologies. EDX confirmed the crystals were ruthenium-, iron-, chromium- and nickel- rich as suggested by XRD. EDX of the needle-like crystals (region A) shows they are Ru- rich while the blocky-like structures (region B) are Mg-, Ni- and Fe- rich. These crystals formed due to incomplete melting of certain elements, eg. Ru, thus leaving crystals on cooling [4].

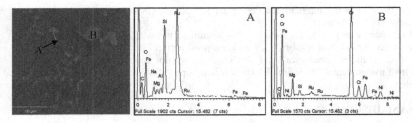

**Figure 2**. SEI of reference sample (left); EDX spectrum of region A (middle) and EDX spectrum of region B (right).

After leach testing, the same features were found as in the reference sample and were confirmed by EDX to be the same crystals. However, cracks were observed in the glass surface as shown in Fig. 3. The cracks appeard to emanate from the crystals and to completely cover the surface. We believe these cracks were generated due to release of mobile ions from the glass creating stress between the glass and crystals [11] although thermal expansion mismatch stress may also play a role. The cracks increased as leaching time increased from 7 to 21 days.

After 28 days of leaching, the surfaces were covered with a completely fractured layer over the entire surface (Fig. 4). This can be due to alkaline and alkaline earth ion depletions and hydration of the layer [12]. The layer was fragile and it lifts out from the surface (Fig. 4). Higher magnification examination of the surface layerrevealed crystals precipitated on it while the bulk glass reveals open pore structures indicating removal of some element species or crystals from the bulk surface.

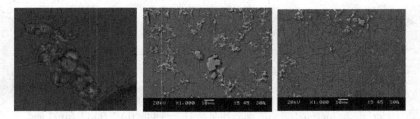

**Figure 3.** SEI of leached samples for 7 days (left), 14 days (middle) and 21 days (right).

To clarify the reason for crack formation in the surface layer, ESEM images were taken under wet conditions on samples that had been leached for 28 days. Fig. 5 (a) shows an image of the surface immediately after being put in the SEM chamber while figure (b) shows the images of the surface after leaving the samples 15mins in the chamber. From this figure we can observe that the cracks occur immediately the surface layers are dried in the ESEM. The ease of crack formation associated with small changes of humidity suggest cracking in repository atmospheres subject to changes in atmospheric conditions could impact radionuclide release.

**Figure 4.** SEI of leached samples for 28 days (centre) with magnified area of the bulk (left) and surface layer (right).

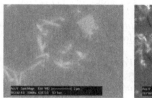 

**Figure 5.** ESEM images of 28 days leached glass taken immediately (left) and after 15mins (right) in the chamber.

Fig. 6 shows the leached ion concentrations found in the leachates were small i.e. less than 4ppm except for Si and B. High Si concentrations are likely due to low network connectivity between Si and mobile elements. When the other ions leached out, Si bonds are broken and so it also is leached. Note that the increment of water pH is small ±0.1 and less than pH 8 indicating that release of mobile ions is small resulting in only small changes in pH [13].

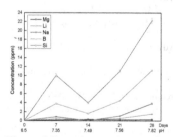

**Figure 6.** Ion concentration as a function of time and pH.

**CONCLUSIONS**

Ion exchange and formation of corrosion layers were observed on simulant UK glasses. The low concentration of leached ions and small pH increase of the adjacent solution indicates reasonable glass durability. However, the presence of additional crystals on the corrosion layer and its propensity to crack on drying indicates further work is needed to understand its role. whether passivating or not. The role of Mg in determining leaching rate of the glass needs further examination.

## ACKNOWLEDGMENTS

The authors would like to thank all who involved in this project, Charlie R. Scales for providing samples, and MOHE and UTM for financial funding.

## REFERENCES

1. N. J. Cassingham, P. A. Bingham and R. J. Hand, Glass Technology, 49 (1), 21-26 (2008).
2. S.W. Swanton and N. Smith, *Experimental studies of the durability of UK HLW and ILW glasses*, Serco Report, Reference Number: RWMD/003/006, Oxfordshire United Kingdom (September 2011).
3. M.I. Ojovan, *Chapter 1 in Handbook of Advanced Radioactive Waste Conditioning Technologies*, (Nova Publishing Co., New York, 2009).
4. V. Jain, Y.M. Pan, *Glass Melt Chemistry and Product Qualification*, Centre for Nuclear Waste Regulatory Analyses (CNWRA) San Antonio, Texas. Nuclear Regulatory Commission Contract NRC-02-97-009 (2000).
5. J.R. Jones, Acta Biomaterialia, 9, 4457-4486 (2013)
6. I. W. Donald, *Waste Immobilisation in Glass and Ceramic Based Hosts*, (A John Wiley & Sons, Ltd., UK, 2010).
7. P. Frugier, S. Gin, Y. Minet, T. Chave, B. Bonin, N. Godon, J. E. Lartigue, P. Jollivet, A. Ayral, L. De Windt and G. Santarini, Journal of Nuclear Materials, 380, 8-21 (2008)
8. S. Swanton, *Geological disposal: experimental studies of the durability of UK HLW and ILW glasses – Roadmap*, Serco Report, Document Number: 15801483, Contract Number: RWM005105, Oxfordshire United Kingdom (March 2012)
9. C. R. Scales, (2011). *Characterisation of simulated vitrified Magnox product manufactured on the VTR*, National Nuclear Laboratory (10) 10929 Report, Issue 6, Work Order No.: 03363.200.
10. W.E. Lee, M.I. Ojovan, M.C. Stennet, and N.C. Hyatt, Adv. in Appl. Ceramics, 105, 1 (2006)
11. P.B. Rose, D.I.W. Woodward, M.I. Ojovan, N.C. Hyatt and W.E. Lee, Journal of Non-Crystalline Solids, 357, 2989-3001 (2011).
12. M.D Bardi, H. Hutter and M. Schreiner, Appl. Surface Science, 282, 195-201 (2013).
13. D.E. Clark and B.K.Zoitos, *Corrosion of Glass, Ceramics and Ceramic Superconductors*, (William Andrew Inc., Florida, 1992)

Mater. Res. Soc. Symp. Proc. Vol. 1665 © 2014 Materials Research Society
DOI: 10.1557/opl.2014.658

# Estimation of the long term helium production in high burn-up spent fuel due to alpha decay and consequences for the canister

Alba Valls[1], Mireia Grivé[1], Olga Riba[1], Maita Morales[1] and Kastriot Spahiu[2]

[1]Amphos 21 Consulting S.L, Passeig de Garcia i Faria, 49-51, 1-1, 08019 Barcelona , Spain
[2]SKB, Swedish Nuclear Fuel and Waste Management Co., Box 250, SE-101 24, Sweden

## ABSTRACT

In the KBS-3 repository concept and safety analysis, the copper container with a cast iron insert plays a central role in assuring isolation of the waste from the surrounding during long periods of time. All processes that affect its stability are thoroughly analysed, including potential detrimental processes inside the canister. For this reason, an estimation of the helium produced during the long term decay of alpha emitters in the spent fuel is necessary to evaluate if the pressures generated inside can have consequences for the canister.

The spent nuclear fuel to be disposed of in Sweden is mainly LWR fuel. The maximum burn-up expected is 60 MWd/kg U for BWR and PWR. A small quantity of BWR MOX is expected to be stored with a maximum burn-up of 50 MWd/kg U.

This work has focused on carrying out calculations of the amounts of He generated during more than 1 million years in Swedish spent nuclear fuels with a benchmarking exercise by using both codes AMBER and Origen-ARP. The performance and agreement of the codes in the He generation from α-decay have been checked and validated against data reported in literature [1].

In the calculation of the maximal pressure inside the canister, the quantity of helium used to pre-pressurise the fuel rods has been accounted for. The pressure inside the canister due to He generation is at all times much lower than the hydrostatic pressure and/or the bentonite swelling pressure outside the canister.

## INTRODUCTION

Performance Assessment (PA) for the deep geological repositories of spent nuclear fuel requires the quantification of the fraction of radionuclides which are expected to be rapidly released when cladding fails and water contacts the fuel matrix. This issue is categorized as a high priority and urgency topic in the Strategic Research Assessment (SRA) of the Implementing Geological Disposal of Radioactive Waste Technology Platform (IGD-TP) published in July 2011.

Johnson and co-workers [2] described the IRF(Instant Release Fraction) as the fraction of the inventory of safety-relevant radionuclides that may be rapidly released from the fuel and fuel assembly materials at the time of canister breaching. This fraction is mainly composed of the inventory located in the grain boundaries and the gap of the fuel rod.

The potential evolution of the IRF with time in an un-breached canister is related to the processes associated with the mechanisms that promote diffusion of radionuclides and/or evolution of pellet microstructure as well as radioactive decay. The possible mechanisms for the evolution of IRF are listed as follows:

- Chemical evolution of the spent fuel with radioactive decay
- Accumulated α-decay damage
- Production of He by α-decay

As was pointed out in [3], neither chemical evolution nor the accumulation of structural damage due to α-decay should alter the pellet microstructure. However, the production and accumulation of helium during the storage period may cause significant effects on the microstructure of spent fuel leading to an increase of the IRF.

The possible microstructural changes due to the helium pressure in intra-granular bubbles were explored by Ferry and co-workers [4, 5]. The authors compared the estimated helium generated with the calculated critical pressure at which micro-cracks could develop in the structure. This process could explain the observations reported in literature showing the development of micro-cracks in the pellet structure [6, 7, 8].

Despite the experimental observations, Ferry and co-workers pointed out that other mechanisms such as exposition to air, bad sintering or sample preparation should also be considered as responsible for the observed micro-cracks [3].

With the aim to quantify the microstructure evolution by formation of micro-cracks due to overpressure in intra-granular He bubbles, the evolution of the critical pressure within pores in grains (defined as the volume of He bubbles in grain) has been calculated in [5]. In the rim region, the critical concentration of He atoms was calculated obtaining a critical concentration of $1.3 \cdot 10^{21}$ at He/cm$^3$. In the central part of the pellet, the critical concentration of He is $4 \cdot 10^{20}$ at He/cm$^3$ considering 10nm bubbles. The calculated amount of helium accumulated in bubbles prior to 10,000 years, based on the conservative assumption that all He atoms can be trapped in bubbles, is much lower than the critical values derived from rupture criteria. Consequently, no evolution of the microstructure of the spent $UO_2$ fuel is expected.

This work focuses on the role of helium generation on the canister stability. For this reason, the entire He generated is assumed to be released from the spent fuel in the free volume of the canister. Three different steps have been followed: (1) Benchmarking exercise of two codes: AMBER and Origen-ARP, (2) Amount of He generated in spent fuel and (3) Build-up of He pressure in the canister.

## BENCHMARKING EXERCISE

Prior to helium generation calculation from fuels used in the Swedish program, a benchmarking exercise has been done for comparing the performance and agreement of two codes (ORIGEN-ARP and AMBER) towards the calculation of He generation from alpha decay. Helium generated from a given spent fuel has been calculated and compared with published results (Roudil et al., 2004).

Roudil et al. (2004) simulated He generation during the storage time for different spent fuels (Figure 1). $UO_2$ fuel with burnup 47 GWd/tU (black squares in Figure 1) and MOX fuel of burnup 47.5 GWd/tU (grey rhombus in Figure 1) are the ones selected for this benchmarking exercise.

The exercise consists on re-calculating the helium generation calculations in Roudil et al (2004) by using the codes AMBER© [9] and ORIGEN-ARP [10] (Figure 2). A brief description of the methodology followed to calculate the amount of He generated with each code is presented below.

- AMBER. Calculations in AMBER have been structured as follows: each single alpha emitting radionuclide generates both its corresponding daughter and an alpha particle. The alpha particle is then assumed to instantaneously transform to helium atom. Helium generated is accumulated with time. Then, the code makes a sum of the overall He generated and accumulated by the alpha emitters in the spent fuel during its storage ($\sum_{i=0}^{n}(He)_i$).

- ORIGEN-ARP. This code provides the evolution of the alpha and beta activity with time, as seen in Figure 3. As expected, the total activity and the alpha activity tend to converge with time, as beta emitters are mostly short-life. Data must be then post processed in order to obtain the amount of helium generated from the alpha emitters. The assumption done is that one atom of helium is produced from every single $\alpha$ decay. The accumulated helium with time might be then obtained from the following expression: $\sum^{n}_{i=1}((A_{\alpha})_i \cdot (\Delta t)_i$, where $A_{\alpha}$ is the alpha activity at the interval of time $\Delta t$.

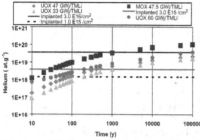

**Figure 1.** Calculated helium generation for different spent fuels during their storage (Figure taken from [1])

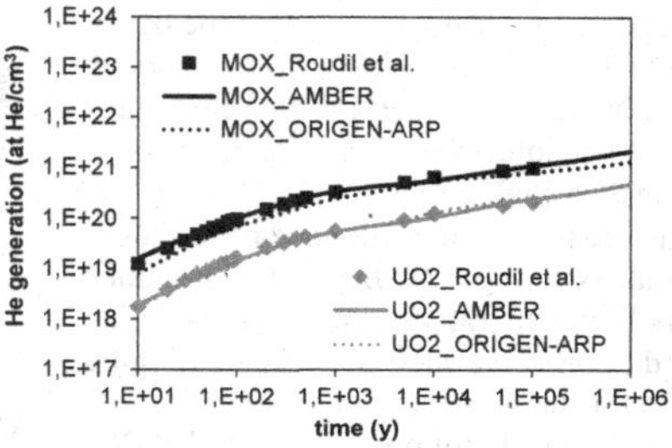

**Figure 2.** Helium generation of $UO_2$ and MOX spent fuel of 47.5 GWd/tU published in literature [1] (black-grey symbol series) and calculated with AMBER (green series) and ORIGEN-ARP (red-orange series) in this work.

Figure 2 shows a comparison of the calculated He amount generated from $UO_2$ and MOX spent fuels of burnup 47.5 GWd/tU by using the different codes:
- Roudil and co-workers [1] (black and grey symbols in Figure 2)
- AMBER (solid lines in Figure 2), in this work
- ORIGEN-ARP (dashed lines in Figure 2), in this work.

As can be seen, both codes (AMBER and ORIGEN) are able to reproduce the He generation previously calculated in [1]. Hence, both codes are appropriate to perform calculations of helium generation during the storage time of spent nuclear fuel. We have selected AMBER in this work, as it was already prepared by Amphos21 to give the needed information directly in the output without a post-treatment.

## HELIUM GENERATION IN SPENT FUEL

This section aims at providing an estimation of helium generated by $\alpha$-decay during the storage of Swedish spent nuclear fuel. Three different types of spent fuel have been taken into account, covering two different types of reactors (PWR and BWR) and two different fuel matrix environments ($UO_2$ and MOX), which characteristics are summarized in Table I. The selection of high burn-up spent fuels has been based on the fact that it is foreseen that burn-up of 60 GWd/tU will generate more He due to the presence of more $\alpha$-decay emitters.

**Table I.** Spent fuel characteristics.

| Spent fuel | Fuel | Reactor | Burn-up | Enrichment |
|---|---|---|---|---|
| $UO_2$ – PWR 60 | $UO_2$ | PWR | 60 GWd/tU | 4.0% $^{235}U$ |
| $UO_2$ – BWR 60 | $UO_2$ | BWR | 60 GWd/tU | 3.6% $^{235}U$ |
| MOX – BWR 50 | MOX | BWR | 50 GWd/tU | 0.2% $^{235}U$ + 4.6% Pufiss |

The corresponding calculated He generation by AMBER from the fuels in Table I is shown in Figure 3. The following information and conclusions can be drawn from this figure:

a) There is no influence of the type of reactor (PWR or BWR) on the amount of helium generated for the same matrix ($UO_2$) (solid black and dashed grey lines in Figure 3).

b) There is a strong influence of the fuel matrix ($UO_2$ or MOX) on the amount of helium generated during its storage (dashed and dotted grey lines in Figure 3) for the same reactor, being the amount of helium produced by MOX about half order of magnitude higher than that generated by $UO_2$ spent fuel.

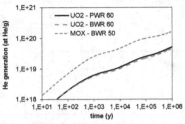

**Figure 3.** Helium generation in $UO_2$ - PWR 60, $UO_2$ - BWR 60 and MOX - BWR 50 spent fuels.

The total amount of helium is the sum of the He generated by each individual alpha emitter. In Figure 4 there is a breakdown of the major contributors on this generation for $UO_2$ – PWR 60 (grey lines) and MOX – BWR 50 (black lines). In both cases, helium is mainly produced by the decay of $^{244}Cm$ during the first hundreds of years. Then, $^{241}Am$ dominates the production of helium up to thousands of years followed by $^{239}Pu$, which becomes the maximum generator of helium at longer time frames.

As seen, the accumulation of helium in $UO_2$ and MOX spent fuel follows the same trend being the main contributors also the same. Differences on the amount of He produced can be explained by the initial inventory after irradiation.

The initial inventory of the main three He producers is detailed in Figure 5. The activity of the three radionuclides is higher for the MOX fuel than in $UO_2$ fuel, which is a direct consequence of the higher initial content of Pu in MOX.

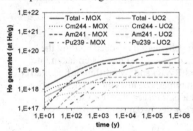

**Figure 4.** Contribution of alpha emitters to total helium generation in $UO_2$ - PWR 60 and MOX - BWR 50 spent fuel.

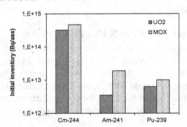

**Figure 5.** Initial inventory of $^{244}Cm$, $^{241}Am$ and $^{239}Pu$ (inventory 2 days after the irradiation phase).

## HELIUM PRESSURE INSIDE A CANISTER

The main objective of this work is to estimate the pressure of He reached in the canister due to He pre-pressurization of rod and He generation by $\alpha$-decay. The calculations have been done for 5 of the 12 BWR and PWR canister configurations considered as reasonable by SKB [11]. Moreover, a complementary case has been selected to be studied where 2 MOX assemblies are disposed in a BWR canister (* in Table II).

**Table II.** Canister configurations considered in this study [11]. The case BWR-MOX-2 has been defined by the authors.

| Canister | N° assembly | Burnup (MWd/kgU) | Age ass. (y) |
|----------|-------------|------------------|--------------|
| **BWR high a** | 12 - UO2 | 47.8 | 48 |
| **BWR high b** | 12 - UO2 | 57 | 60 |
| **BWR-MOX** | 11 - UO2 | 37.7 | 43 |
| | 1 - MOX | 50 | 50 |
| **BWR-MOX-2\*** | 10 - UO2 | 37.7 | 43 |
| | 2- MOX | 50 | 50 |
| **PWR high** | 4 - UO2 | 57 | 55 |
| **PWR-MOX** | 3 - UO2 | 44.8 | 32 |
| | 1 - MOX | 34.8 | 57 |

The pressure of helium reached in the canister as a function of time only due to He generation by $\alpha$-decay is shown in Figure 6. BWR canisters which contain MOX assemblies generate higher helium pressures; 0.35 and 0.43 MPa depending on the number of MOX-assemblies in the canister (one or two respectively). However, no significant differences are observed.

It is important to consider the amount of He, and therefore the pressure that this gas exerts within the canister, due to the rod pre-pressurization. Depending on the type of fuel the pressure of He introduced is different. In this work, it has been considered that PWR rods are pre-pressurized with 2.1 MPa while BWR rods with 0.7 MPa He (average from various fuel manufacturers). Figure 7 shows the total He pressure inside the canister considering both rod pre-pressurization and generation by $\alpha$-decay.

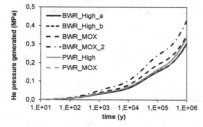 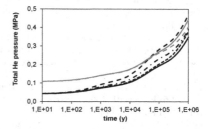

**Figure 6.** He pressure generated in the canister by decay of $\alpha$-emitters.

**Figure 7.** Total He pressure ($P_{He}$ due to rod pre-pressurization + $P_{He}$ due to $\alpha$-decay)

Even considering the quantity of helium used to pre-pressurize the fuel rods, the maximum He pressure is achieved in 2ass.-MOX-BWR fuels with a pressure of 0.47 MPa. Nevertheless, no significant differences are observed between the studied cases, thus the He pressure ranges from 0.35 to 0.47 MPa (BWR_High_a and BWR_MOX_2, respectively).

## CONCLUSIONS

After a benchmarking exercise with Roudil and co-workers data [1], Amber has been selected for performing He generation calculations. The amount of helium produced by α-emitters form high burn-up spent fuels of different matrix and from different reactors has been calculated. At the end, the He pressure reached in the canister was calculated considering both He from rod pre-pressurization and He generated from the α-decay.

One of the conclusions obtained from this work is that, as expected, the amount of He generated in MOX spent fuels is much higher than in $UO_2$ spent fuels. However, when estimating the pressure of He reached in BWR and PWR canisters with different configurations (only $UO_2$ spent fuel assemblies or combination of $UO_2$ and MOX spent fuel assemblies) no significant differences are observed. That is the results of the combination of a small amount of MOX fuel with $UO_2$ fuel of low or medium burn-up.

To conclude, the estimated maximal He pressure reached inside a KBS-3 canister is much lower than the external hydrostatic pressure (about 5 MPa at 500 m depth) and/or the bentonite swelling pressure(5-15 MPa).

## ACKNOWLEDGMENTS

We want to acknowledge Swedish Nuclear Fuel and Waste Management Co. (SKB) for funding this work.

## REFERENCES

1. Roudil, D., Deschanels, X., Trocellier, P. Jégou, C., Peuget, S., Bart, J-M. (2004) Helium thermal diffusion in a uranium dioxide matrix. J. Nucl. Mater., 325, 148-158.
2. Johnson, L., Ferry, C., Poinssot, C., Lovera, P. (2005) Spent fuel radionuclide source-term model for assessing spent fuel performance in geological disposal. Part I: Assessment of the instant release fraction. J. Nucl. Mater., 346, 56-65.
3. Ferry, C., Piron, J.P., Poinssot, C. (2006) Evolution of the spent nuclear fuel during the confinement phase in repository conditions: Major outcomes of the French research. MRS Symp.Proceedings, 932,pp. 513-520
4. Ferry, C., Piron, J.P., Poulesquen, A., Poinssot, C. (2008) Radionuclides release from the spent fuel under disposal conditions: re-evaluation of the instant release fraction. MRS Symp.Proc., 1107.pp. 447–454
5. Ferry, C., Piron, J.P., Ambard, A. (2010) Effect of helium on the microstructure of spent duel in a repository: An operational approach. J Nucl Mater, 407, 100-109.
6. Ronchi, C., Hiernaut, J. (2004) Helium diffusion in uranium and plutonium oxides. J. Nucl. Mater, 325, 1-12.
7. Roudil, D., Jegou, C., Deschanels, X., P., Peuget, S., Raepseat, C., Gallien, J-P., Broudic, V. (2006) Effects of alpha self-irradiation on actinide-doped spent fuel surrogate matrix. MRS Symp.Proceedings, 932, 529–536.
8. Guilbert, S., Sauvage, T., Erramli, H., Barthe, M.-F., Desgardin, P., Blondiaux, G., Corbel, C., Piron, J.P. (2003) Helium behaviour in $UO_2$ polycrystalline disks. J. Nucl. Mater, 321, 121-128.
9. Quintessa Limited (2012) AMBER 5.6 Reference Guide. QE-AMBER-1, Version 5.6.
10. Oak Ridge National Laboratory (2001) Scale: A Comprehensive Modeling and Simulation Suite for Nuclear Safety Analysis and Design, ORNL/TM-2005/39, Version 6.1.
11. SKB (2010) Spent nuclear fuel for disposal in the KBS-3 repository. SKB-TR-10-13.

Mater. Res. Soc. Symp. Proc. Vol. 1665 © 2014 Materials Research Society
DOI: 10.1557/opl.2014.659

# Design of a New Reactor to Work at Low Volume Liquid/Surface Solid Ratio and High Pressure and Temperature: Dissolution Rate Studies of UO2 Under Both Anoxic and Reducing Conditions.

A. Martínez-Torrents[1,2], J.Giménez[2], I.Casas[2], J. de Pablo[1,2].

[1]CTM Centre Tecnològic, Plaça de la ciència 2, 08243 Manresa, Spain
[2]Departament of Chemical Engineering, Universitat Politècnica de Catalunya, Diagonal 647 H-4, 08028 Barcelona, Spain.

## ABSTRACT

A flow-through experimental reactor has been designed in order to perform studies at both high pressure and high temperature conditions. A chromatographic pump is used to impulse the leachant throughout the reactor in order to work at very low flows but high pressures. Therefore, high surface solid to volume leachant ratios, similar to the ones predicted in the final repository, can be obtained. The reactor allows working at different atmospheres at pressures up to 50 bars. The temperature inside the reactor can be set using a jacket.

Using this new reactor the evolution of uranium concentrations released from an UO2 sample was studied at different conditions.

The results show that at hydrogen pressures between 5 and 7 bars, hydrogen peroxide does not seem to significantly oxidize the uranium (IV) oxide. Uranium concentrations in those experiments remain between $10^{-8}$ mol·l$^{-1}$ and $10^{-9}$ mol·l$^{-1}$.

## INTRODUCTION

The study of the behavior of the Spent Nuclear Fuel Matrix is critical for the deep geological repository safety assessment. As an approach to the chemical behavior of the Spent Nuclear Fuel Matrix, it is possible to use unirradiated UO2 as an analogue. Casas et al. [1] studied the dissolution kinetics of UO2 under oxidizing conditions proposing a first mechanism of oxidation-dissolution of UO2. Later the same group studied the role of pe, pH and carbonate on the solubility of UO2 at reducing conditions [2]. The oxidative dissolution mechanism was improved by De Pablo et al. [3,4,5] adding the effect of Temperature, pH, carbonate concentration and oxygen partial pressure and a mechanism of the dissolution of UO2 due to the uranium-carbonate complexation was proposed[5]. Due to the radiolysis of water, species like $H_2$, $O_2$ and $H_2O_2$ are formed. Clarens et al. [6] added the concentration of hydrogen peroxide as a new parameter to the oxidative dissolution mechanism studies and studied the effect of the pH in the dissolution on UO2 in $H_2O_2$ solutions. Precipitation of the uranium peroxide Studtite was observed at different concentrations of hydrogen peroxide adding complexity to the dissolution mechanism. Casas et al. [7] added a new parameter to the dissolution experiments: pressure. They designed a reactor in order to perform experiments up to 100 bars and temperatures up to 100 °C. The reactor was continuously stirred and they determined UO2(s) dissolution rates in a hydrogen peroxide and carbonate media as a function of pressure and temperature. Lately, Casas et al. [8] improved the knowledge in the effect of carbonate and hydrogen peroxide concentration in UO2 dissolution observing and increase in the dissolution when both the concentration of carbonate and hydrogen peroxide increased.

Flow through reactors were also used to determine matrix dissolution rates of nuclear spent fuel at very different conditions. Earlier studies are collected in Gray and Wilson [9], the effect of carbonate concentration, oxygen pressure and temperature is well documented. In Röllin et al. [10] matrix dissolution rate were determined at oxidizing, anoxic and reducing conditions. Recently, dissolution rates of high burn-up spent fuels have been determined by Serrano-Purroy et al. [11,12].

All these experiments have something in common. In all of them the solid was saturated with the leachant. This situation does not reproduce what it is expected in a deep geologic repository. Wronkiekicz et al. [13] take this fact in consideration and used an experimental set-up with a very low flow (leachant drops through the solid) of leaching solution during 10 years at 90°C. They observed a decrease in the release rate after the first two years produced by a dense mat of alteration phases that trap the loose particles of $UO_2$. They also observed precipitation of secondary uranium phases, reducing the concentration of $UO_2$ in solution. Following this trend a new reactor has been designed where the $UO_2$ solid is not saturated with the leachant. Leachant drops pass through the solid particle at a certain flow rate. Reactor can be operated at different pressures, temperatures, atmosphere composition and leaching composition. One of the aims of this set-up is to simulate the experimental conditions that could be found in a deep geologic repository.

## EXPERIMENTAL

### Experimental set-up

The main idea of the experimental set-up is to have a solid in powder or in pellet in contact with drops of leachant in a very low flow. The contact would be at a controlled atmosphere, pressure and temperature. The leachate is afterwards collected at the bottom of the reactor.

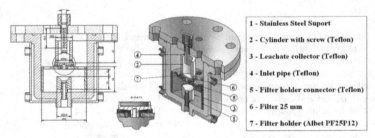

1 - Stainless Steel Suport

2 - Cylinder with screw (Teflon)

3 - Leachate collector (Teflon)

4 - Inlet pipe (Teflon)

5 - Filter holder connector (Teflon)

6 - Filter 25 mm

7 - Filter holder (Albet PF25P12)

**Figure 1.** Design of the stainless steel reactor.

1 - Stainless Steel Suport

2 - Cylinder with screw (Teflon)

3 - Leachate collector (Teflon)

4 - Inlet pipe (Teflon)

5 - Filter holder connector (Teflon)

6 - Filter 25 mm

7 - Filter holder (Albet PF25P12)

**Figure 2.** Three of the four parts of the Teflon structure inside the reactor. The absent part is the Teflon pipe that is connected to the upper part of the reactor.

In order to fulfill this idea, a stainless-steel (AISI 316) reactor was designed (Figure 1). This reactor allows working at inner gas pressures up to 50 bars, even with hydrogen gas. The reactor is equipped with a manometer and a thermometer to measure the pressure and the temperature inside the reactor. It also has a security valve to avoid overpressure and two inlets, one for gas and the other for liquid. The inlet of gas connects through different valves and manometers the inside of the reactor with the selected gas cylinder. The leachant is stored in a bottle continuously purged with the selected gas. From this bottle, a chromatographic pump (Knauer, Smartline Pump 100) impulses the leachant into the reactor directly into the Teflon structure inside the reactor. The use of a chromatographic pump allows working at higher pressures and low flow. The Teflon structure (Figure 1 and 2) guides the leaching solution and stores the solid avoiding any contact of the solid with the steel. It is formed by 4 pieces. First a Teflon cylinder that guides the leachant into the inside of the reactor in order to assure that the leachant drops fall vertically into the top of the solid and from a close distance. The second Teflon piece is a larger cylinder with a screw at the end. This piece assures that all the leachant falls into the solid and it is connected to a commercial filter holder (Albet PF25P12). The filter holder contains a filter with the solid as powder or as a pellet. This second piece also contains the solid into the Teflon structure avoiding dispersion into the insides of the steel reactor, especially when venting or pressurizing the reactor. The third piece is connected with the bottom of the filter holder allowing that the leachate drops into the bottom of the reactor where the last Teflon piece recovers the leachate, and guides it into the outlet in the bottom of the steel reactor. The leachate then can be recovered using a pair of valves in a closed bottle with a septum in order to avoid air contamination. The reactor has a jacket in order to change the temperature inside the reactor by circulating some fluid at a different temperature (Figure 3).

**Figure 3.** Experimental Set-up

## Materials and Methods

Crystalline Uranium (IV) dioxide was obtained from a $UO_2$ pellet supplied by ENUSA (Empresa Nacional del Uranio S.A.). X-ray powder diffraction (XRD) analysis showed the bulk of the sample to correspond to $UO_{2.01}$ [2]. For the experiments two different particles sizes were used, one with particles larger than 500 μm and the other with particles between 150 and 500 μm (surface areas of 0.005 and 0.01 $m^2 g^{-1}$, respectively) [16,17]. The hydrogen peroxide solutions were obtained from the same initial solution (Merck) and the concentration was periodically standardized with thiosulfate (Scharlau) in $H_2SO_4$. Cylinders of 99.99% hydrogen gas and 99.99% nitrogen gas were provided by Abelló Linde S.A.

Two kinds of leaching solutions were prepared. One was prepared with $10^{-3}$ mol·l$^{-1}$ $Na_2CO_3$ and $19·10^{-3}$ mol·l$^{-1}$ $NaClO_4$. The second one was prepared simulating cement pore water diluted ten times to avoid precipitation problems. The composition after the dilution is shown in Table I. The uranium composition was determined by means of a Time Resolved Laser Fluorescence Spectroscopy and adding FLURAN.® to increase Uranium (VI) fluorescence [14]. Some samples were analyzed by ICP-MS technique to assure that the concentration results were correct. Experiments were made at room temperature.

A Platinum mesh with Palladium electrodeposition was used to catalyse the hydrogen reaction with oxygen in order to eliminate oxygen traces and assure better reducing conditions. The Platinum mesh was used to avoid the solid dispersion in the filter due to the mechanical effects of the falling drops.

**Table I.** Cement pore water composition.

| Element | Composition (M) |
|---------|-----------------|
| pH | 12.2 |
| [Al] | $2.6·10^{-6}$ |
| [K] | $1.8·10^{-3}$ |
| [Na] | $4.9·10^{-3}$ |
| [Ca] | $7.1·10^{-3}$ |
| [SO$_4^{2-}$] | $4.3·10^{-5}$ |
| [Si] | $5.3·10^{-4}$ |

## RESULTS AND DISCUSSION

The experiments carried out in this work are collected in Table II. In experiments 1 and 2, the flow rate was varied during the experiment, from $3.34·10^{-7}$ L/s to $10^{-5}$ L/s. The concentration of Uranium in mol·l$^{-1}$·m$^{-2}$ was represented versus the inverse of the flow in s·l$^{-1}$. The points follow a straight line, showing that a stationary state was reached. In figure 4 the concentration of uranium versus the inverse of the leaching flow, in experiment 1, is shown as an example.

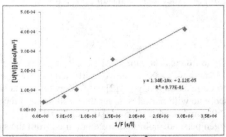

**Figure 4.** Uranium (VI) concentration (mol l$^{-1}$ m$^{-2}$) versus the inverse of flow rate in experiment 1.

Experiments performed at different conditions have been compared by using the dissolution rate, using the following equation:

$$r\left(\frac{mol}{m^2s}\right) = \frac{[U(VI)]\left(\frac{mol}{l}\right)*Q\left(\frac{l}{s}\right)}{m_{particle}\ (g)*A_{sup}\cdot\left(\frac{m^2}{g}\right)}$$

(eq. 1)

In equation 1, r is the dissolution rate in $mol\cdot m^{-2}\cdot s^{-1}$, [U(VI)] is the concentration of uranium (VI) found in solution in $mol\cdot l^{-1}$, Q is the flow rate in $l\cdot s^{-1}$, $m_{particle}$, is the mass of the $UO_2$ particles in grams and $A_{sup}$ is the specific surface area of the particles in $m^2\cdot g^{-1}$.

**Table II.** Experiments performed in this work.

| EXP. | LEACHANT | pH | [H₂O₂] (mol·l⁻¹) | FLOW RATE (l s⁻¹) | GAS (5-7 bars) | PARTICLE SIZE (μm) | MASS (g) |
|------|----------|-----|------------------|-------------------|----------------|--------------------|----------|
| 1 | $1\cdot10^{-3}$ mol·l⁻¹ HCO₃⁻ $19\cdot10^{-3}$ mol·l⁻¹ NaClO₄ | 8 | $10^{-3}$ | variable | N₂ | 150-500 | 0.2021 |
| 2 | $1\cdot10^{-3}$ mol·l⁻¹ HCO₃⁻ $19\cdot10^{-3}$ mol·l⁻¹ NaClO₄ | 8 | $10^{-3}$ | variable | H₂ | 150-500 | 0.2021 |
| 3 | $1\cdot10^{-3}$ mol·l⁻¹ HCO₃⁻ $19\cdot10^{-3}$ mol·l⁻¹ NaClO₄ | 8 | $10^{-3}$ | $1.67\cdot10^{-5}$ | N₂ | ~500 | 0.0233 |
| 4 | $1\cdot10^{-3}$ mol·l⁻¹ HCO₃⁻ $19\cdot10^{-3}$ mol·l⁻¹ NaClO₄ | 8 | $10^{-3}$ | $1.67\cdot10^{-5}$ | H₂ | ~500 | 0.0243 |
| 5 | Pore water | 12.2 | $10^{-3}$ | $1.67\cdot10^{-5}$ | N₂ | ~500 | 0.0243 |
| 6 | Pore water | 12.2 | $10^{-3}$ | $1.67\cdot10^{-5}$ | H₂ | ~500 | 0.0243 |

The dissolution rate for each experiment in Table II is shown in Figure 5 and 6. In figure 5 the experiments at variable flow rate, a particle size of 150-500 μm and 0,2021 g of $UO_2$ are shown, it is clear that dissolution rate is slower in hydrogen atmosphere. Once the steady state was demonstrated experiments at only one flow rate were performed. In these experiments another particle size was used with 10 times less mass. At these high flow rate the concentrations of uranium obtained are in the range between $10^{-9}$ and $5\cdot10^{-9}$ mol·l⁻¹, similar to the steady-state values found in the literature [10, 18, 19]. However these low concentrations are near the detection limit of the technique and therefore the error in the measurements is higher. Nevertheless, in general, it is clear that rates are slower in hydrogen than in nitrogen, this can be due to the reaction of hydrogen peroxide with hydrogen, recently pointed out by Gimenez et al. [15]. The decrease on the dissolution rate due to the presence of hydrogen is of about one third in all the experiments.

Dissolution rates are higher when bicarbonate is present in the solution in both atmospheres nitrogen and hydrogen. Scanning Electron Microscopy SEM was made after the experiments and no secondary phases of uranium were found. So the lower dissolution rate in pore water is not due to the precipitation of secondary phases like calcium uranates.

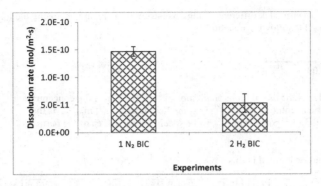

**Figure 5.** Dissolution rates of the experiments 1 and 2.

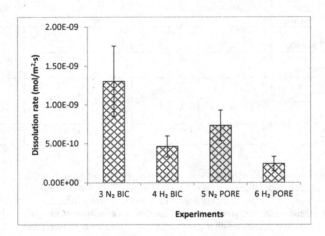

**Figure 6.** Dissolution rates of the experiments 3,4,5 and 6.

An extra experiment was made without uranium, and the hydrogen peroxide concentration was measured by iodometry with standardized thiosulfate. The concentration of hydrogen peroxide did not change due to its pass throw the experimental set-up, so it is possible to say that hydrogen peroxide did not react with any of the materials of the experimental set-up.

**Future Studies: The reactor as a multipurpose tool**

This reactor can also be used to perform:
- Secondary phase formation in long-term experiments.
- RedOx studies on solid surfaces.
- Sorption research.

- Experiments where three different phases solid-liquid-gas interact.
- Studies on the effect of high pressure and temperature above 298K.

## ACKNOWLEDGEMENTS

The authors want to thank Aurora Martínez-Esparza for her valuable comments and support. This work has been supported by ENRESA and Ministerio de Economia y Competitividad (Spain) project CTM2008-06662-C02-01, CTM2011-27680-C02-01 and the associated scholarship (A.M.-T.).

## REFERENCES.

[1] I. Casas, J. Giménez, V. Martí, M.E. Torrero and J. de Pablo. Radiochim. Acta 66/67 23-27 (1994).

[2] I. Casas, J. de Pablo, J. Giménez, M.E. Torrero, J. Bruno, E. Cera, R.J. Finch and R.C. Ewing. Geochim. Cosmochim. Acta 62 13 2223-2231 (1998).

[3] J. de Pablo, I. Casas, J. Giménez, M. Molera, M. Rovira, L. Duro and J. Bruno. Geochim. Cosmochim. Acta 63 19/20 3097-3103 (1999).

[4] J. de Pablo, I. Casas, J. Giménez, F. Clarens, L. Duro and J. Bruno. Mater. Res. Soc. Symp. Proc. 807, 83 (2003).

[5] J. Giménez, F. Clarens, I. Casas, M. Rovira, J. de Pablo and J. Bruno. J. Nucl. Mater. 345 232-238 (2005).

[6] F. Clarens, J. de Pablo, I. Casas, J. Giménez, M. Rovira, J. Merino, E. Cera, J. Bruno, J. Quiñones and A. Martínez-Esparza. J. Nucl. Mater. 345, 225-231 (2005).

[7] I. Casas, M. Borrell, L. Sánchez, J. de Pablo, J. Giménez and F. Clarens. J. Nucl. Mater. 375, 151-156 (2008).

[8] I. Casas, J. de Pablo, F. Clarens, J. Gimenez, J. Merino, J. Bruno and A. Martinez-Esparza. Radiochim. Acta 97, 9, 485-490 (2009).

[9] Gray W. J. and Wilson C. N. (1995) Spent fuel dissolution studies: FY1991 to 1994. Report PNL-10540 (USA).

[10] S.Röllin, K.Spahiu and U.-B. Eklund. J. Nucl. Mater. 297 231-243 (2001).

[11] D. Serrano-Purroy, F. Clarens, J.-P- Glatz, B. Christiansen, J. de Pablo, J. Giménez, I. Casas, A. Martínez-Esparza. Radiochim. Acta 97, 491-496 (2009).

[12] D. Serrano-Purroy, I. Casas, E. González-Robles, J.P. Glatz, D.H. Wegen, F. Clarens, J. Giménez, J. de Pablo, A. Martínez-Esparza. Journal Nuclear Materials 434, 451-460 (2013).

[13] D.J. Wronkiewicz, J.K. Bates, S.F. Wolf and E.C. Buck. J. Nucl. Mater. 78-95 (1996).

[14] K.W. Jung, J.M. Kim, C.J. Kim and J.M. Lee. J. Korean Nucl. Soc. 19, 4, 242-248, (1987).

[15] J. Giménez, I. Casas, R. Sureda and J. de Pablo. Radiochim. Acta 100 445-448 (2012).

[16] J. de Pablo, I. Casas, F. Clarens, J. Giménez, and M. Rovira (2003). Vol.1. Publicaciones técnicas, ENRESA, SPAIN.

[17] E. Iglesias, J. Quiñones, S. Pérez de Andrés, J. M. Cobo and J. Alcaide. (2005) DFN/RR-02/SP-05, CIEMAT.

[18] K. Spahiu, L. Werme and U.-B. Eklund. Radiochim. Acta 88 507-511 (2000).

[19] V. Neck and J.I. Kim. Radiochim. Acta 89 1-16 (2001).

# Ceramic and advanced materials

Mater. Res. Soc. Symp. Proc. Vol. 1665 © 2014 Materials Research Society
DOI: 10.1557/opl.2014.660

# A Study of Natural Metamict Yttrium Niobate as Analogue of Actinide Ceramic Waste Form

Cao Qiuxiang[1, 2], Anton I. Isakov[2, 3], Liu Xiaodong[1], Sergey V. Krivovichev[2], Boris E. Burakov[3]
[1]East China Institute of Technology, Guanglan Road, 418, 330013, Nanchang, Jiangxi, China
[2]Saint-Petersburg State University, Universitetskaya emb. 7/9, 199034, St.-Petersburg, Russia
[3]V.G. Khlopin Radium Institute, 2-nd Murinskiy Ave. 28, 194021, St.-Petersburg, Russia

## ABSTRACT

Natural metamict mineral found as large (1-3 cm in size) homogeneous grains (as assumed, former single crystals), was investigated by X-ray powder diffraction (pXRD), high-temperature pXRD, scanning electron microscopy (SEM) and electron microprobe analysis (EMPA). The average chemical composition obtained by EMPA is (wt. %): $Nb_2O_5$ – 42.6; $Ta_2O_5$ – 4.4; $TiO_2$ – 9.2; $UO_3$ – 4.4; $ThO_2$ – 1.0; $MnO$ – 1.3; $FeO$ – 19.4; $Y_2O_3$ – 16.6. The untreated (original) sample is X-ray amorphous. The sample remained amorphous after annealing at 400 °C for 1 hour. The sample became almost fully crystalline after annealing at 700 °C for 1 hour with an X-ray diffraction pattern similar to that of Fe-columbite (ICCD: 01-074-7356). Further annealing at 1000 °C and higher temperatures caused changes in the phase composition of the sample. It was proposed that under self-irradiation a single-phase U-Th-bearing solid solution, based on monocrystalline Y-niobate, became metamict but remained homogeneous without evidence of solid solution destruction. However, this metamict solid solution is unstable under thermal treatment and recrystallization.

## INTRODUCTION

Since the term "metamict" was defined in 1893, the words "metamict" or "metamictization" have been increasingly synonymous for "amorphous" or "amorphization". Minerals considered to be metamict were judged amorphous because of their conchoidal fracture and isotropic optical properties; however, well-developed crystal faces evidenced the prior crystalline state [1,2].

Metamict minerals may contain significant quantities of U and Th that lead to the transition from the crystalline to the aperiodic, amorphous state. Amorphization occurs mainly due to the effects of heavy-particle irradiation and their crystalline structures can be reconstituted by heating [3,4]. It was also surprising that the metastable, amorphous state could be preserved in a material for long periods of geological time (up to ~$10^9$ years) [1]. Consequently, metamict minerals can serve as natural analogues of ceramic nuclear waste-forms regarding their response to radiation damage [5-7]. The metamict state raised important fundamental questions concerning the stability of different structure types in a radiation field [8].

In this article, we report on the occurrence and study of a metamict mineral found in granite pegmatites from Karelia, Russia.

## EXPERIMENTAL DETAILS

(1) EMPA analyses and secondary electron (SE) SEM imaging: Camscan electron microscope; 30 kV, 100 nA. Investigations were carried out in V.G. Khlopin Radium Institute.

(2) XRD: Bruker D2 Phaser diffractometer; $Cu_{K\alpha}$ radiation; $2\theta = 10-70°$.

(3) High-temperature XRD: Rigaku Ultra IV diffractometer; $Cu_{K\alpha}$ radiation; $2\theta = 6-70°$; the temperature variation from 20 to 1050 °C; the temperature step 100 °C (20–500 °C) and 25 °C (500–1050 °C).

Analyses of the X-ray diffraction patterns were performed using a PDXL-2 software package and the PDF database (Release 2011 RDB) at the X-ray Diffraction Centre of Saint-Petersburg State University.

## RESULTS AND DISCUSSION

Separate pieces of metamict mineral untreated and annealed at 400 and 700 °C are dense, homogeneous, conchoidal-fractured and have no faces (figure 1). This could be the evidence for their amorphous state.

**Figure 1.** The samples of metamict Y-niobate untreated (a), annealed at 400 (b) and 700 °C (c).

Untreated separate pieces of metamict mineral were investigated by means of SEM and EMPA. Generally a homogeneous, dense matrix with insignificant amount of cracks and inclusions was observed at low magnification (figure 2, a). The included (secondary) phase was observed at high magnification - the dark area in the picture (figure 2, b). The chemical compositions of the matrix and secondary phase were determined using EMPA (Table 1). According to the chemical composition, the main phase (matrix) was designated as yttrium niobate. Precise mineralogical identification of this metamict material was not possible because the phase composition of initial crystal (before metamictization) is still unclear. It can be seen that the composition of the secondary phase (Table 1, **Inc**; figure 2, point 17) differs from the composition of the matrix (Table 1, **Mx**; figure 2, points 8-15). It mainly consists of calcium oxide; the concentrations of oxides of niobium, yttrium and iron are significantly lower. The contents of U and Th oxides in the disseminated phase are increased in comparison with the matrix phase.

**Table I.** The chemical composition obtained by EMPA (wt. %) of matrix phase (Mx) and included phase (Inc) of metamict Y-niobate.

|     | $Nb_2O_5$ | FeO  | $Y_2O_3$ | $TiO_2$ | $Ta_2O_5$ | $UO_3$ | $ThO_2$ | MnO  | CaO  | $Nd_2O_3$ | $Sm_2O_3$ | $Gd_2O_3$ |
|-----|-----------|------|----------|---------|-----------|--------|---------|------|------|-----------|-----------|-----------|
| **Mx**  | 42.6  | 19.4 | 16.6     | 9.2     | 4.4       | 4.4    | 1.0     | 1.3  | 0.3  | 0.1       | 0.2       | 0.4       |
| **Inc** | 32.0  | 11.2 | n.d.     | 9.3     | 7.6       | 8.9    | 2.4     | n.d. | 28.6 | n.d.      | n.d.      | n.d.      |

n.d. – not detected; errors of measurement for Ca, Mn and Ti are $\pm$ 0.3 %, for others $\pm$ 0.1 %

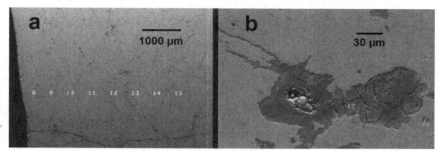

**Figure 2.** SEM image of untreated sample of Y-niobate (SE). General view (a) and inclusion (b). The digits denote the points analyzed.

Separate pieces of the same grain of metamict Y-niobate were annealed in air at different temperatures: 400, 700, 1000 and 1200 °C and, together with the untreated sample, studied by XRD.

The results demonstrate that the untreated sample and the sample annealed at 400 °C are amorphous (figure 3). It may indicate that the metamict mineral was initially crystalline, but, due to the irradiation damage caused by the α-decay of $^{238}$U and $^{232}$Th, it became fully amorphous.

The sample annealed at 700 °C is almost fully crystalline. The diffraction pattern (figure 3) is similar to that of Fe-columbite $FeNb_2O_6$ («columbite-like phase»). According to the previous R. Ewing's research of metamict columbite "the recrystallized phase may or may not be the original pre-metamict phase" [9].

The samples annealed at 1000 and 1200 °C were completely crystalline, but the phase composition differs. The phase composition of the sample annealed at 1000 °C is not identified yet.

The sample annealed at 1200 °C consists of the following main phases: $Ta_2O_5$, $FeNbO_4$, $FeTaO_4$ and $YNbO_4$. Hereafter: it should be noted that the phases are named conventionally because of the width of isomorphic substitutions; it is felt as more correct to use «$FeNbO_4$–like phase» etc. The peaks of few other phases are there, but these phases are not recognized yet. $ThO_2$ and $UO_2$ phases were not observed, but their presence should not be excluded. It is because they would be below the limit of detection.

It can be concluded that annealing could cause not only recrystallization, but also decomposition of an initially homogeneous columbite-like phase and the formation of new phases through reaction with oxygen at high temperature with the possibility of partial release of

U and Th in separate phases.

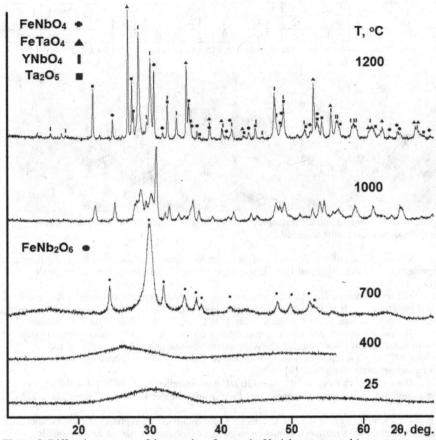

**Figure 3.** Diffraction patterns of the samples of metamict Y-niobate untreated (room temperature - 25 °C) and annealed (400, 700, 1000 and 1200 °C). The dots denote peak positions of the corresponding phases.

An untreated sample of metamict Y-niobate was investigated by mean of high-temperature XRD. When heated in air, the untreated sample undergoes two solid-state transitions (figure 4).

In the first stage, when the temperature is in the range of 400-700 °C, the amorphous sample undergoes recrystallization and forms a columbite-like phase. This result is in agreement with the annealing experiments.

In the second stage, when the temperature ranges from 900 to 1050 °C, the newly formed columbite-like phase undergoes decomposition, also in agreement with the results of annealing experiments. But, in contrast to the annealing experiments, it can be seen that the final product represents a mixture of phases with the structure types of $YTaO_4$, $FeNb_2O_6$, $Nb_2O_5$ and, probably, $TiTaO_4$. $ThO_2$ and $UO_2$ were not observed. The differences between the annealing and high-temperature XRD experiments were the heating rate and the exposure time. For annealing experiments the heating rate was 3 hours (25-1200 °C) and the exposure was 1 hour. For high-temperature XRD experiments the heating rate was 15 hours (25-1050 °C) and the exposure was 0.5 hour.

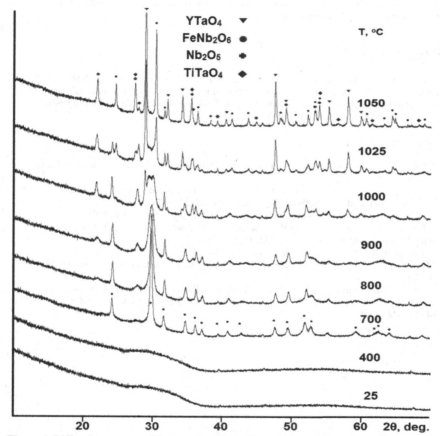

**Figure 4.** Diffraction patterns of the sample of metamict Y-niobate observed at different temperatures. The dots denote peak positions of the corresponding phases.

## CONCLUSIONS

The results obtained demonstrate very complex behavior of single-phase Th-U-bearing solid solution based on Y-niobate under self-irradiation and recrystallization. Long-term radiation damage may cause metamictization of single-phase minerals without destruction of the solid solution. However, recrystallization caused by heating of the metamict mineral may cause formation of new separate crystalline phases resulting in destruction of the initially homogeneous solid solution. This process is a subject of potential serious concern because it may be accompanied by segregation of U, Th (and other actinides for artificial minerals) into separate phases. This phenomenon should be taken into account in order to develop optimal ceramic waste-forms for actinide immobilization.

## ACKNOWLEDGMENTS

We would like to thank M.G. Krzhizhanovskaya, N.V. Platonova and D.V. Spiridonova (Crystallography department, Saint-Petersburg State University) for assistance in XRD data collection and phase analyses. Also we would like to thank Yu. Kretser (V.G. Khlopin Radium Institute) for assistance in SEM and EMPA investigations.

## REFERENCES

1. R.C. Ewing, B.C. Chakoumakos, G.R. Lumpkin, T. Murakami. Mater. Res. Soc. Bull. 12(4). 58–66 (1987).
2. D. Malczewski, J.E. Frackowiak, E.V. Galuskin. Hyperfine Interact. 166:529–536 (2005).
3. R.C. Ewing. Nucl. Instrum. Methods Phys. Res. B91, 22–29 (1994).
4. R.C. Ewing and R.F. Haaker, Nucl, and Chem. Waste Management 1, 51 (1980).
5. R.C. Ewing, B.C. Chakoumakos, G.R. Lumpkin, T. Murakami, R.B. Greegor, F.W. Lytle. Nucl. Instrum. Methods Phys. Res. B32, 487–497 (1988).
6. G.R. Lumpkin, R.C. Ewing. Phys Chem Minerals. 16:2–20 (1988).
7. A. Meldrum, L.A. Boatner, W.J. Weber, R.C. Ewing, Geochim. Cosmochim. Acta. 62, 2509 (1998).
8. W.J. Weber, R.C. Ewing, C.R.A. Catlow, T. Diaz de la Rubia, L.W. Hobbs, C. Kinoshita, Hj. Matzke, A.T. Motta, M. Nastasi, E.H.K. Salje, E.R. Vance and S.J. Zinkle. J. Mat. Res. 13, 1434–1484 (1998).
9. R.C. Ewing. Mineral. Mag. 40:89–899 (1976).

Mater. Res. Soc. Symp. Proc. Vol. 1665 © 2014 Materials Research Society
DOI: 10.1557/opl.2014.661

# Calcium Vanadinite: An Alternative Apatite Host for Cl-rich Wastes

M. R. Gilbert
AWE, Aldermaston, Reading, RG7 4PR, UK.

## ABSTRACT

Apatites are often seen as good potential candidates for the immobilization of halide-rich wastes and, in particular, chlorapatite ($Ca_5(PO_4)_3Cl$) has received much attention in recent years. However, synthesis of chlorapatite waste-forms can produce a complicated multi-phase system, with a number of secondary phases forming, including $\beta$-TCP ($Ca_3(PO_4)_2$), spodiosite ($Ca_2(PO_4)Cl$) and pyrophosphate ($Ca_2P_2O_7$), many of which require elevated temperatures and extended calcinations times to reduce. Calcium vanadinite ($Ca_5(VO_4)_3Cl$) demonstrates a much simpler phase system, with calcination at 750 °C yielding $Ca_5(VO_4)_3Cl$ together a small quantity of a $Ca_2V_2O_7$ secondary phase, the formation of which can be retarded by the addition of excess $CaCl_2$. Characterization of compositions doped with $SmCl_3$ as an inactive analogue for $AnCl_3$ show the Cl to be immobilized in the vanadinite whilst the Sm forms a wakefieldite ($SmVO_4$) phase.

## INTRODUCTION

Chlorapatite ($Ca_5(PO_4)_3Cl$) has been of interest as a waste-form due to its relatively high natural Cl content (6.8 wt. %), which makes it a potential candidate for the immobilization of Cl-rich wastes arising from the pyrochemical reprocessing of Pu [1,2]. Chlorapatite can be synthesized via conventional solid state synthesis from stoichiometric quantities of $CaHPO_4$, $CaCO_3$, and $CaCl_2$. However, the product formed is not single-phase, but a mix of chlorapatite with a number of secondary phases, including $\beta$-TCP ($\beta$-$Ca_3(PO_4)_2$), spodiosite ($Ca_2(PO_4)Cl$) and pyrophosphate ($Ca_2P_2O_7$) [3,4]. These require elevated temperatures and extended calcinations times to reduce, however, to prevent excess volatilization of Cl the maximum calcination temperature possible is 800 °C, and even after 8 hours calcination at this temperature the spodiosite, $\beta$-TCP and pyrophosphate phases are still present at levels of approximately 3, 4 and 11 wt. % respectively (as determined by Rietveld refinement) [1,4-5]. Such a varied multi-phase mix presents issues regarding the partitioning of species between different phases and deconvolution of their long-term durability [5]. A simplified system would therefore be beneficial in assessing both its chemical and long-term behavior.

Calcium vanadinite ($Ca_5(VO_4)_3Cl$) is a chlorapatite isomorph, crystallizing in the hexagonal $P6_3/m$ system with the Cl⁻ anions lying in the [00$z$] channels. Vanadinites have been viewed as prospective waste-forms for a number of years, with lead iodo-vanadinite ($Pb_5(VO_4)_3I$) proposed as a potential host for [129]I, the iodine able to be immobilized within the channels of the z-axis [6,7]. Synthesis of single-phase $Ca_5(VO_4)_3Cl$ is documented in the literature using a two-step process, firstly reacting $CaCO_3$ and $V_2O_5$ in a 3:1 molar ratio at 1100 °C for 10 h to form $Ca_3(VO_4)_2$, then reacting this with $CaCl_2$ in a 3:1 molar ratio at 900 °C repeatedly for 1 h [8,9]. These processing conditions are too extreme to meet AWE's requirements (limit of 800 °C), so a way to reduce the calcination temperature and time is required. One option is to switch from a $CaCO_3$ to CaO reagent, as this can potentially reduce the required calcination temperature by up to 200 – 250 °C, as has been demonstrated in the synthesis of calcium chlorosilicates [10-13]. The increased hygroscopic nature of CaO was not deemed to be an issue as controls would already be in place to limit absorption by the highly hygroscopic $CaCl_2$. It was therefore decided to attempt the synthesis of $Ca_5(VO_4)_3Cl$ in a one-step reaction using a mix of CaO, $CaCl_2$ and $V_2O_5$ reagents.

## EXPERIMENT

Calcium vanadinite was fabricated via a solid state synthesis reaction as shown in Equation 1.

$$4\tfrac{1}{2}CaO + 1\tfrac{1}{2}V_2O_5 + \tfrac{1}{2}CaCl_2 \rightarrow Ca_5(VO_4)_3Cl \quad \cdot \tag{1}$$

Stoichiometric quantities of CaO, $CaCl_2$ and $V_2O_5$ were placed into a Nalgene mill pot together with yttria-stabilized zirconia (YSZ) milling media and dry milled overnight. The resulting powder mix was passed through a 250 μm sieve mesh, placed in an alumina crucible, heated to 300 °C at 3.4 °C/min and held for 1 h to drive off any absorbed moisture. This dried powder was then heated at 3.4 °C/min to 650, 700, 750 or 800 °C and calcined in air for 5 h in order to assess the effect of calcination temperature on the final product.

In order to reduce the amount of $Ca_2V_2O_7$ secondary phase being formed an excess of $CaCl_2$ was added to drive the reaction toward the formation of $Ca_5(VO_4)_3Cl$. $CaCl_2$:$V_2O_5$:CaO in a 1:1:3 molar ratio were used following the same experimental conditions outline previously.

$Sm^{3+}$ was used as an inactive analogue for $Pu^{3+}$ and $Am^{3+}$. Four compositions of the solid solution $(Ca_{(5-1.5x)}Sm_x)(VO_4)_3Cl$ were synthesized, where $x = 0.05, 0.1, 0.25$ and $0.5$. Synthesis followed the same experimental conditions outlined above using a 1 mol excess of $CaCl_2$.

The calcined plugs were broken up, passed through a 250 μm sieve mesh and characterized by X-ray diffraction (XRD). Powder XRD was carried out using a Bruker D8 Advance diffractometer operating in Bragg-Brantano flat plane geometry using Cu $K_{\alpha1}$ radiation ($\lambda = 1.54056$ Å). Diffraction patterns were measured over a 2θ range of 10 – 90 ° using a step size of 0.025 ° and a collection time of 3.4 s per step. The phases present were identified by pattern matching using the Bruker Eva software and PDF database. Refinement of the phase assemblage and unit cell parameters was carried out using the GSAS suite of software [14].

## RESULTS

### Ca5(VO4)3Cl

Figure 1 shows the powder XRD patterns from the solid state synthesis of $Ca_5(VO_4)_3Cl$ with the phase assemblages of the respective products detailed in Table 1. As can be seen, two secondary phases are formed alongside the $Ca_5(VO_4)_3Cl$; a $Ca_2V_2O_7$ pyrophosphate isomorph and a $Ca_2(VO_4)Cl$ spodiosite isomorph. Formation of the $Ca_2(VO_4)Cl$ phase can successfully be eliminated by increasing the calcination temperature, such that at ≥ 750 °C only the $Ca_5(VO_4)_3Cl$ and $Ca_2V_2O_7$ phases are formed. Although increasing the calcination temperature further does begin to reduce the amount of $Ca_2V_2O_7$ formed, at 800 °C (the maximum potential operating temperature for this process) it still represents 28 wt. % of the final product.

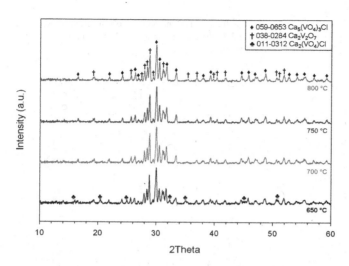

**Figure 1.** XRD of Ca$_5$(VO$_4$)$_3$Cl calcined at 650 – 800 °C.

**Table I.** Phase assemblage (wt. %) of products.

| Synthesis temp. (°C) | Ca$_5$(VO$_4$)$_3$Cl | Ca$_2$V$_2$O$_7$ | Ca$_2$(VO$_4$)Cl | $R_{wp}$ |
|---|---|---|---|---|
| 650 | 54.4 % | 33.4 % | 12.2 % | 5.70 |
| 700 | 62.2 % | 35.8 % | 2.1 % | 2.66 |
| 750 | 63.7 % | 36.3 % | 0.0 % | 1.36 |
| 800 | 71.6 % | 28.4 % | 0.0 % | 5.57 |

The Ca$_2$V$_2$O$_7$ secondary phase is formed via a side reaction of CaO and V$_2$O$_5$, as shown in Equation 2.

$$2CaO + V_2O_5 \rightarrow Ca_2V_2O_7 \tag{2}$$

In order to reduce the amount of Ca$_2$V$_2$O$_7$ secondary phase being formed an excess of CaCl$_2$ was added to drive the reaction toward the formation of Ca$_5$(VO$_4$)$_3$Cl. The resulting XRD patterns are shown in Figure 2 with the phase assemblages of the respective products detailed in Table 2. Initially, at 650 °C a much greater amount of Ca$_2$(VO$_4$)Cl phase is present than seen previously, 32.9 wt. % compared to 12.2 wt. % when not using excess CaCl$_2$. However, as with before, increasing the calcination temperature successfully eliminates the formation of this phase. The addition of 1 mol excess CaCl$_2$ is highly effective at retarding the formation of the

$Ca_2V_2O_7$ secondary phase. At 650 °C less than half of the amount observed previously is present, decreasing still further with increasing calcination temperature, such that at temperatures $\geq$ 750 °C near single-phase $Ca_5(VO_4)_3Cl$ is formed.

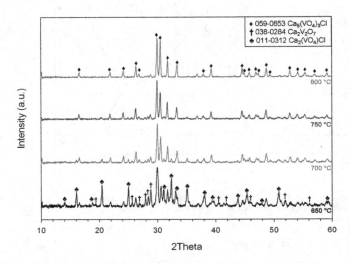

**Figure 2.** XRD of $Ca_5(VO_4)_3Cl$ calcined with excess $CaCl_2$ at 650 – 800 °C.

**Table II.** Phase assemblage (wt. %) of products from calcination with excess $CaCl_2$.

| Synthesis temp. (°C) | $Ca_5(VO_4)_3Cl$ | $Ca_2V_2O_7$ | $Ca_2(VO_4)Cl$ | $R_{wp}$ |
|---|---|---|---|---|
| 650 | 51.1 % | 16.0 % | 32.9 % | 5.24 |
| 700 | 89.5 % | 7.6 % | 3.0 % | 2.33 |
| 750 | 93.3 % | 5.1 % | 1.6 % | 1.20 |
| 800 | 96.7 % | 3.3 % | 0.0 % | 0.97 |

## $(Ca_{(5-1.5x)}Sm_x)(VO_4)_3Cl$

Figure 3 shows the powder XRD patterns from the solid state synthesis of $(Ca_{(5-1.5x)}Sm_x)(VO_4)_3Cl$ with the phase assemblages of the respective products detailed in Table 3. As can be seen from the XRD data, the product is not a $(Ca_{(5-1.5x)}Sm_x)(VO_4)_3Cl$ solid solution but rather a mixture of separate phases. Whilst the Cl continues to be immobilized within the $Ca_5(VO_4)_3Cl$ phase, the Sm instead forms a tetragonal $Sm(VO_4)$ wakefieldite phase (a vanadate analogue of xenotime). This $Sm(VO_4)$ phase is the primary Sm host, growing steadily with increasing Sm content as can be seen by the increasing 101, 200, 112 and 312 reflexions at 18.5,

24.6, 33.1 and 48.9 ° 2θ respectively. No incorporation of Sm into the $Ca_5(VO_4)_3Cl$ phase with increasing Sm content is observed, with no evidence of unit cell contraction that would be caused by the substitution of Ca for Sm. The unit cell parameters for $Ca_5(VO_4)_3Cl$ and $Sm(VO_4)$ phases are shown in Table 4.

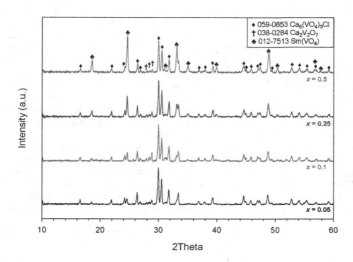

**Figure 3.** XRD of $(Ca_{(5-1.5x)}Sm_x)(VO_4)_3Cl$ where $x = 0.05, 0.1, 0.25$ and $0.5$.

**Table III.** Phase assemblage (wt. %) of products.

| $x$ | $Ca_5(VO_4)_3Cl$ | $Ca_2V_2O_7$ | $Sm(VO_4)$ | $R_{wp}$ |
|------|------|------|------|------|
| 0.05 | 92.8 % | 5.2 % | 2.0 % | 6.35 |
| 0.1 | 87.1 % | 7.8 % | 5.0 % | 2.17 |
| 0.25 | 85.5 % | 5.3 % | 9.1 % | 1.17 |
| 0.5 | 75.3 % | 3.0 % | 21.7 % | 4.10 |

**Table IV.** Unit cell parameters for $Ca_5(VO_4)_3Cl$ and $Sm(VO_4)$ phases formed.

| $Ca_5(VO_4)_3Cl$ | | $Sm(VO_4)$ | |
|------|------|------|------|
| Space Group | $P6_3/m$ | Space Group | $I4_1/amd$ |
| $a = b$ | 10.143(7) | $a = b$ | 7.249(7) |
| c | 6.778(3) | c | 6.377(5) |
| $\alpha = \beta$ | 90 ° | $\alpha = \beta = \gamma$ | 90 ° |
| $\gamma$ | 120 ° | | |

## CONCLUSIONS

Switching from a $CaCO_3$ to $CaO$ reagent enables the synthesis of calcium vanadinite ($Ca_5(VO_4)_3Cl$) via a single-step solid state reaction. By eliminating the need to form a $Ca_3(VO_4)_2$ intermediate, requiring calcination at 1100 °C for at least 10 h [8,9], synthesis of $Ca_5(VO_4)_3Cl$ has been achieved at much lower temperatures and in a much shorter timescale than has been previously reported in the literature

Initially, synthesis of calcium vanadinite yields $Ca_5(VO_4)_3Cl$ together with two secondary phases; $Ca_2(VO_4)Cl$ and $Ca_2V_2O_7$, isomorphous to the spodiosite and pyrophosphate secondary phases seen in chlorapatite synthesis [1,3-4]. The $Ca_2(VO_4)Cl$ secondary phase can successfully be removed by increasing the calcination temperature to $\geq$ 750 °C, whilst the formation of the $Ca_2V_2O_7$ secondary phase can be successfully retarded via the addition of a 1 mol excess of $CaCl_2$, such that for calcination at 800 °C near single-phase $Ca_5(VO_4)_3Cl$ is formed, with only 3.3 wt. % of the $Ca_2V_2O_7$ secondary phase present. This represents a markedly simpler phase assemblage than currently seen in the synthesis of chlorapatite under similar conditions [1,4-5].

Incorporation of $SmCl_3$ as an inactive analogue does not produce a $(Ca_{(5-1.5x)}Sm_x)(VO_4)_3Cl$ solid solution at any level of Sm-doping. In previous work with chlorapatite, the Cl is immobilized within the [00$z$] channels of the apatite, whilst the Sm is initially incorporated within the $Ca_3(PO_4)_2$ secondary phase, before forming a $Sm(PO_4)$ phase at higher levels of Sm-doping [5]. In the case of $Ca_5(VO_4)_3Cl$, the Cl is similarly contained within the [00$z$] channels of the vanadinite lattice, however, without the presence of an analogous $\beta$-$Ca_3(PO_4)$ phase, the Sm is immobilized within a $Sm(VO_4)$ phase at all levels of Sm-doping.

## REFERENCES

1. B. L. Metcalfe, I. W. Donald, S. K. Fong, L. A. Gerrard, D. M. Strachan, R. D. Scheele, *Mat. Res. Soc. Symp. Proc.*, **1124**, 207 (2009).
2. W. E. Lee, M. R. Gilbert, S. T. Murphy, R. W. Grimes, *J. Am. Ceram. Soc.*, **96**, 2005 (2013).
3. S. K. Fong, I. W. Donald, B. L. Metcalfe, *J. Alloys Comp.*, **444/445**, 424 (2007).
4. B. L. Metcalfe, I. W. Donald, S. K. Fong, L. A. Gerrard, D. M. Strachan, R. D. Scheele, *Mat. Res. Soc. Symp. Proc.*, **985**, 157 (2007).
5. S. K. Fong, B. L. Metcalfe, D. Strachan, R. Scheele, *Mat. Res. Soc. Symp. Proc.*, **1518**, in press (2013).
6. F. Audubert, J. Carpena, J. L. Lacout, F. Tetard, *Solid Sate Ionics*, **95**, 113 (1997).
7. M. C. Stennett, I. J. Pinnock, N. C. Hyatt, *J. Nucl. Mater.*, **414**, 352 (2011).
8. H. Kreidler, *Am. Mineral.*, **55**, 180 (1970)
9. H. P. Beck, M. Douiheche, R. Haberkorn, H. Kohlmann, *Solid State Sci.*, **8**, 64 (2006)
10. M. R. Gilbert, *Mat. Res. Soc. Symp. Proc.*, **1518**, in press (2013).
11. J. Liu, H. Lian, C. Shi, J. Sun, *J. Electrochem. Soc.*, **152**, 880 (2005).
12. W. Ding, J. Wang, M. Zhang, Q. Zhang, Q. Su, *J. Solid State Chem.*, **179**, 3582 (2006).
13. N. Karkada, D. Porob, P. Kumar, *ECS Trans.*, **33**, 39 (2010).
14. H. M. Rietveld, *Acta Cryst.*, **22**, 151 (1967).

Mater. Res. Soc. Symp. Proc. Vol. 1665 © 2014 Materials Research Society
DOI: 10.1557/opl.2014.662

# Molten Salt Synthesis of Zirconolite Polytypes

M. R. Gilbert
AWE, Aldermaston, Reading, RG7 4PR, UK.

## ABSTRACT

Zirconolite ($CaZrTi_2O_7$), a durable and compositionally flexible titanate ceramic for the immobilization of separated actinides, is currently the UK's preferred candidate phase for the immobilization of plutonium dioxide arising from aqueous reprocessing. Here, its suitability as a waste-form for actinide chlorides arising from pyrochemical reprocessing is investigated through synthesis via a molten salt mediated reaction using a number of different salt eutectics ($MgCl_2$:NaCl, $CaCl_2$:NaCl and KCl:NaCl). It is found that the effectiveness of the molten salt synthesis of zirconolite is governed by the solubility of $ZrO_2$ in the salt medium used; the synthesis proceeding via the formation of a perovskite ($CaTiO_3$) intermediate which then reacts with $ZrO_2$ to form zirconolite via a solution-diffusion mechanism. Most notably, in the KCl:NaCl eutectic different zirconolite polytypes are formed at different synthesis temperatures, with zirconolite-3T forming at 900 °C, giving way to zirconolite-2M at 1200 °C.

## INTRODUCTION

Pyrochemical reprocessing techniques enable the recovery of Pu metal from spent nuclear material without the need to convert it to $PuO_2$ and back [1]. These methods utilise an electrorefining process, where the Pu is separated from the impurities in a molten chloride salt, most typically either $CaCl_2$ or an equimolar mixture of NaCl-KCl, at temperatures of between 750 – 850 °C [2]. Post-reprocessing, this chloride salt must be replaced, as it now contains a number of different waste streams which will contaminate the cathode and affect the properties of the molten salt. This contaminated salt must be disposed of in such a way as to immobilize the radionuclides contained within. However, halide-rich wastes such as these can be problematic to immobilize, as not only are their solubilities in melts very low, but even in small quantities they can seriously affect the properties of the waste-form [3,4]. In addition, processing temperatures are often severely limited in order to prevent the volatilisation of the halides.

One approach is to selectively immobilize the actinide cations from the chloride media in titanate ceramics, using the salt as a reaction medium for molten salt synthesis (MSS). In this method, a salt or eutectic mix of salts with low melting point is added to the reactants (typically 80-100 wt. % of the reactant mixture) and heated to above the melting point of the salt. The molten salt medium essentially acts as a solvent, assisting the diffusion of reactant species and thus reducing the temperature and time required for synthesis. There are two generally accepted mechanisms of product formation through MSS: solution-precipitation and solution-diffusion (sometimes referred to as "templating"). Which mechanism is dominant depends upon the solubility and relative dissolution rates of the reactants in the molten salt [5-7].

Nominally $CaZrTi_2O_7$, zirconolite exhibits a wide range of stoichiometries and polytypes (including 2M, 4M, 3O, 3T and 6T [8]) of general formula $CaZr_xTi_{(3-x)}O_7$ where $0.8 < x < 1.35$ [9,10]. The structures of all zirconolite polytypes are derived from an anion-deficient fluorite archetype. The most common polytype is that of zirconolite-2M, a lamellar monocline structure with space group C2/c consisting of two distinct layers of cations: a hexagonal-tungsten-bronze (HTB) layer and a layer of Ca and Zr cation polyhedra.

## EXPERIMENT

As the waste salt in question is composed largely of $CaCl_2$ and $MgCl_2$, $CaCl_2:NaCl$ and $MgCl_2:NaCl$ eutectics, together with the $KCl:NaCl$ eutectic demonstrated to be successful in the MSS of pyrochlore [11,12], have been investigated over a range of synthesis temperatures, as displayed in Table 1.

**Table I.** Molten salt media and synthesis temperatures for zirconolite MSS.

| Eutectic | Melting Point (°C) | Synthesis Temp (°C) |
|---|---|---|
| NaCl:KCl | 658 | 700 – 1200 |
| $CaCl_2:NaCl$ | 505 | 500 – 1000 |
| $MgCl_2:NaCl$ | 465 | 500 – 1000 |

A 7:1 molar ratio of molten salt to zirconolite product ($MS:CaZrTi_2O_7$) was used for each eutectic. Stoichiometric quantities of CaO, $ZrO_2$, $TiO_2$ (anatase) and salt medium were placed into a Nalgene mill pot together with yttria-stabilised zirconia (YSZ) milling media. 5 wt. % excess $ZrO_2$ and $TiO_2$ were added to drive the reaction to formation of zirconolite and the powders were dry milled overnight. The resulting powder mix was passed through a 250 µm sieve mesh and pressed at 101.5 MPa for 1 min into pellets. The green pellets were placed into a closed alumina crucible, heated to 300 °C at 5 °C/min and held for 1 h to drive off any absorbed moisture, before being fired at 500/700–1000 °C (5 °C/min) for 2 h. Upon cooling to room temperature the fired pellets were broken up in a pestle and mortar, washed with hot deionised water to remove the molten salt medium and the product collected via vacuum filtration.

Powder XRD was carried out using a Bruker D8 Advance diffractometer operating in Bragg-Brantano flat plane geometry using Cu $K_{\alpha 1}$ radiation ($\lambda = 1.54056$ Å). Diffraction patterns were measured over a $2\theta$ range of 10 – 90 ° using a step size of 0.025 ° and a collection time of 3.4 s per step.

The phases present were identified by pattern matching using the Bruker Eva software and PDF database. Refinement of the unit cell parameters and phase assemblage was carried out using the GSAS suite of software [13,14].

## RESULTS

### MgCl₂:NaCl Eutectic

Figure 1 shows the powder XRD patterns from the attempted MSS of zirconolite at temperatures from 500 – 1000 °C. As can be seen from the XRD data, the only reaction occurring in the $MgCl_2:NaCl$ eutectic is a reaction between the $TiO_2$ and the $MgCl_2$ present in the salt, forming a geikielite ($MgTiO_3$) phase. Small amounts of unreacted $TiO_2$ are present at lower temperatures (500 – 700 °C), however, by 800 °C the remaining $TiO_2$ is consumed, leaving the geikielite together with the unreacted $ZrO_2$.

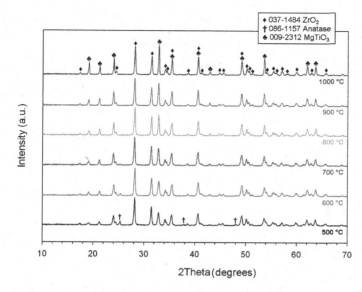

**Figure 1.** XRD of CaZrTi$_2$O$_7$ MSS in MgCl$_2$:NaCl salt medium.


## CaCl$_2$:NaCl Eutectic

Figure 2 shows the powder XRD patterns from the MSS of zirconolite at temperatures from 500 – 1000 °C. At temperatures < 800 °C no reaction takes place, the XRD patterns showing the products to be a mixture of unreacted TiO$_2$ and ZrO$_2$. At 800 °C the first evidence of zirconolite formation can be observed with the appearance of the major 202 reflexion of the hexagonal zirconolite-3T polytype at 30.4 ° 2θ. However, at this stage zirconolite-3T is only present as a minor phase, accompanied by a perovskite (CaTiO$_3$) secondary phase. The major phase at this point is Ca$_2$Ti$_5$O$_{12}$, together with unreacted ZrO$_2$ and TiO$_2$, now present as both anatase and rutile polymorphs (it should be noted that TiO$_2$ converts from the anatase to rutile polymorph at temperatures ≥ 800 °C). At temperatures ≥ 900 °C the Ca$_2$Ti$_5$O$_{12}$ is no longer present, leaving zirconolite-3T and perovskite phases of approximately equal amounts, together with unreacted ZrO$_2$ and TiO$_2$ (rutile). The phase assemblage of the products can be seen in Table 2.

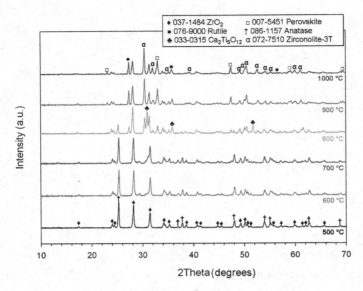

**Figure 2.** XRD of CaZrTi$_2$O$_7$ MSS in CaCl$_2$:NaCl salt medium.

**Table II.** Phase assemblage of products by MSS in CaCl$_2$:NaCl eutectic.

| Synthesis temp. | Zirconolite-3T | Perovskite | Ca$_2$Ti$_5$O$_{12}$ | ZrO$_2$ | TiO$_2$ |
|---|---|---|---|---|---|
| 800 °C | 6.7 wt. % | 3.2 wt. % | 48.2 wt. % | 32.0 wt. % | 9.9 wt. % |
| 900 °C | 21.1 wt. % | 21.0 wt. % | 0.0 wt. % | 46.3 wt. % | 11.7 wt. % |
| 1000 °C | 23.9 wt. % | 23.5 wt. % | 0.0 wt. % | 39.7 wt. % | 12.9 wt. % |

## KCl:NaCl Eutectic

Figure 3 shows the powder XRD patterns from the MSS of zirconolite at temperatures from 700 – 1200 °C. At temperatures < 900 °C the products are mainly unreacted ZrO$_2$ and TiO$_2$, although a small amount of perovskite is also formed. At 900 °C clear evidence of zirconolite-3T formation can be observed, together with the perovskite secondary phase and unreacted ZrO$_2$ and TiO$_2$, which are now starting to be consumed. As the synthesis temperature increases the perovskite secondary phase gradually disappears and more of the ZrO$_2$ and TiO$_2$ reactants are consumed, until at 1100 °C zirconolite-3T is the major phase formed, before undergoing a phase change between 1100 – 1200 °C and converting to near single-phase zirconolite-2M. The phase assemblage of the products can be seen in Table 3.

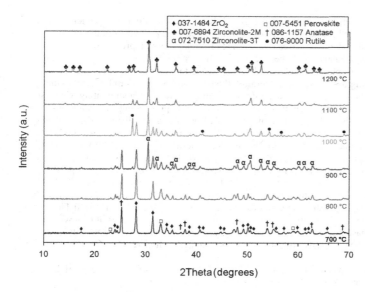

**Figure 3.** XRD of CaZrTi$_2$O$_7$ MSS in KCl:NaCl salt medium.

**Table III.** Phase assemblage of products formed by MSS in KCl:NaCl eutectic.

| Synthesis temp. | Zirconolite-3T | Zirconolite-2M | Perovskite | ZrO$_2$ | TiO$_2$ |
|---|---|---|---|---|---|
| 900 °C | 19.0 wt. % | 0.0 wt. % | 11.9 wt. % | 42.0 wt. % | 27.1 wt. % |
| 1000 °C | 24.1 wt. % | 0.0 wt. % | 12.1 wt. % | 40.2 wt. % | 23.7 wt. % |
| 1100 °C | 55.3 wt. % | 0.0 wt. % | 0.0 wt. % | 24.1 wt. % | 20.6 wt. % |
| 1200 °C | 0.0 wt. % | 80.6 wt. % | 0.0 wt. % | 5.5 wt. % | 13.9 wt. % |

## DISCUSSION AND CONCLUSIONS

In the MgCl$_2$:NaCl eutectic, the low solubility of the ZrO$_2$ in the salt means that it remains completely unreacted. Instead, the CaO and TiO$_2$ react with the large excess of MgCl$_2$ present in the salt medium to form MgTiO$_3$. By 800 °C all of the CaO and TiO$_2$ reactants have been consumed, leaving a mix of essentially single-phase MgTiO$_3$ together with unreacted ZrO$_2$.

Replacing the MgCl$_2$ in the salt with CaCl$_2$ to form a CaCl$_2$:NaCl eutectic eliminates this issue. However, the low solubility of the ZrO$_2$ in the salt again makes it very unreactive. As a result the TiO$_2$ and CaO, as well as the large excess of Ca now present in the salt, react to form the calcium titanate phases Ca$_2$Ti$_5$O$_{12}$ and CaTiO$_3$ (perovskite) at 800 °C. As the temperature increases the Ca$_2$Ti$_5$O$_{12}$ decomposes to form CaTiO$_3$ and TiO$_2$. At 900 °C the first substantial evidence of zirconolite formation can be observed, a result of the CaTiO$_3$ templating onto the surface of the $ZrO_2$ particles to form the 3T polytype. This increases with increasing temperature, together with the reaction of the remaining $TiO_2$ with the excess Ca present from the salt to form more perovskite, such that the two phases remain in approximately equal proportions.

Using a KCl:NaCl eutectic removes the excess Ca from the system and hence stops the formation of the $Ca_2Ti_5O_{12}$ phase, the CaO and $TiO_2$ reacting together to form only the $CaTiO_3$ perovskite. On increasing the synthesis temperature the reaction follows a similar trend to that seen in the $CaCl_2$:NaCl eutectic, with zirconolite-3T forming at 900 °C via templating. Given the low solubility of the $ZrO_2$ at this stage, the reaction is most likely proceeding via a solution-diffusion mechanism. Without the presence of excess Ca from the salt medium, as the synthesis temperature increases further the perovskite is eventually consumed, resulting in, at 1100 °C, the zirconolite-3T becoming the major phase for the first time. Between 1100 and 1200 °C a rapid increase in the consumption of the $ZrO_2$ and $TiO_2$ reactants is observed, potentially indicating an increase in the solubility of the $ZrO_2$ at these elevated temperatures leading to a solution-precipitation reaction mechanism. The zirconolite itself also undergoes a phase change, converting to the more common natural 2M polytype originally targeted, such that at 1200 °C the product is a mix of near single-phase zirconolite-2M together with the excess $ZrO_2$ and $TiO_2$. However, by having to raise the synthesis temperature to such an elevated level, it is now in the region where single-phase zirconolite-2M can readily be fabricated by conventional solid state synthesis [15,16], thus negating any advantage of using a molten salt process.

## REFERENCES

1. T. Nishimure, T. Koyama, M. Iizuka, H. Tanaka, *Prog. Nucl. Energy*, **32**, 381 (1998).
2. I. N. Taylor, M. L. Thompson, T. R. Johnson, *Proceedings of the International Conference and Technology Exposition on Future Nuclear*, **1**, 690 (1993).
3. W. E. Lee, R. W. Grimes, *Energy Materials*, **1**, 22 (2006).
4. T. O. Sandland, L.-S. Du, J.F. Stebbins, J.D. Webster, *Geochim. Cosmochim. Acta*, **68**, 5059 (2004).
5. D. Segal, *J. Mater. Chem.*, **7**, 1297 (1997).
6. S. Zhang, *Pak. Mater. Soc.*, **1**, 49 (2007).
7. T. Kimura, in: *Advances in Ceramics – Synthesis and Characterization, Processing and Specific Applications*, ed. C. Sikalidis (InTech, 2011).
8. K. L. Smith, G. R. Lumpkin, *Defects and Processes in the Solid State: Geoscience Applications*, eds. J. N. Boland and J. D. Fitzgerald (Elsevier, 1993).
9. H. J. Rossell, *Nature*, **283**, 282 (1980).
10. B. M. Gatehouse, I. E. Grey, R. J. Hill, H. J. Rosell, *Acta Cryst. B*, **37**, 306 (1981).
11. M. L. Hand, M. C. Stennett, N. C. Hyatt, *J. Eur. Ceram. Soc.*, **32**, 3211 (2012).
12. M. C. Stennett, M. L. Hand, N. C. Hyatt, *Mater. Res. Soc. Symp. Proc.*, **1518**, 97 (2013).
13. H. M. Rietveld, *Acta Cryst.*, **22**, 151 (1967).
14. H. M. Rietveld, *J. Appl. Cryst.*, **2**, 65 (1969).
15. M. C. Stennett, N. C. Hyatt, M. R. Gilbert, F. R. Livens, E. R. Maddrell, *Mater. Res. Soc. Symp. Proc.*, **1107**, 413 (2008).
16. M. R. Gilbert, C. Selfslag, M. Walter, M. C. Stennett, J. Somers, N. C. Hyatt, F. R. Livens, *IOP Conf. Ser.: Mater. Sci. Eng.*, **9**, 12007 (2010).

# AUTHOR INDEX

# SUBJECT INDEX

Printed in the United States
By Bookmasters